CABINETS AND COUNTERTOPS

Charles Self

McGraw-Hill

New York   San Francisco   Washington, D.C.   Auckland   Bogotá
Caracas   Lisbon   London   Madrid   Mexico City   Milan
Montreal   New Delhi   San Juan   Singapore
Sydney   Tokyo   Toronto

To Frances, with thanks for having the patience to see me through this one.

Library of Congress Cataloging-in-Publication Data

Self, Charles R.
　　Cabinets and countertops / Charles Self.
　　　　p.　　cm.
　　Includes index.
　　ISBN 0-07-134899-9
　　1. Kitchen cabinets.　2. Counter tops.　3. Cabinetwork.　I. Title.
TT197.5.K57S45　　2000
684.1'6—dc21　　　　　　　　　　　　　　　　00-28353
　　　　　　　　　　　　　　　　　　　　　　　CIP

McGraw-Hill

A Division of The McGraw·Hill Companies

1 2 3 4 5 6 7 8 9 0　DOC/DOC　0 6 5 4 3 2 1 0

ISBN 0-07-134899-9

The sponsoring editor for this book was Zoe G. Foundotos, the editing supervisor was Stephen M. Smith, and the production supervisor was Sherri Souffrance. It was set in Melior by Michele Pridmore of McGraw-Hill's Hightstown, N.J., Professional Book Group composition unit.

Printed and bound by R. R. Donnelley & Sons Company.

McGraw-Hill books are available at special quantity discounts to use as premiums and sales promotions, or for use in corporate training programs. For more information, please write to the Director of Special Sales, Professional Publishing, McGraw-Hill, Two Penn Plaza, New York, NY 10121-2298. Or contact your local bookstore.

 This book is printed on recycled, acid-free paper containing a minimum of 50% recycled de-inked fiber.

CONTENTS

The purpose of *Cabinets and Countertops* is to provide you with enough information to make cabinets in what might be considered a small way—that is, to set up a cabinetmaking shop with one, two, or three other people, either employees or partners, and to go on and do an excellent job of cabinetmaking, while also making a decent living. To that purpose, you'll find extensive information on business formation and general business procedures in the appendices. Otherwise, *Cabinets and Countertops* is devoted to setting up the shop, filling it with tools, and using those tools to produce cabinets that please your customers. You won't find any information on the more sophisticated, and far more costly, tools; nor is there any detail on CNC—computer numeric control—machining of wood. All that has value, but not at the start of a career in cabinetmaking, unless one has immense (by my standards) capital.

The variety of cabinetry available in stock and custom factory-produced styles today is greater than ever; that presents a challenge to a small custom shop, but there are numerous ways around the problem. While catalog cabinet styles cover most of the needs that the designers can imagine, the needs that actually crop up often require special additions or changes to those stock styles. Large manufacturers are limited by context. Although the small cabinetry shop may not be able to do some huge jobs as quickly, smaller jobs that require more creativity and freedom of operation are the stock in trade of small shops, and it should

> The variety of cabinetry available in stock and custom factory-produced styles today is greater than ever before. That presents a challenge to a small custom shop, but there are numerous ways around the problem. Large manufacturers are limited by context; while the small cabinetry shop may not be able to do some huge jobs as quickly, smaller jobs that require more creativity and freedom of operation are the stock in trade of small shops, and should always be strongly emphasized: You can fit any style, any area, any odd shape or size.

always be strongly emphasized: You can fit any style, any area, any odd shape or size, that stock cabinetmakers cannot.

You can't compete with Home Depot, so don't even try. Be careful to pre-qualify your customers at the outset. As soon as clients say, "But we don't want to spend a lot of money," let them know that they are in the wrong place and send them to Home Depot. You need to make sure that what you offer is a significant step up in quality; but at the same time, remember that most customers are not experienced with the level of quality and style you will be offering. You won't make it by trying to offer the best price in town, but you might make it if your quality and service are above reproach.

Almost all the big cabinetry suppliers are weak in custom work. Their design work usually is mediocre as well. And often their materials and workmanship, including installation, are far from perfect. And that's particularly true when compared to what a top small custom shop can supply.

Here is what you should offer:

- *Top-notch design.*   If you can't do it, better hire a designer-sales-person who can.

- *Top-notch quality.*   That means special ¾-inch carcass sides, ½-inch backs, mortise-and-tenon face frames, and beautiful stains and finishes. It also means handmade moldings and three-piece crowns, real dentil molding, and on. You know what it all means: no compromises. The folks who buy luxury cars and $700,000 homes want to know that they are getting the absolute best that money can buy. (Some may also ask if they can avoid paying sales tax by paying with cash. Even if you lose that job, refuse. Avoiding taxes is legal; evading them is a felony.)

- *Top-notch service.*   If the customer says there is a problem, get there and fix it right away. No waiting. Don't even stammer. Just do it.

- *Service to high-end clientele who want to know they are getting something special.*   Be sure to casually mention any celebrity clients you may have, any design degrees, any awards, and similar things. Make them believe how fortunate they are to have the opportunity to

> **If the customer says there is a problem, get there and fix it right away. No waiting. Don't even stammer. Just do it.**

hire a craftsperson as highly skilled as yourself. Your top-notch clients will love it—but it had better be true.

Many Routes to a Fine Job

In fact, there is seldom one way to do any cabinetmaking job, and often more than a single tool works for each job. A simple job like straight raised panels can be done in 2 standard ways on a router table; they could also be made on a shaper; and a table saw also produces a fine raised panel. Of course, for the purist, raised panels may also be made by hand, using special planes. And sticking a curve or arch into the panel doesn't always create a major limit in the tools that are needed or useful. Templates that work on a shaper can be adapted for routers, and other methods and jigs can be readily developed to use most tools for the purposes desired. Too much handwork, of course, can price a job, even for a $700,000 home, out of the market; so make the needed adjustments between your craft and art.

Thus, what I'm trying to do here is not to lock you into any one style or method of cabinetmaking, but to meld the types that seem to work best for me and for those I know. If another style, learned elsewhere, works well for you, use it. Don't change just to be changing. Any technique you're currently using that is effective, efficient, and safe is probably the one you should stick with. Much of cabinetmaking is fairly standard after all the years of experience in the craft by many thousands of people. And much of what I show here is standard, useful in almost all situations. We aim to please as many people as possible. The nuances of any craft are learned on site, and over time, and only a few bits and pieces of such nuances can be gleaned from a book. Even those need extensive practice for real ease of use. You can use a level to position lines to set holes in countertops with one look at the drawings. However, you need to spend a half-dozen or more attempts to learn to make dovetails for drawers (except with the Keller jig, which usually works exactly right either the first or second time.)

Generally, we recommend reading and then trying—assuming you have a need for the procedure. Ignore all procedures that you have no need for, as you ignore procedures that make you uneasy from a safety standpoint. Regardless of who is telling you how to do something, if the method seems to *you* at all unsafe, don't use it. Period. Woodworking

> When a woodworking method seems to *you* at all unsafe, don't use it. Period. Woodworking in any form can be a dangerous enterprise. If instructions for safety for each tool are followed, then the danger is appreciably reduced. If you follow your own instincts, you reduce the danger even more.

in any form can turn into a dangerous enterprise. If instructions for safety for each tool are followed, then the danger is appreciably reduced. If you follow your own instincts, you will reduce the danger even more. I would suggest that if you're far enough along in woodworking and cabinetry to consider setting up your own small shop, and you have no such instincts, then you need to find a different way of thinking about the things you do, or you need to find a different way of making a living.

The most effective cabinetmaker is also the safest.

There is probably the widest variety of opportunities available today, because we're in the midst, as I write *Cabinets and Countertops,* of the longest building boom in history, especially considering the size, complexity, and price of today's new homes. Starter homes today are larger than final homes of just a few years ago, and they have many more expensive features (central air conditioning, superb wood cabinetry in several rooms, hardwood floors, and so on). After World War II, Levittown was begun as a way to help veterans afford homes. Homes less than 1000 square feet, on concrete pads, were erected by the thousands, at the lowest possible cost (prices averaged $6000 for the first 17,000 or so houses erected, in two-bedroom single-bath styles on 60-foot by 100-foot lots). The market boomed for years as returning veterans, most of whom left home as youngsters who had been living with their parents, came home needing living space of their own.

Today, the building boom is driven by similar market needs. A huge number of the children of those World War II veterans need, and desire, larger homes for themselves, while the next two generations are unwilling to start where today's grandparents and great-grandparents began, in a small home. Thus the sale of huge (by 1960 standards) homes is still rising. The opportunity for cabinetmakers is obvious, but the effort to remodel older housing is almost as strong, and that market may be even stronger, and for the small cabinetmaker more lucrative, than the new-home market. It is often in older homes, with oddly shaped rooms that are also often oddly placed, that the residential cabinetmaker comes into his or her own.

The opportunities exist, and if you have most of the skills, many of the tools, and a few dollars in the bank (or backing—see the appendices on setting up and running a cabinetmaking business), this is probably one of the best times in history to start out as a custom cabinetmaker.

Today, there are more reasons than ever for starting your own business: First, and probably foremost, is the flexibility that working for yourself gives you. Working as a small business created some communications problems in the past, but with today's electronic machines, there is not much reason to ever miss a step, regardless of whether a helper is free to tend the phone when you're busy with other things. There are other considerations: Many companies are downsizing, and the economy is in a current long-term up cycle. But building starts are a shade down as I write this, but are still very high, just below current record levels, and appear aimed at remaining high for some time to come. That means there is another good reason for jumping off the precipice at this time: plenty of work for those who know how to get it. As a craftsman or craftswoman, you're already more independent than most. It probably hasn't been all that long since you loaded your tools and took your skills to another outfit when you were displeased with something from company tool policy to the condition of the toilets. Or the next guy down the block may just represent "greener grass" for you, a more complex, more interesting set of cabinets to design, build, and install.

So we've got computers and other machines to tend the office "fires" while you're busy sawing wood and building doors and drawers. And we've got a scarcity of good jobs on any long-term basis. Many larger companies today are doing their hiring on a temporary-to-hire basis, with some companies leaving good workers hanging for as long as 2 years before making the final decision. There are plenty of $7 per hour jobs, and the number is apt to grow, but those don't support a family or provide much in the way of a real living. We've got plenty of residential and other small building construction work, so the demand for cabinetry is high. Remodeling work is also another strong source of cabinetry work, and it remains strong.

On the opposing side, the Internal Revenue Service is always forming ever more restrictive regulations in regard to the small workplace. The cost of keeping decent books is a reasonable cost of doing business in today's world, the cost of complying with some IRS regulations is going to make you flinch, but must be borne. A good accountant is an essential, but make absolutely sure your accountant understands the necessities of the cabinetmaking business.

It is possible to get a job, though you probably won't make as much as you'll make in a good year, working out of your home, with a small cabinetry business. And you'll probably have much more fun and may well turn out better work, working from your own shop, for yourself.

If you do it right, have a little luck, and just generally pay attention and are willing to work very hard, then working out of a small shop will do it for you and will pay good dividends. Those dividends can be large. You are going to build a business based on your reputation, but you will end up with a business that has equipment, maybe buildings, a stable customer list, and lots of other desirable facets that can make a healthy retirement even happier when the business is sold.

Check it out. If your personality is right, you are energetic and involved and enjoy woodworking, and you think out what you're doing, you should do well, and provide a valuable product for many people.

Charles Self

Planning Kitchens

Obviously, the use of cabinetry in a kitchen is exceptionally important in the planning of that kitchen. Some buyers have special needs, and that's why a custom cabinetmaker is really in business, to meet those needs by building cabinets or adapting stock cabinets to fit. Naturally, appliance size and human size figure strongly into both overall kitchen design and cabinet design. Cabinets must be placed at the correct height and have the correct depth, and most of the size needs are spelled out fairly carefully in design books (and here shortly). Today, larger homes with larger appliances are the rule, rather than the exception, so it is often necessary to add to some of the basic sizes. Although cabinet depth is seldom more than 25 inches and never less than 12 inches, height can and must vary to suit the height of the users. If you're a drafter, today's computer-assisted design (CAD) programs can help, and we'll look at a couple of the simpler ones (and a couple that are even free, at least at the outset).

The first important things are the locations of plumbing setups, doors, and windows. Everything moves from there. It is a nuisance for the cabinetmaker to have to place a sink on one side of the room when plumbing is already installed on the other side, although it can be done by moving the plumbing, which is a job for a qualified

> The first imperatives are the locations of plumbing setups, doors, and windows. Everything moves from there.

Sienna cabinets from Amera by Merillat, here in cherry in the Garnet finish, feature a flat, monolithic door style with radiused corners and edges that is at home in contemporary settings and will complement retro, deco, and eclectic designs as well. *(Merillat)*

plumber. The choice of styles is a function of the desires of the client and the realities of the space available: You can't install an island in a tiny kitchen, and small kitchens don't do too well with the country look that includes a table and chairs. Otherwise, most cabinetry works with reasonable simplicity to fit into the size and overall design of the room, including ceiling and floor and wall treatments.

Start by taking careful overall measurements of the room, beginning with standard length and width measurements. I then like to take diagonal measurements of the room to get an idea of just how far out of square the room is—this is effective even in L-shaped rooms, if you take separate diagonal measurements for each rectangular room segment.

It's unlikely you'll actually need to build cabinets that are out of square to fit into a kitchen or bath. But there is a possibility that you may need some extrawide molding or filler strips at some point, and it's helpful to know ahead of time that you may need them, because it simplifies the ordering of materials for special jobs.

Frameless construction cabinets
Raised panel doors; arch doors for wall cabinets
Solid cherry doors, countertop edges, and molding
Hunter green laminate on counter, 4" solid back splash

Frameless cabinets. *(KCDw Cabinetmakers Software)*

The three basic patterns for kitchen layout—U, straight, and L—can be easily and often modified, but they still seem to work better than anything else as starting points. Equally important is the layout of work surfaces to allow easy steps from one station to another while preparing meals, stashing groceries, or cleaning up after meals. Remember, though, that these are basics, and in today's "anything goes" home-building atmosphere, you will almost certainly have to come up with additions of utility and style if you're going to be successful as a custom cabinetmaker.

Reception area
White laminate frameless cabinets
Formica Fog 961 Countertops
Blum 170° hinges

White laminate cabinets. *(KCDw Cabinetmakers Software)*

Designing for Use

Kitchen use varies a lot according to lifestyle. My wife, Frances, is a country cook, and as such she doesn't do what others consider gourmet cookery. But she does do a fair amount of baking of cakes, biscuits, and similar foods, so she needs space for laying out dough, kneading it, cutting it, letting cake pans cool, and letting bread rise. Also there is always the need to store larger quantities of flour and other staples than many kitchens require. Frances also likes to use

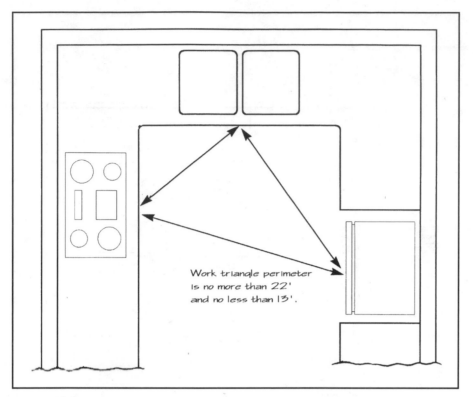

U-shaped kitchen.

Work triangle perimeter is no more than 22' and no less than 13'.

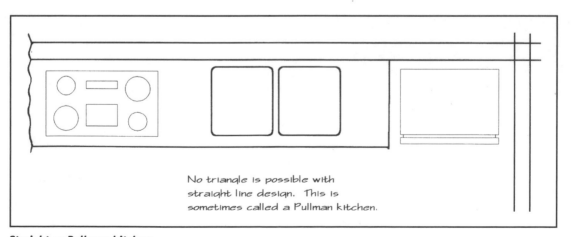

Straight or Pullman kitchen.

No triangle is possible with straight line design. This is sometimes called a Pullman kitchen.

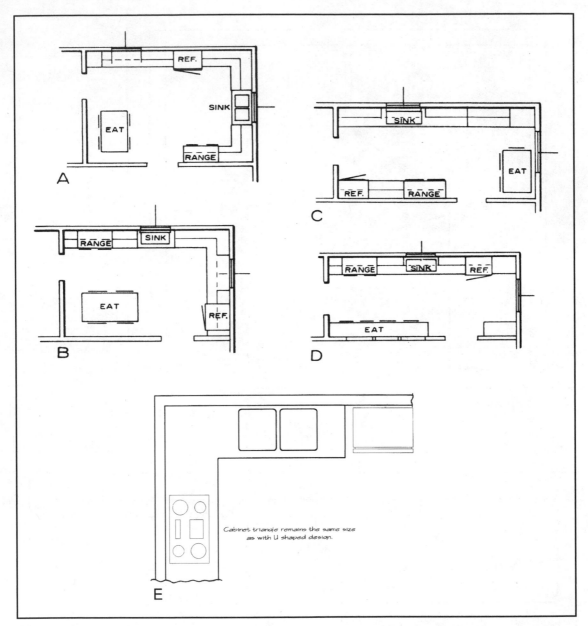

Kitchen layouts. (*A*) U-shaped; (*B*) L-shaped; (*C*) "parallel wall" type; (*D*) sidewall type; (*E*) L-shaped.

the telephone while she cooks, cleans up, and generally prepares food for the family. Another cook, one who doesn't bake or bakes infrequently, needs a different space allotment for those things, possibly with more preparation room for meats, less for vegetables, and perhaps different storage emphasis. That person may prefer

> If you're doing a kitchen for a new home, or a remodel on an older home up for sale, usually the kitchen has to be generic, designed to suit most lifestyles without emphasizing any one too much.

to use the telephone only while sitting. Except when family is present, we tend to eat informally, so quick and easy serving areas are essential. Others prefer formality for each meal and would be lost in an eat-in kitchen, preferring to always eat in the dining room or a specific dining alcove.

If you're designing a kitchen to fit a specific family's lifestyle, adapting and making needed changes are almost elemental. If you're doing a common kitchen for a new home, or are remodeling an older home up for sale, things are less interesting and probably a little less difficult, because the kitchen has to be generic, designed to suit lifestyles across the board.

It pays to remember that kitchens do a lot of jobs. Most families store foods, both packaged and unpackaged, in one area or another of the kitchen (here, I count the pantry as part of the kitchen, because it is essentially the kitchen "closet"). Families prepare those stored foods for cooking, or direct serving, in the kitchen. The prepared foods are then cooked, when needed, and readied for serving. In some kitchens, they are served and eaten in the same room. Once the food has been eaten, storage and cleanup come back into the picture, so leftovers are wrapped and properly stored and dirty dishes and utensils are washed and returned to their stored positions.

Waste disposal is a part of kitchen life, and it needs to be considered in the cabinetry. Some hiding spot for garbage and trash is a nice touch. Often today, the kitchen becomes at least the first-stage family recycling center, so room and holders for those items are essential.

In addition, the kitchen needs to have ready access to the dining room and fairly quick access to other rooms: Snacks in the living room (or recreation or family room) are a feature of life for many of us today, as is the cocktail party with drinks often made in the kitchen and served elsewhere in the house.

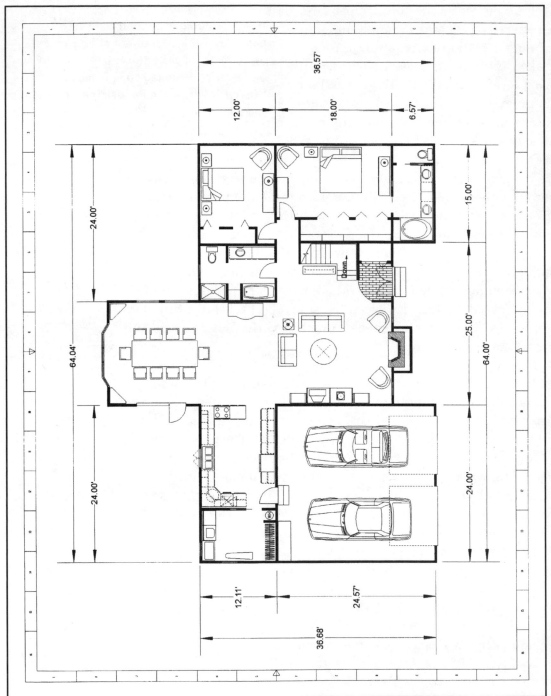

Note the relationship of the kitchen to the rest of the house, and the ease of getting trash to the garage or back door and of serving in the dining room. (*AutoDesk*)

Meeting these needs is not particularly difficult in today's cabinet-making world, but keeping these needs in mind, and speaking with your clients about them, is a great idea whose time really has come. See what clients would like emphasized, and do it.

Styles change, too. Today, the kitchen island is more popular than ever, and as this book is being written, stainless-steel cabinets are available from at least two companies. Brushed stainless surfaces go well with the many commercial-looking stainless appliances that are popular at the moment. I suggest you avoid this trendy, difficult-to-handle material. Currently, concrete countertops are in vogue, but their design and installation are almost a specialty itself, something most readers aren't going to want to fool with. In my experience, cabinetmakers are woodworkers, who sometimes handle plastics and stone for countertops and some facings, but becoming a sheet-metal worker is out of bounds for most of us.

Eat-In Kitchens

Grandma's kitchen was big enough to have a sizable table dropped down in its center, where many of the family's meals were taken. Some

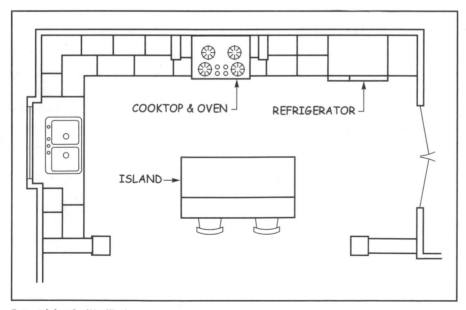

Eat-at island. *(Merillat)*

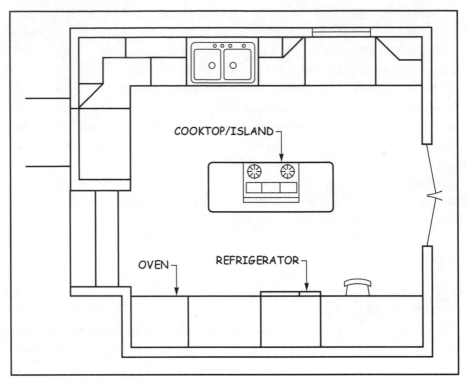

Cooktop island. *(Merillat)*

kitchens today are also large enough to encompass such a table. And similar tables, with the porcelainized metal tops, are available today, so that's a good solution for one eat-in design, when size and overall design permit. One of my favorite kitchens (built in the 1950s) was in an old farmhouse I rented years ago. The kitchen had a pantry larger than many kitchens, which looked strange with a single guy's food stock on the shelves, but the kitchen itself had a 12-foot-long peninsula that served for 90 percent of my meals. Cabinets were under a 6-inch overhang under the plywood and linoleum top (with stainless-steel nail-on edging). Shelves were located at the curved end, but still set under enough for knee room off stools. The peninsula held a microwave oven, a toaster oven, a coffee maker, a toaster, a set of canisters, some odds and ends, and enough space for four or five people to eat with no hassle and no setup. There was also room in that kitchen for a table large enough to seat six, and one corner had its own built-in corner china cabinet. It was a true marvel, with a huge, old cast-iron sinktop on a sheet-metal (white and rust) cabinet, with some wood cabinets to the sides and above, two huge windows, across-the-hall access to the dining room (which became my office), direct access to one living room (this old house had two living rooms, or, if you prefer, a drawing room and a living room), and, if memory serves, was at least 22 feet by 24 feet.

All country kitchens should aspire to a model similar to that, or so I feel.

Fortunately, or unfortunately, not all of us have the space or desire for a country kitchen. My present kitchen is about 11 feet by 11 feet, and it barely has space for two people to turn around, with all cabinets

Amera Manor legged island with square, raised-panel doors. Pigmented stains offer a colorful alternative to natural wood tones, while allowing the fine maple grain to show through. They are suitable for use in entire projects, such as shown here, or they may be used selectively to highlight one part of a room or to create the generations-old look of an unfitted kitchen. *(Merillat)*

A simple frame and panel styling combined with a dynamic use of space gives this country kitchen visual appeal. Numerous accessories add interest and functional storage. Used throughout the kitchen, spice drawer organizers are perfect for small items. The Dorchester cabinets are from Amera Fine cabinetry by Merillat. *(Merillat)*

and appliances in place. But it has direct access to the dining room on one side and, catty-cornered, direct access to the living room, a window over the sink, and other pleasing touches.

Safety: A Part of the Design

Safety needs to be designed into kitchens. Any place where people work with hot utensils, hot meats, hot oils, and similar hazardous items is going to become the site of accidents. Toss children into the mix—and that's sometimes how it seems on a busy day at home—and the chances increase quickly. Start with adequate lighting, although that's not the cabinetmaker's concern. Make sure there is a decent overall light and good task lighting at all working spaces. Obviously, task lighting needs to be placed under cabinets when they're installed or, as is popular now, in the ceiling with individual porthole lights, switched either in banks or individually. Task lights under cabinets are usually reserved for those wall cabinets raised more than 15 inches off the countertop, though not always. These are frequently fluorescent.

Safety needs to be designed into kitchens. Anyplace where people work with hot utensils, hot meats, hot oils, and similar hazardous items is going to become the site of accidents. Toss children into the mix—and that's sometimes how it seems on a busy day at home—and the chances go up further and faster.

The cooktop—or range—location is of great importance. Most kitchen accidents happen around the cooking area, so being able to prevent some problems at the outset is a big help. Start by locating the cooktop or range away from opening windows. Drafts can blow curtains onto the top when burners are lighted or, with a gas range or top, blow out burners. Keep the sides of the cooktop at least 1 foot away from any wall. This allows the cook to step away to either side in case of a pan fire.

Ranges or cooktops need countertops at least 1 foot long (some say 15 inches) to each side. Ranges and cooktops don't go on the ends of counter runs. Make the surface of one or both (preferred) of these countertops heat-resistant.

The overall work triangle size is important to both safety and efficiency in a kitchen where one person will be doing most of the preparation and cooking. Where chores such as cleaning vegetables, mixing batters, and so on are shared, the triangle may be larger. For a minimum, in the work triangle, the distance from the refrigerator to the sink to the stove and back to the refrigerator needs to be no less than 13 feet total (perimeter). For a maximum, most sources list 22 feet. The 13-foot minimum allows a careful person plenty of space to move foods from refrigerator to preparation center (sink) to the rangetop for cooking, at which point a return to the refrigerator enables one to get more food for preparation, and so on. The 22-foot size presumably helps keep down excess walking. Although the legs of the triangle will vary in size, hold to those total lengths to help in designing the efficient kitchen. For very large kitchens, placing a cooktop in an island, or on a peninsula, can reduce distance traveled, as can placing a sink there. Some modern kitchens even have two sinks, one in a long countertop run and the other placed in an island to reduce travel between food preparation steps. Of course, steps to clean up are also reduced, so overall effort is reduced.

Overall work triangle size is important to both safety and efficiency.

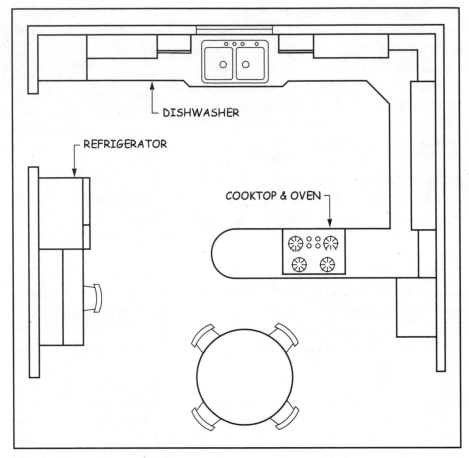

Space U-shaped layout. *(Merillat)*

Variations on the Themes

There is so much possible variation in the design of kitchens that we should never see two alike. But similar, yes. In fact, in one way or another, all kitchens are similar; but even if all kitchens were 16 feet by 16 feet, there'd be no reason to make them all alike. Some people prefer a galley-style kitchen, while others prefer a one-wall kitchen (the least efficient of all, because no work triangle is possible, but often a feature of small apartments), and others demand a U-shaped or L-shaped kitchen.

Galley kitchens are often small, and are featured in smaller apartments (and some larger ones: I lived for some time in a large, airy two-bedroom apartment in New York that had a galley-style kitchen that was barely large enough for me to sidle in sideways, but included every needed device, plus a dumbwaiter). Efficiency tends to be good, because of the compressed work triangle. In some galley kitchens, it's actually difficult to find enough space to get the 13-foot minimum. Often, galley kitchens are designed as pass-throughs and suffer from traffic while cooking is going on, although the ones I've seen didn't have this feature.

One-wall kitchens tend to be long, drawn-out affairs that are placed so traffic flows through, often interfering with work. Where possible, closing off one end of the kitchen can help, as may turning cabinet ends into the room. Those, then, create a U-shaped kitchen, one of the two most efficient floor layouts.

The L-shaped kitchen can be located in a corner of a larger dining-kitchen area and still have the efficient work triangle, because it has the two walls needed to allow the third triangle leg to form. Like the U-shaped kitchen, this one has corners, and that creates some extra work for the designer, because corner cabinets need more layout space (and you'll probably want to provide extras, such as lazy susans). As noted above, you need to make sure there is at least 1 foot on the corner side of the range or cooktop, so as to increase safety and reduce that crowded feeling for the cook.

Along about here, you need to be sure you know all the components in a basic kitchen area. Range, cooktop, oven, sink, dishwasher, microwave, and other work elements are obvious, but if you're going to build or install cabinetry, know that in stock catalogs from many companies there is a great variety of different types of cabinets available. Semicustom units from larger cabinetmakers give you even greater competition for fit. Also, hinges can be installed either way on almost all stock cabinetry sold today. Wall cabinets generally are available in 12- and 24-inch depths. The 24-inch depths are generally meant for use over refrigerators, and some custom and semicustom lines adapt sizes, but with longer waiting periods and higher cost, to any dimension desired between 6 and 24 inches.

Most companies will not work to the half inch, but only to full inch sizes, so you have an advantage here. This holds true for any stock cabinet line that goes to semicustom. Also, such changes are generally only available for specific lines and specific models of cabinet within

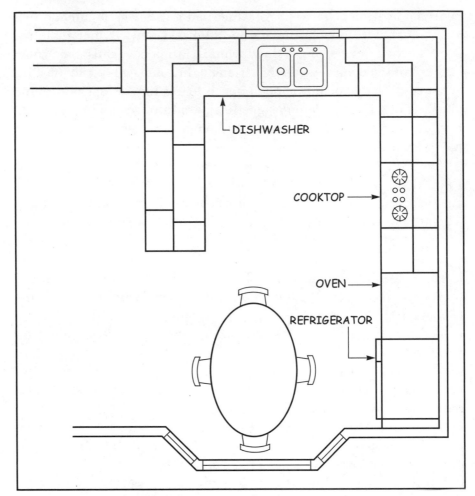

Kitchen components. *(Merillat)*

those lines. The small cabinetmaker can do it all, making the reductions to any real size wanted, or increasing depth, within practical limits, to any size wanted or needed. In general, stock companies offer wall cabinets 12 inches deep, 42 inches high, in single-door units 9, 12, 15, 18, and 21 inches wide (the actual size may vary slightly from company to company). Two-door cabinets without center mullions come in 24- and 27-inch widths, while two-door cabinets with center mullions are usually available in 30-, 33-, and 36-inch widths. Three-foot-high single-door cabinets come in the same range of widths, for

WALL CABINETS

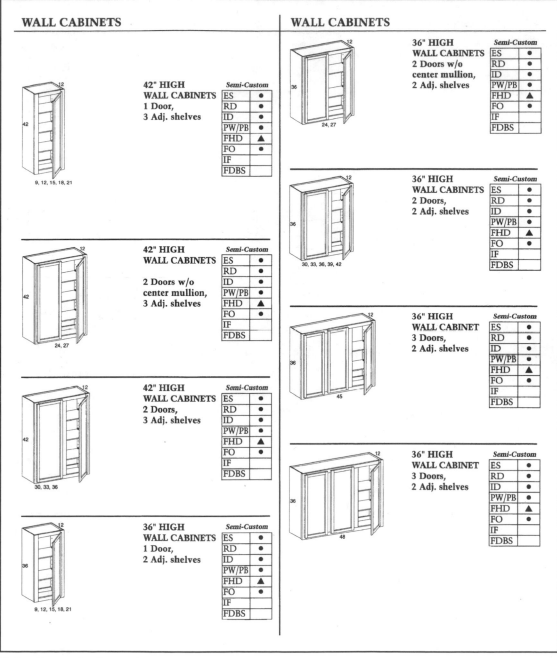

**42" HIGH
WALL CABINETS
1 Door,
3 Adj. shelves**

Semi-Custom	
ES	●
RD	●
ID	●
PW/PB	●
FHD	▲
FO	●
IF	
FDBS	

9, 12, 15, 18, 21

**42" HIGH
WALL CABINETS

2 Doors w/o
center mullion,
3 Adj. shelves**

Semi-Custom	
ES	●
RD	●
ID	●
PW/PB	●
FHD	▲
FO	●
IF	
FDBS	

24, 27

**42" HIGH
WALL CABINETS
2 Doors,
3 Adj. shelves**

Semi-Custom	
ES	●
RD	●
ID	●
PW/PB	●
FHD	▲
FO	●
IF	
FDBS	

30, 33, 36

**36" HIGH
WALL CABINETS
1 Door,
2 Adj. shelves**

Semi-Custom	
ES	●
RD	●
ID	●
PW/PB	●
FHD	▲
FO	●
IF	
FDBS	

9, 12, 15, 18, 21

WALL CABINETS

**36" HIGH
WALL CABINETS
2 Doors w/o
center mullion,
2 Adj. shelves**

Semi-Custom	
ES	●
RD	●
ID	●
PW/PB	●
FHD	▲
FO	●
IF	
FDBS	

24, 27

**36" HIGH
WALL CABINETS
2 Doors,
2 Adj. shelves**

Semi-Custom	
ES	●
RD	●
ID	●
PW/PB	●
FHD	▲
FO	●
IF	
FDBS	

30, 33, 36, 39, 42

**36" HIGH
WALL CABINET
3 Doors,
2 Adj. shelves**

Semi-Custom	
ES	●
RD	●
ID	●
PW/PB	●
FHD	▲
FO	●
IF	
FDBS	

45

**36" HIGH
WALL CABINET
3 Doors,
2 Adj. shelves**

Semi-Custom	
ES	●
RD	●
ID	●
PW/PB	●
FHD	▲
FO	●
IF	
FDBS	

48

Wall cabinets. *(Wellborn Cabinet, Inc.)*

WALL CABINETS

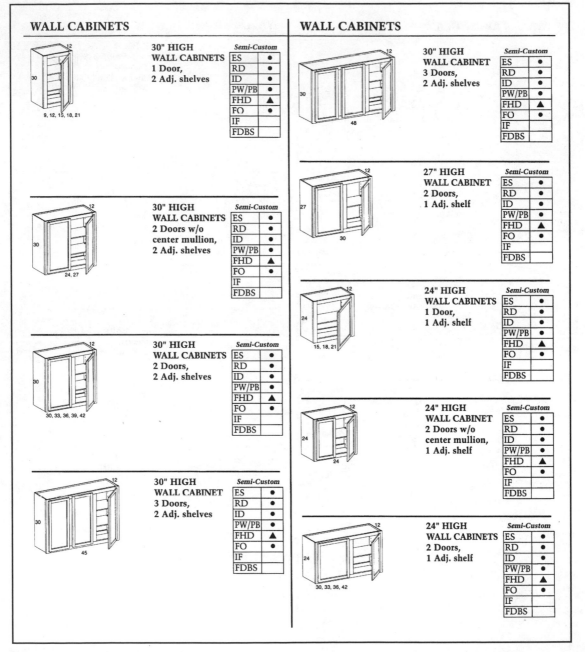

30" HIGH WALL CABINETS
1 Door,
2 Adj. shelves

9, 12, 15, 18, 21

Semi-Custom	
ES	●
RD	●
ID	●
PW/PB	●
FHD	▲
FO	●
IF	
FDBS	

30" HIGH WALL CABINETS
2 Doors w/o center mullion,
2 Adj. shelves

24, 27

Semi-Custom	
ES	●
RD	●
ID	●
PW/PB	●
FHD	▲
FO	●
IF	
FDBS	

30" HIGH WALL CABINETS
2 Doors,
2 Adj. shelves

30, 33, 36, 39, 42

Semi-Custom	
ES	●
RD	●
ID	●
PW/PB	●
FHD	▲
FO	●
IF	
FDBS	

30" HIGH WALL CABINET
3 Doors,
2 Adj. shelves

45

Semi-Custom	
ES	●
RD	●
ID	●
PW/PB	●
FHD	▲
FO	●
IF	
FDBS	

WALL CABINETS

30" HIGH WALL CABINET
3 Doors,
2 Adj. shelves

48

Semi-Custom	
ES	●
RD	●
ID	●
PW/PB	●
FHD	▲
FO	●
IF	
FDBS	

27" HIGH WALL CABINET
2 Doors,
1 Adj. shelf

30

Semi-Custom	
ES	●
RD	●
ID	●
PW/PB	●
FHD	▲
FO	●
IF	
FDBS	

24" HIGH WALL CABINETS
1 Door,
1 Adj. shelf

15, 18, 21

Semi-Custom	
ES	●
RD	●
ID	●
PW/PB	●
FHD	▲
FO	●
IF	
FDBS	

24" HIGH WALL CABINET
2 Doors w/o center mullion,
1 Adj. shelf

24

Semi-Custom	
ES	●
RD	●
ID	●
PW/PB	●
FHD	▲
FO	●
IF	
FDBS	

24" HIGH WALL CABINETS
2 Doors,
1 Adj. shelf

30, 33, 36, 42

Semi-Custom	
ES	●
RD	●
ID	●
PW/PB	●
FHD	▲
FO	●
IF	
FDBS	

Wall cabinets. *(Wellborn Cabinet, Inc.)*

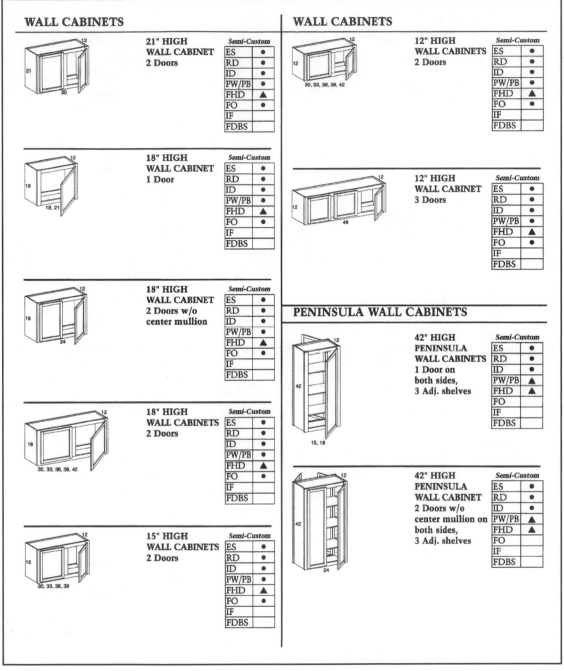

WALL CABINETS

21" HIGH WALL CABINET 2 Doors

Semi-Custom	
ES	●
RD	●
ID	●
PW/PB	●
FHD	▲
FO	●
IF	
FDBS	

18" HIGH WALL CABINET 1 Door

Semi-Custom	
ES	●
RD	●
ID	●
PW/PB	●
FHD	▲
FO	●
IF	
FDBS	

18" HIGH WALL CABINET 2 Doors w/o center mullion

Semi-Custom	
ES	●
RD	●
ID	●
PW/PB	●
FHD	▲
FO	●
IF	
FDBS	

18" HIGH WALL CABINETS 2 Doors

Semi-Custom	
ES	●
RD	●
ID	●
PW/PB	●
FHD	▲
FO	●
IF	
FDBS	

15" HIGH WALL CABINETS 2 Doors

Semi-Custom	
ES	●
RD	●
ID	●
PW/PB	●
FHD	▲
FO	●
IF	
FDBS	

WALL CABINETS

12" HIGH WALL CABINETS 2 Doors

Semi-Custom	
ES	●
RD	●
ID	●
PW/PB	●
FHD	▲
FO	●
IF	
FDBS	

12" HIGH WALL CABINET 3 Doors

Semi-Custom	
ES	●
RD	●
ID	●
PW/PB	●
FHD	▲
FO	●
IF	
FDBS	

PENINSULA WALL CABINETS

42" HIGH PENINSULA WALL CABINETS 1 Door on both sides, 3 Adj. shelves

Semi-Custom	
ES	●
RD	●
ID	●
PW/PB	▲
FHD	▲
FO	
IF	
FDBS	

42" HIGH PENINSULA WALL CABINET 2 Doors w/o center mullion on both sides, 3 Adj. shelves

Semi-Custom	
ES	●
RD	●
ID	●
PW/PB	▲
FHD	▲
FO	
IF	
FDBS	

Wall cabinets. *(Wellborn Cabinet, Inc.)*

PENINSULA WALL CABINETS

42" HIGH PENINSULA WALL CABINETS
2 Doors on both sides,
3 Adj. shelves

Semi-Custom	
ES	•
RD	•
ID	•
PW/PB	▲
FHD	▲
FO	
IF	
FDBS	

36" HIGH PENINSULA WALL CABINETS
1 Door on both sides,
2 Adj. shelves

Semi-Custom	
ES	•
RD	•
ID	•
PW/PB	▲
FHD	▲
FO	
IF	
FDBS	

36" HIGH PENINSULA WALL CABINET
2 Doors w/o center mullion on both sides,
2 Adj. shelves

Semi-Custom	
ES	•
RD	•
ID	•
PW/PB	▲
FHD	▲
FO	
IF	
FDBS	

36" HIGH PENINSULA WALL CABINETS
2 Doors on both sides,
2 Adj. shelves

Semi-Custom	
ES	•
RD	•
ID	•
PW/PB	▲
FHD	▲
FO	
IF	
FDBS	

PENINSULA WALL CABINETS

30" HIGH CORNER PENINSULA WALL CABINET
1 Door,
2 Adj. shelves

Semi-Custom	
ES	•
RD	•
ID	•
PW/PB	▲
FHD	▲
FO	
IF	
FDBS	

No door on back.

FRONT   BACK

30" HIGH PENINSULA WALL CABINETS
1 Door on both sides,
2 Adj. shelves

Semi-Custom	
ES	•
RD	•
ID	•
PW/PB	▲
FHD	▲
FO	
IF	
FDBS	

30" HIGH PENINSULA WALL CABINET
2 Doors w/o center mullion on both sides,
2 Adj. shelves

Semi-Custom	
ES	•
RD	•
ID	•
PW/PB	▲
FHD	▲
FO	
IF	
FDBS	

30" HIGH PENINSULA WALL CABINETS
2 Doors on both sides,
2 Adj. shelves

Semi-Custom	
ES	•
RD	•
ID	•
PW/PB	▲
FHD	▲
FO	
IF	
FDBS	

Wall cabinets. *(Wellborn Cabinet, Inc.)*

one and two doors, but extend out to three-door models in 45- and 48-inch widths. Shorter 30-inch-tall wall cabinets again come in 9-, 12-, 15-, 18-, and 21-inch widths in single-door; in 24- and 27-inch widths in double-door models without a center mullion; and in 30-, 33-, 36-, 39-, and 42-inch widths in two-door mullion models. Three-door models come in 45- and 48-inch heights as well. Normally, 27-inch-high cabinets come in only single-door, 15-, 18-, and 21-inch-wide models; while 24-inch-high cabinets come in 24-inch widths with double doors, then 30-, 33-, 36-, and 42-inch-wide mullion double doors. Many companies offer variations on the theme from that point, with possibly one or more 21-inch-tall, 12-inch-deep wall cabinets, and several 18-inch-high cabinets (most in two-door models), with widths of 18, 21, 30, 33, 36, 39, and 42-inches. The standard 15-inch-high cabinet comes with two doors, generally in 30-, 33-, 36-, and 39-inch widths, as does the 12-inch-high model, which adds a 42-inch width in many lists and sometimes offers a three-door, 48-inch-wide model.

For peninsula areas, you'll find stock cabinets in 42-inch heights and 12-inch depths with doors opening from both sides, in 15- and 18-inch-widths, for the single door, and 24-inch width for the double door, without a center mullion. Center-mullion two-door peninsula wall cabinets come in 30- and 36-inch widths, and the variations tend to hold true through 36- and 30-inch heights, and there is even an 18-inch height available from some companies.

Further wall cabinet styles are found in 24-inch-deep refrigerator cabinets: These are usually available in 27-, 24-, 21-, 18-, 15-, and 12-inch heights, in widths from 30 to 42 inches depending on height and company, and are two-door, mullion cabinets. This is a cabinet you probably won't have much occasion to make, and one you may want to spend some time talking your customer out of, should the customer request it. Placing a 24-inch-deep cabinet above normal reach for most people tends to make about 10 inches, and often more, of the back area of the cabinet unreachable without some form of step stool or at least a two-step ladder.

Finished interior cabinets are those with open shelves, or with glass doors of one kind or another, and the variety is exceptionally wide, including corner cabinets, with both angled and curved shelves, and with and without doors. It is possible to build a reasonably sophisticated

REFRIGERATOR CABINETS

24" DEEP REFRIGERATOR CABINETS 2 Doors

	Semi-Custom
ES	•
RD	•
ID	
PW/PB	
FHD	▲
FO	•
IF	
FDBS	

33, 36, 39, 42

FINISHED INTERIOR CABINETS

36" HIGH FINISHED INTERIOR CABINETS No door, 2 Adj. matching shelves

	Semi-Custom
ES	•
RD	•
ID	•
PW/PB	•
FHD	
FO	•
IF	
FDBS	

9, 12, 15, 18, 21

FINISHED INTERIOR CABINETS

42" HIGH FINISHED INTERIOR CABINETS No door, 3 Adj. matching shelves

	Semi-Custom
ES	•
RD	•
ID	•
PW/PB	
FHD	
FO	•
IF	
FDBS	

9, 12, 15, 18, 21

36" HIGH FINISHED INTERIOR CABINETS No door, 2 Adj. matching shelves

	Semi-Custom
ES	•
RD	•
ID	•
PW/PB	•
FHD	
FO	•
IF	
FDBS	

24, 27

42" HIGH FINISHED INTERIOR CABINETS No door, 3 Adj. matching shelves

	Semi-Custom
ES	•
RD	•
ID	•
PW/PB	
FHD	
FO	•
IF	
FDBS	

24, 27

36" HIGH FINISHED INTERIOR CABINETS No doors, 2 Adj. matching shelves

	Semi-Custom
ES	•
RD	•
ID	•
PW/PB	•
FHD	
FO	•
IF	
FDBS	

30, 33, 36, 39, 42

42" HIGH FINISHED INTERIOR CABINETS No doors, 3 Adj. matching shelves

	Semi-Custom
ES	•
RD	•
ID	•
PW/PB	•
FHD	
FO	•
IF	
FDBS	

30, 33, 36

36" HIGH FINISHED INTERIOR CABINET No doors, 2 Adj. matching shelves

	Semi-Custom
ES	•
RD	•
ID	•
PW/PB	•
FHD	
FO	•
IF	
FDBS	

45

Finished interior cabinets. *(Wellborn Cabinet, Inc.)*

PENINSULA WALL FINISHED INTERIOR CABINETS

30" HIGH PENINSULA WALL FINISHED INTERIOR CABINET
No door,
2 Adj. matching shelves

	Semi-Custom	
ES	●	
RD	●	
ID	●	
PW/PB	▲	
FHD		
FO		
IF		
FDBS		

30" HIGH PENINSULA WALL FINISHED INTERIOR CABINETS
No doors,
2 Adj. matching shelves

	Semi-Custom	
ES	●	
RD	●	
ID	●	
PW/PB	▲	
FHD		
FO		
IF		
FDBS		

24" HIGH PENINSULA WALL FINISHED INTERIOR CABINETS
No doors,
1 Adj. matching shelf

	Semi-Custom	
ES	●	
RD	●	
ID	●	
PW/PB	▲	
FHD		
FO		
IF		
FDBS		

DIAGONAL CORNER FINISHED INTERIOR CABINETS

42" HIGH DIAGONAL CORNER FINISHED INTERIOR CABINET
No door,
3 Adj. matching shelves

	Semi-Custom	
ES		
RD		
ID		
PW/PB		
FHD		
FO	●	
IF		
FDBS		

DIAGONAL CORNER FINISHED INTERIOR CABINETS

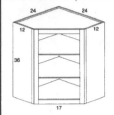

36" HIGH DIAGONAL CORNER FINISHED INTERIOR CABINET
No door,
2 Adj. matching shelves

	Semi-Custom	
ES		
RD		
ID		
PW/PB		
FHD		
FO	●	
IF		
FDBS		

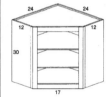

30" HIGH DIAGONAL CORNER FINISHED INTERIOR CABINET
No door,
2 Adj. matching shelves

	Semi-Custom	
ES		
RD		
ID		
PW/PB		
FHD		
FO	●	
IF		
FDBS		

INSIDE CURVED CORNER WALL FINISHED INTERIOR CABINETS

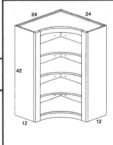

42" HIGH INSIDE CURVED CORNER WALL FINISHED INTERIOR CABINET
No door,
3 Adj. matching shelves

Peninsula finished interior cabinets. *(Wellborn Cabinet, Inc.)*

FINISHED INTERIOR MULLION DOOR CABINETS

30" HIGH FINISHED INTERIOR MULLION DOOR CABINET
3 Mullion doors,
2 Adj. matching shelves

	Semi-Custom
ES	●
RD	●
ID	●
PW/PB	●
FHD	
FO	●
IF	
FDBS	

30" HIGH FINISHED INTERIOR MULLION DOOR CABINET
3 Mullion doors,
2 Adj. matching shelves

	Semi-Custom
ES	●
RD	●
ID	●
PW/PB	●
FHD	
FO	●
IF	
FDBS	

24" HIGH FINISHED INTERIOR MULLION DOOR CABINET
2 Mullion doors
w/o center mullion,
1 Adj. matching shelf

	Semi-Custom
ES	●
RD	●
ID	●
PW/PB	●
FHD	
FO	●
IF	
FDBS	

24" HIGH FINISHED INTERIOR MULLION DOOR CABINETS
2 Mullion doors,
1 Adj. matching shelf

	Semi-Custom
ES	●
RD	●
ID	●
PW/PB	●
FHD	
FO	●
IF	
FDBS	

PENINSULA WALL FINISHED INTERIOR MULLION DOOR CABINETS

30" HIGH PENINSULA WALL FINISHED INTERIOR MULLION DOOR CABINET
4 Mullion doors
w/o center mullion,
2 Adj. matching shelves

	Semi-Custom
ES	●
RD	●
ID	●
PW/PB	▲
FHD	
FO	
IF	
FDBS	

30" HIGH PENINSULA WALL FINISHED INTERIOR MULLION DOOR CABINETS
4 Mullion doors,
2 Adj. matching shelves

	Semi-Custom
ES	●
RD	●
ID	●
PW/PB	▲
FHD	
FO	
IF	
FDBS	

24" HIGH PENINSULA WALL FINISHED INTERIOR MULLION DOOR CABINETS
4 Mullion doors,
1 Adj. matching shelf

	Semi-Custom
ES	●
RD	●
ID	●
PW/PB	▲
FHD	
FO	
IF	
FDBS	

DIAGONAL CORNER FINISHED INTERIOR MULLION DOOR CABINETS

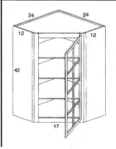

42" HIGH DIAGONAL CORNER FINISHED INTERIOR MULLION DOOR CABINET
1 Mullion door,
3 Adj. matching shelves

	Semi-Custom
ES	
RD	
ID	
PW/PB	
FHD	
FO	●
IF	
FDBS	

Finished interior mullion door cabinets. *(Wellborn Cabinet, Inc.)*

DIAGONAL CORNER FINISHED INTERIOR MULLION DOOR CABINETS

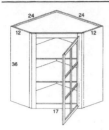

36" HIGH DIAGONAL CORNER FINISHED INTERIOR MULLION DOOR CABINET
1 Mullion door, 2 Adj. matching shelves

Semi-Custom	
ES	
RD	
ID	
PW/PB	
FHD	
FO	●
IF	
FDBS	

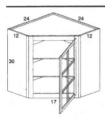

30" HIGH DIAGONAL CORNER FINISHED INTERIOR MULLION DOOR CABINET
1 Mullion door, 2 Adj. matching shelves

Semi-Custom	
ES	
RD	
ID	
PW/PB	
FHD	
FO	●
IF	
FDBS	

ANGLE WALL FINISHED INTERIOR MULLION DOOR CABINETS

42" HIGH ANGLE WALL FINISHED INTERIOR MULLION DOOR CABINET
1 Mullion door, 3 Adj. matching shelves

Semi-Custom	
ES	
RD	
ID	
PW/PB	
FHD	
FO	●
IF	
FDBS	

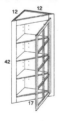

36" HIGH ANGLE WALL FINISHED INTERIOR MULLION DOOR CABINET
1 Mullion door, 2 Adj. matching shelves

Semi-Custom	
ES	
RD	
ID	
PW/PB	
FHD	
FO	●
IF	
FDBS	

ANGLE WALL FINISHED INTERIOR MULLION DOOR CABINETS

30" HIGH ANGLE WALL FINISHED INTERIOR MULLION DOOR CABINET
1 Mullion door, 2 Adj. matching shelves

Semi-Custom	
ES	
RD	
ID	
PW/PB	
FHD	
FO	●
IF	
FDBS	

CORNER WALL CABINETS

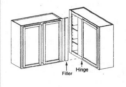

42" HIGH CORNER WALL CABINETS
1 Door, 3 Adj. shelves

Semi-Custom	
ES	●
RD	●
ID	●
PW/PB	●
FHD	▲
FO	●
IF	
FDBS	

Diagonal corner wall cabinets. *(Wellborn Cabinet, Inc.)*

corner cabinet—built-in—of stock components with these units. And, of course, any sort of dining room cabinetry may be built of similar straight-line and corner units, too, if your customer doesn't want to use such cabinetry in the kitchen.

Standard nonangled corner wall units allow a run of cabinets to butt against an open end (minus a door or filler) placed against the adjacent wall. These will be matching units for a wide variety of any company's styles and sizes, and they will come in the needed heights, with widths varying according to styles. Easy-reach corner cabinets are characterized by a flow of shelving, usually in a single piece of shelving per shelf, around the 90° corner, but with the back of the corner clipped off and the front rounded, so that it is easy to reach all the way into the unit. A single door may swing out to one side, with two-door units solidly fixed to each other at 90°; or two doors, for wider, easy-reach cabinets, may swing the same way and may not be solidly fixed (accordion style). Of course, corner cabinets may be bought open, and rotating shelves added, or may be bought with rotating shelves. These come in varying sizes with usually two or three round shelves on the rotating pole. Similar shelves may be placed in corners with the bottom designed as an appliance garage, using a tambour door. Such units are higher than most because they're meant to extend down to the countertop. So generally they're available in 60-, 54-, and 48-inch shelf heights, with corner units designed to take 24 inches of wall space on each side. Tambour cabinets for flat walls are also available in similar heights, without the lazy susan rotating shelves.

From this point, we continue to specialty cabinets, including microwave cabinets. These also tend to have added height, many being 48 inches tall, in widths of 27 and 30 inches. Bottom shelves are deeper than the basic 12-inch depth of these cabinets—the shelves are usually 17 inches or more deep—so the kitchen TV set might well fit on such a cabinet, too.

Cabinet companies today offer a wide variety of open-display cabinets, both in corner and straight wall styles, with different kinds of trim and different intended users. Those with a diagonally inset inside grid are useful only as racks for wine bottles, while plate racks tend to have vertical dowels. Bookcases are available in a wide range of styles from most cabinet companies, and they are a good sideline for custom cabinetmakers. People tend to moan about the prices of top-quality

DIAGONAL CORNER WALL CABINETS

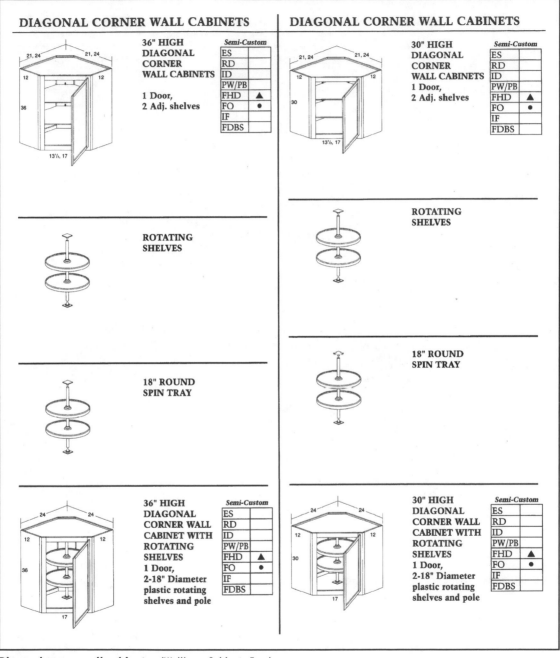

36" HIGH DIAGONAL CORNER WALL CABINETS

1 Door,
2 Adj. shelves

Semi-Custom	
ES	
RD	
ID	
PW/PB	
FHD	▲
FO	●
IF	
FDBS	

ROTATING SHELVES

18" ROUND SPIN TRAY

36" HIGH DIAGONAL CORNER WALL CABINET WITH ROTATING SHELVES

1 Door,
2-18" Diameter plastic rotating shelves and pole

Semi-Custom	
ES	
RD	
ID	
PW/PB	
FHD	▲
FO	●
IF	
FDBS	

DIAGONAL CORNER WALL CABINETS

30" HIGH DIAGONAL CORNER WALL CABINETS

1 Door,
2 Adj. shelves

Semi-Custom	
ES	
RD	
ID	
PW/PB	
FHD	▲
FO	●
IF	
FDBS	

ROTATING SHELVES

18" ROUND SPIN TRAY

30" HIGH DIAGONAL CORNER WALL CABINET WITH ROTATING SHELVES

1 Door,
2-18" Diameter plastic rotating shelves and pole

Semi-Custom	
ES	
RD	
ID	
PW/PB	
FHD	▲
FO	●
IF	
FDBS	

Diagonal corner wall cabinets. *(Wellborn Cabinet, Inc.)*

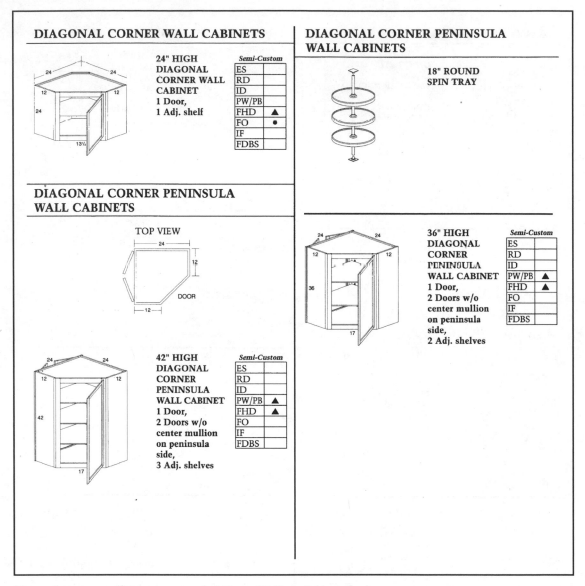

DIAGONAL CORNER WALL CABINETS

24" HIGH DIAGONAL CORNER WALL CABINET
1 Door,
1 Adj. shelf

Semi-Custom	
ES	
RD	
ID	
PW/PB	
FHD	▲
FO	●
IF	
FDBS	

DIAGONAL CORNER PENINSULA WALL CABINETS

TOP VIEW

DOOR

42" HIGH DIAGONAL CORNER PENINSULA WALL CABINET
1 Door,
2 Doors w/o center mullion on peninsula side,
3 Adj. shelves

Semi-Custom	
ES	
RD	
ID	
PW/PB	▲
FHD	▲
FO	
IF	
FDBS	

DIAGONAL CORNER PENINSULA WALL CABINETS

18" ROUND SPIN TRAY

36" HIGH DIAGONAL CORNER PENINSULA WALL CABINET
1 Door,
2 Doors w/o center mullion on peninsula side,
2 Adj. shelves

Semi-Custom	
ES	
RD	
ID	
PW/PB	▲
FHD	▲
FO	
IF	
FDBS	

Diagonal corner appliance garage and shelves. *(Wellborn Cabinet, Inc.)*

bookcases, but when you offer some special molding features, special sizes, greater range of adjustments, different kinds of adjustments, and possible other custom variations, they do tend to pay the higher prices. Most people don't really enjoy the stapled-together bookcases of pressed wood available at chain stores. Solid wood, or top-grade

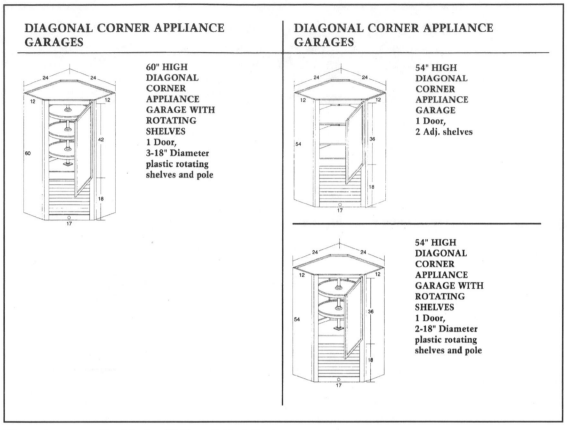

DIAGONAL CORNER APPLIANCE GARAGES

60" HIGH DIAGONAL CORNER APPLIANCE GARAGE WITH ROTATING SHELVES
1 Door,
3-18" Diameter plastic rotating shelves and pole

DIAGONAL CORNER APPLIANCE GARAGES

54" HIGH DIAGONAL CORNER APPLIANCE GARAGE
1 Door,
2 Adj. shelves

54" HIGH DIAGONAL CORNER APPLIANCE GARAGE WITH ROTATING SHELVES
1 Door,
2-18" Diameter plastic rotating shelves and pole

Diagonal corner wall cabinets. *(Wellborn Cabinet, Inc.)*

veneer plywood, with care taken in joinery and design, sells almost any good wood product these days, as it always has.

Spice drawers and small specialty wall cabinets abound, as do whatnot shelves that can almost redesign a kitchen when fitted on the end of a cabinet run. There are many kinds of hood accessories of wood, as well as hanging pot racks and light boxes and single- and double-shelf units, cassette drawer units, condiment shelves, organizer units, mug holders, stemmed-glass holders, mixer garages, tambour storage units (to form appliance garages), and valances. And we're still in wall-hung cabinetry. You can accessorize wall cabinets with leaded-glass doors, stained-glass doors, all in many sizes and patterns and, of course, shapes to suit those cabinet styles that use crowned doors as well as flat panel doors.

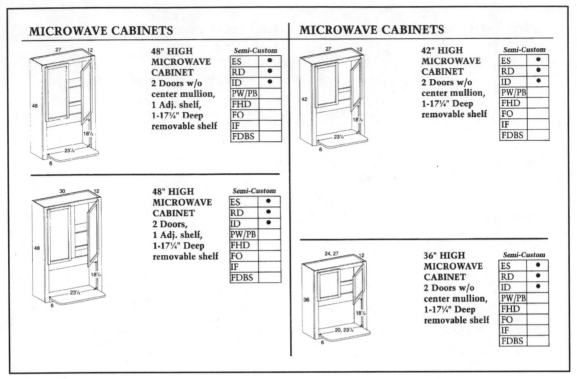

MICROWAVE CABINETS

48" HIGH MICROWAVE CABINET
2 Doors w/o center mullion,
1 Adj. shelf,
1-17¼" Deep removable shelf

	Semi-Custom
ES	●
RD	●
ID	●
PW/PB	
FHD	
FO	
IF	
FDBS	

48" HIGH MICROWAVE CABINET
2 Doors,
1 Adj. shelf,
1-17¼" Deep removable shelf

	Semi-Custom
ES	●
RD	●
ID	●
PW/PB	
FHD	
FO	
IF	
FDBS	

MICROWAVE CABINETS

42" HIGH MICROWAVE CABINET
2 Doors w/o center mullion,
1-17¼" Deep removable shelf

	Semi-Custom
ES	●
RD	●
ID	●
PW/PB	
FHD	
FO	
IF	
FDBS	

36" HIGH MICROWAVE CABINET
2 Doors w/o center mullion,
1-17¼" Deep removable shelf

	Semi-Custom
ES	●
RD	●
ID	●
PW/PB	
FHD	
FO	
IF	
FDBS	

Microwave cabinets. *(Wellborn Cabinet, Inc.)*

Base Cabinet Styles

There aren't quite as many base cabinet styles available for kitchens, for which we can breathe a slight sigh of relief, but there are plenty. The lack of variety comes in height: Standard base cabinets are 34½ inches, and variants, usually to 32½ inches, are considered "accessible" cabinetry for use by those in wheelchairs. There are companies with slightly different accessible heights, but few stock houses offer kitchen cabinets for the taller people who are now everywhere around us. Growing up, I felt tall. Today, I feel not much more than average height many days, as the youngsters around me surpass my 6 feet 2 inches by several inches with some frequency. This shows up when people are working in kitchens, and 34½ inches simply may not be tall enough for such people. At this point, the custom cabinetmaker is coming into his or her own.

(*Text continues on p. 37.*)

MICROWAVE CABINETS

36" HIGH MICROWAVE CABINET 2 Doors, 1-17¼" Deep removable shelf	Semi-Custom	
	ES	●
	RD	●
	ID	●
	PW/PB	
	FHD	
	FO	
	IF	
	FDBS	

RETURN ANGLE WALL CABINETS

BASE/WALL RETURN ANGLE CABINET 2 Doors, 2 Adj. shelves	Semi-Custom	
	ES	
	RD	
	ID	
	PW/PB	
	FHD	▲
	FO	●
	IF	
	FDBS	▲

Door Sizes
½ Overlay: 15½ x 27½
Full Overlay: 17⅝ x 29⅝

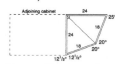

Degrees shown on diagrams are for use when cutting molding.

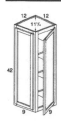

42" HIGH RETURN ANGLE WALL CABINET 2 Doors, 3 Adj. shelves	Semi-Custom	
	ES	
	RD	
	ID	
	PW/PB	
	FHD	▲
	FO	●
	IF	
	FDBS	

RETURN ANGLE WALL CABINETS

36" HIGH RETURN ANGLE WALL CABINET 2 Doors, 2 Adj. shelves	Semi-Custom	
	ES	
	RD	
	ID	
	PW/PB	
	FHD	▲
	FO	●
	IF	
	FDBS	

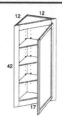

30" HIGH RETURN ANGLE WALL CABINET 2 Doors, 2 Adj. shelves	Semi-Custom	
	ES	
	RD	
	ID	
	PW/PB	
	FHD	▲
	FO	●
	IF	
	FDBS	

ANGLE WALL CABINETS

42" HIGH ANGLE WALL CABINET 1 Door, 3 Adj. shelves	Semi-Custom	
	ES	
	RD	
	ID	
	PW/PB	
	FHD	▲
	FO	●
	IF	
	FDBS	

36" HIGH ANGLE WALL CABINET 1 Door, 2 Adj. shelves	Semi-Custom	
	ES	
	RD	
	ID	
	PW/PB	
	FHD	▲
	FO	●
	IF	
	FDBS	

Angle wall cabinets. *(Wellborn Cabinet, Inc.)*

ANGLE WALL CABINETS

30" HIGH
ANGLE WALL
CABINET
1 Door,
2 Adj. shelves

	Semi-Custom
ES	
RD	
ID	
PW/PB	
FHD	▲
FO	●
IF	
FDBS	

CURVED CORNER WALL CABINETS

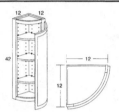

42" HIGH
CURVED
CORNER WALL
CABINET
1 Door,
3 Adj. shelves

	Semi-Custom
ES	
RD	
ID	
PW/PB	
FHD	▲
FO	
IF	
FDBS	

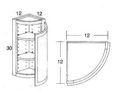

30" HIGH
CURVED
CORNER WALL
CABINET
1 Door,
2 Adj. shelves

	Semi-Custom
ES	
RD	
ID	
PW/PB	
FHD	▲
FO	
IF	
FDBS	

RETURN ANGLE WALL CABINETS

Degrees shown on diagrams are for use when cutting molding.

RETURN ANGLE WALL CABINETS

42" HIGH
RETURN
ANGLE WALL
CABINET
2 Doors,
3 Adj. shelves

	Semi-Custom
ES	●
RD	
ID	
PW/PB	
FHD	▲
FO	●
IF	
FDBS	

36" HIGH
RETURN
ANGLE WALL
CABINET
2 Doors,
2 Adj. shelves

	Semi-Custom
ES	●
RD	
ID	
PW/PB	
FHD	▲
FO	●
IF	
FDBS	

30" HIGH
RETURN
ANGLE WALL
CABINET
2 Doors,
2 Adj. shelves

	Semi-Custom
ES	●
RD	
ID	
PW/PB	
FHD	▲
FO	●
IF	
FDBS	

WALL ROTATING SHELF UNIT CABINET

WALL
ROTATING
SHELF UNIT
CABINET
1 Door,
1 Rotating shelf
unit

	Semi-Custom
ES	●
RD	
ID	
PW/PB	●
FHD	▲
FO	●
IF	
FDBS	

Curved corner wall cabinets. *(Wellborn Cabinet, Inc.)*

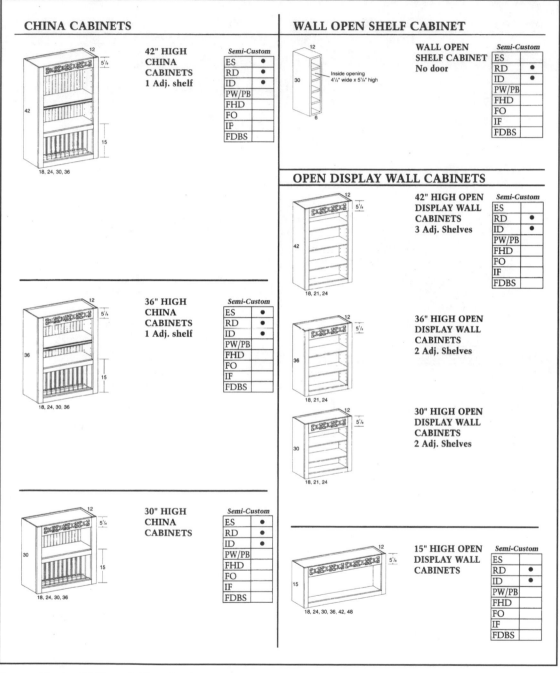

CHINA CABINETS

42" HIGH CHINA CABINETS 1 Adj. shelf

Semi-Custom	
ES	●
RD	●
ID	●
PW/PB	
FHD	
FO	
IP	
FDBS	

18, 24, 30, 36

36" HIGH CHINA CABINETS 1 Adj. shelf

Semi-Custom	
ES	●
RD	●
ID	●
PW/PB	
FHD	
FO	
IF	
FDBS	

18, 24, 30, 36

30" HIGH CHINA CABINETS

Semi-Custom	
ES	●
RD	●
ID	●
PW/PB	
FHD	
FO	
IF	
FDBS	

18, 24, 30, 36

WALL OPEN SHELF CABINET

Inside opening 4¼" wide x 5¼" high

WALL OPEN SHELF CABINET No door

Semi-Custom	
ES	
RD	●
ID	●
PW/PB	
FHD	
FO	
IF	
FDBS	

OPEN DISPLAY WALL CABINETS

42" HIGH OPEN DISPLAY WALL CABINETS 3 Adj. Shelves

Semi-Custom	
ES	
RD	●
ID	●
PW/PB	
FHD	
FO	
IF	
FDBS	

18, 21, 24

36" HIGH OPEN DISPLAY WALL CABINETS 2 Adj. Shelves

18, 21, 24

30" HIGH OPEN DISPLAY WALL CABINETS 2 Adj. Shelves

18, 21, 24

15" HIGH OPEN DISPLAY WALL CABINETS

Semi-Custom	
ES	
RD	●
ID	●
PW/PB	
FHD	
FO	
IF	
FDBS	

18, 24, 30, 36, 42, 48

Open cabinets. *(Wellborn Cabinet, Inc.)*

OPEN FRONT DISPLAY CORNER WALL CABINETS

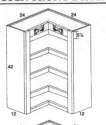

42" HIGH OPEN FRONT DISPLAY CORNER WALL CABINET
3 Adj. shelves

	Semi-Custom
ES	
RD	●
ID	●
PW/PB	
FHD	
FO	
IF	
FDBS	

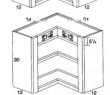

36" HIGH OPEN FRONT DISPLAY CORNER WALL CABINET
2 Adj. shelves

30" HIGH OPEN FRONT DISPLAY CORNER WALL CABINET
2 Adj. shelves

ARCHED FRET OPEN DISPLAY CORNER WALL CABINETS

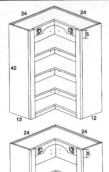

42" HIGH ARCHED FRET OPEN DISPLAY CORNER WALL CABINET
3 Adj. shelves

	Semi-Custom
ES	
RD	●
ID	
PW/PB	
FHD	
FO	
IF	
FDBS	

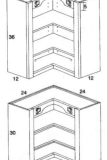

36" HIGH ARCHED FRET OPEN DISPLAY CORNER WALL CABINET
2 Adj. shelves

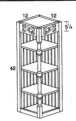

30" HIGH ARCHED FRET OPEN DISPLAY CORNER WALL CABINET
2 Adj. shelves

ARCHED FRET OPEN DISPLAY WALL CABINETS

42" HIGH ARCHED FRET OPEN DISPLAY WALL CABINETS
3 Adj. shelves

	Semi-Custom
ES	
RD	●
ID	●
PW/PB	
FHD	
FO	
IF	
FDBS	

36" HIGH ARCHED FRET OPEN DISPLAY WALL CABINETS
2 Adj. shelves

30" HIGH ARCHED FRET OPEN DISPLAY WALL CABINETS
2 Adj. shelves

OPEN FRONT DISPLAY WALL CABINETS

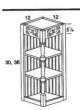

OPEN FRONT DISPLAY WALL CABINET
3 Fixed shelves

OPEN FRONT DISPLAY WALL CABINETS
2 Fixed Shelves

Open cabinets. *(Wellborn Cabinet, Inc.)*

WALL ACCESSORIES

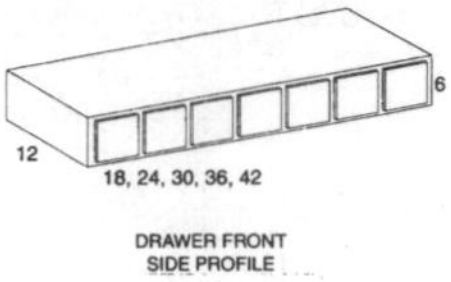

DRAWER UNIT

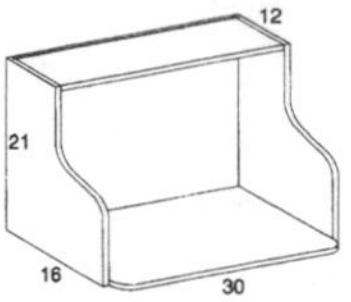

12

18, 24, 30, 36, 42

6

DRAWER FRONT
SIDE PROFILE

**APPLIANCE
CABINET**

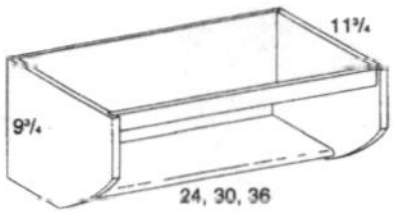

21

16   30

12

**CONTEMPORARY
SHELVES**

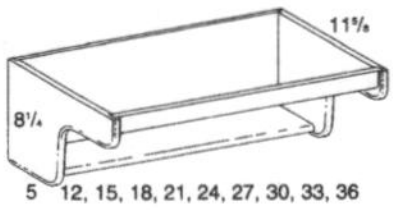

$9^{3}/_{4}$

$11^{3}/_{4}$

24, 30, 36

**CONTEMPORARY
CONDIMENT
SHELVES**

$8^{1}/_{4}$

$11^{3}/_{4}$

5   12, 15, 18, 21, 24, 27, 30, 33, 36

WALL ACCESSORIES

21, 24   21, 24

$8^{1}/_{4}$ $11^{3}/_{4}$

5

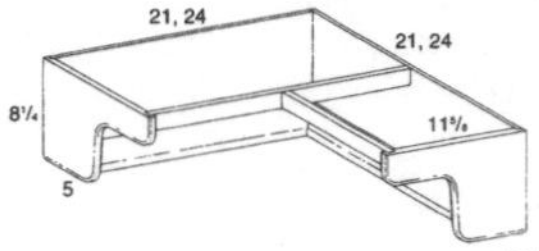

**CORNER
CONTEMPORARY
CONDIMENT
SHELVES**

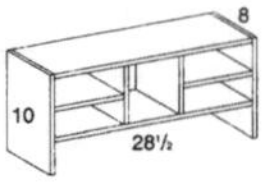

8

10    $28^{1}/_{2}$

**ORGANIZER
UNIT**

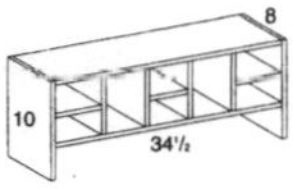

8

10    $34^{1}/_{2}$

**ORGANIZER
UNIT**

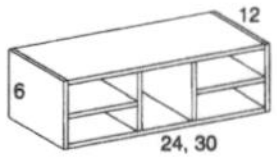

12

6   24, 30

**WALL
ORGANIZER
UNITS**

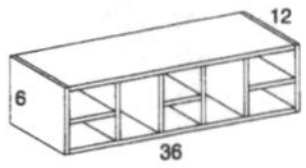

12

6   36

**WALL
ORGANIZER
UNIT**

Accessories. *(Wellborn Cabinet, Inc.)*

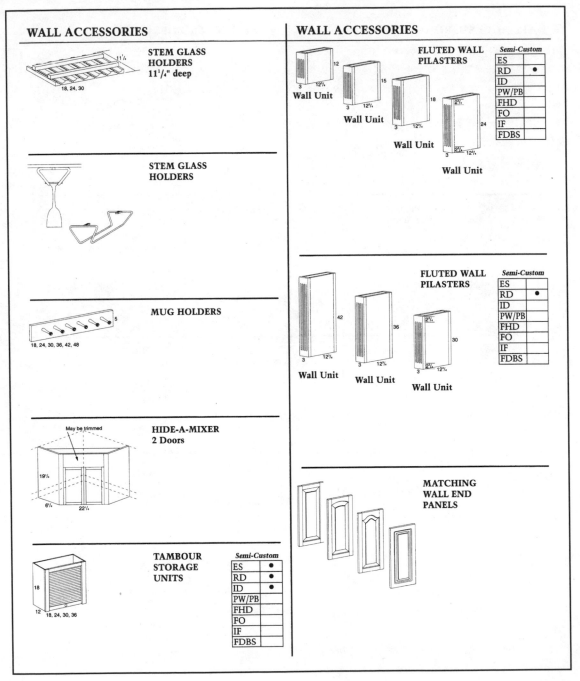

WALL ACCESSORIES

STEM GLASS HOLDERS 11¹/₄" deep

18, 24, 30

STEM GLASS HOLDERS

MUG HOLDERS

5

18, 24, 30, 36, 42, 48

HIDE-A-MIXER 2 Doors

May be trimmed

19¹/₄

6¹/₄   22¹/₄

TAMBOUR STORAGE UNITS

18

12   18, 24, 30, 36

	Semi-Custom
ES	●
RD	●
ID	●
PW/PB	
FHD	
FO	
IF	
FDBS	

WALL ACCESSORIES

FLUTED WALL PILASTERS

Wall Unit   12   3   12¹/₄

Wall Unit   15   3   12¹/₄

Wall Unit   18   3   12¹/₄

Wall Unit   24   2³/₄   3   12¹/₄   2³/₄   3

	Semi-Custom
ES	
RD	●
ID	
PW/PB	
FHD	
FO	
IF	
FDBS	

FLUTED WALL PILASTERS

Wall Unit   42   3   12¹/₄

Wall Unit   36   3   12¹/₄

Wall Unit   30   2³/₄   3   12¹/₄

	Semi-Custom
ES	
RD	●
ID	
PW/PB	
FHD	
FO	
IF	
FDBS	

MATCHING WALL END PANELS

Accessories. *(Wellborn Cabinet, Inc.)*

Also, there are still a good number of people who extend less distance up from the ground (and who pay less for clothing, and have an easier time finding styles and sizes, and often, if not usually, live longer). These people are forced to almost stand on tiptoe to use 34^{1}/$_{2}$-inch base cabinets, or to stand on a work step stool. Again, the custom cabinetmaker has a distinct advantage in remodels and custom houses that are made for both above- and below-average-height couples and families. Of course, in the average family, there is some compromise: my wife is 5 feet 5 inches, so the kitchen is sized for her.

For base cabinets at 34^{1}/$_{2}$-inch height, widths for single-door units start at 9 inches, and styles include a swinging door or a pull-out spice rack. Single-drawer, single-door units then go from 12 to 15 to 18 to 21 inches wide. Double-door, single-drawer units can usually be found 24 and 27 inches wide, while two-drawer two-door units come in widths of 30, 33, 36, 39, and 42 inches. Three-door, three-drawer units come in 45- and 48-inch widths, and next come those units with a shelf, but no drawer, in the same height ranges. Some companies, probably most, these days offer lower shelves that slide out, in pretty much the same widths as standard base cabinets, up to 45 inches. Dishwasher bases are available, as are breadbox drawer base cabinets, range base cabinets, sink and range base cabinets, wastebasket cabinets, island base cabinets, curved-corner base cabinets, desk file drawer base cabinets, angled-back base cabinets, curved base cabinet fronts for peninsulas, curved base cabinets with two doors (curved), and an array of corner and straight sink base cabinets, display base cabinets, open-shelf base cabinets, and so on. Some of the availability depends on the manufacturer, but you will find yourself having to compete with each and every one of these styles and sizes at some point in your career as a custom cabinetmaker. The competition is strong, but it shouldn't be daunting, because in most cases it is reasonably simple to get a beat on, with your own designs and patterns and ideas. Curved doors are pretty much out of the question for the small start-up cabinetmaker (lots of gear, some waste of wood). Otherwise, there isn't a thing here you can't meet or beat in your shop, given time and tools. What you need to determine is the cost-effectiveness. Buying and installing top-of-the-line manufactured cabinetry for your customer in some cases may make more sense than building your own to fit the spaces available.

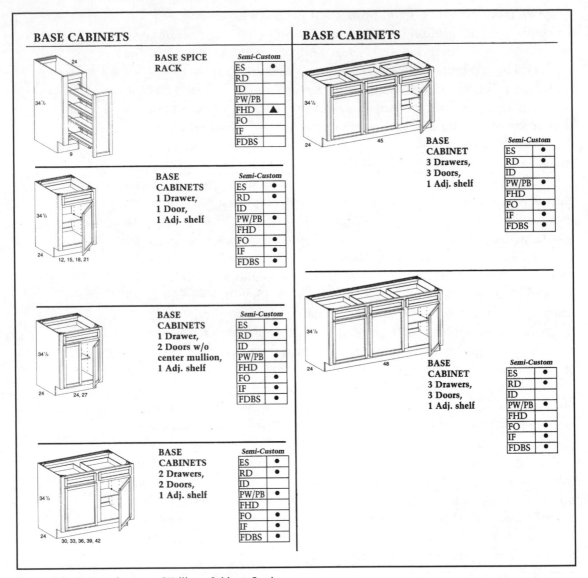

BASE CABINETS

BASE SPICE RACK

	Semi-Custom
ES	●
RD	
ID	
PW/PB	
FHD	▲
FO	
IF	
FDBS	

BASE CABINETS
1 Drawer,
1 Door,
1 Adj. shelf

	Semi-Custom
ES	●
RD	●
ID	
PW/PB	●
FHD	
FO	●
IF	●
FDBS	●

BASE CABINETS
1 Drawer,
2 Doors w/o
center mullion,
1 Adj. shelf

	Semi-Custom
ES	●
RD	●
ID	
PW/PB	●
FHD	
FO	●
IF	●
FDBS	●

BASE CABINETS
2 Drawers,
2 Doors,
1 Adj. shelf

	Semi-Custom
ES	●
RD	●
ID	
PW/PB	●
FHD	
FO	●
IF	●
FDBS	●

BASE CABINETS

BASE CABINET
3 Drawers,
3 Doors,
1 Adj. shelf

	Semi-Custom
ES	●
RD	●
ID	
PW/PB	●
FHD	
FO	●
IF	●
FDBS	●

BASE CABINET
3 Drawers,
3 Doors,
1 Adj. shelf

	Semi-Custom
ES	●
RD	●
ID	
PW/PB	●
FHD	
FO	●
IF	●
FDBS	●

Base cabinets, top drawers. *(Wellborn Cabinet, Inc.)*

Spacing the Parts

Kitchen layout comes next. You must make sure each and every job the client wants to do in the new kitchen can be done. That means checking to see that you're not shortening up countertop areas where flat space is needed in order to slip in an extra section of pantry or an

BASE CABINETS WITH FULL HEIGHT DOORS

BASE CABINETS WITH FULL HEIGHT DOORS	Semi-Custom	
2 Full height doors w/o center mullion, 1 Adj. shelf	ES	●
	RD	●
	ID	
	PW/PB	●
	FHD	▲
	FO	●
	IF	
	FDBS	●

BASE CABINETS WITH FULL HEIGHT DOORS	Semi-Custom	
2 Full height doors, 1 Adj. shelf	ES	●
	RD	●
	ID	
	PW/PB	●
	FHD	▲
	FO	●
	IF	
	FDBS	●

BASE CABINETS WITH FULL HEIGHT DOORS	Semi-Custom	
3 Full height doors, 1 Adj. shelf	ES	●
	RD	●
	ID	
	PW/PB	●
	FHD	▲
	FO	●
	IF	
	FDBS	●

BASE CABINETS WITH SLIDING SHELVES INSTALLED

BASE CABINETS WITH SLIDING SHELF	Semi-Custom	
1 Drawer, 1 Door, 1 Adj. shelf, 1 Sliding shelf	ES	●
	RD	
	ID	
	PW/PB	●
	FHD	
	FO	●
	IF	●
	FDBS	●

BASE CABINETS WITH SLIDING SHELVES INSTALLED

BASE CABINETS WITH SLIDING SHELF	Semi-Custom	
1 Drawer, 2 Doors w/o center mullion, 1 Adj. shelf, 1 Sliding shelf	ES	●
	RD	
	ID	
	PW/PB	●
	FHD	●
	FO	●
	IF	●
	FDBS	●

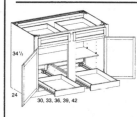

BASE CABINETS WITH SLIDING SHELVES	Semi-Custom	
2 Drawers, 2 Doors, 1 Adj. shelf, 2 Sliding shelves	ES	●
	RD	
	ID	
	PW/PB	●
	FHD	●
	FO	●
	IF	●
	FDBS	●

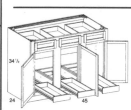

BASE CABINET WITH SLIDING SHELVES	Semi-Custom	
3 Drawers, 3 Doors, 1 Adj. shelf, 3 Sliding shelves	ES	●
	RD	
	ID	
	PW/PB	●
	FHD	●
	FO	●
	IF	●
	FDBS	●

Base cabinets, full height doors. *(Wellborn Cabinet, Inc.)*

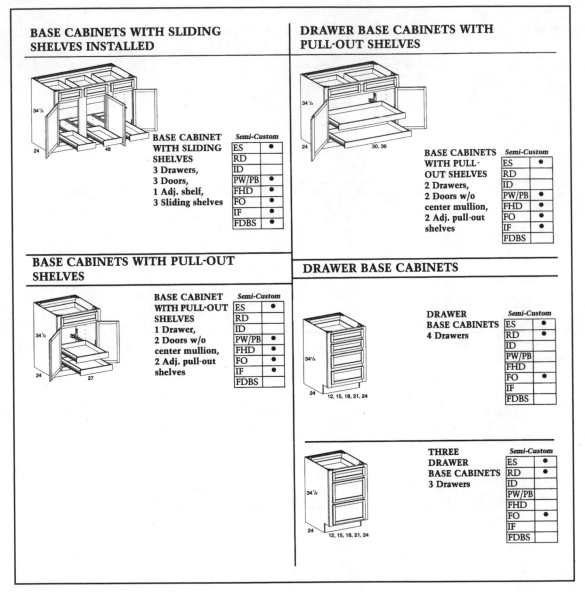

BASE CABINETS WITH SLIDING SHELVES INSTALLED

BASE CABINET WITH SLIDING SHELVES
3 Drawers,
3 Doors,
1 Adj. shelf,
3 Sliding shelves

	Semi-Custom
ES	●
RD	
ID	
PW/PB	●
FHD	●
FO	●
IF	●
FDBS	●

DRAWER BASE CABINETS WITH PULL-OUT SHELVES

BASE CABINETS WITH PULL-OUT SHELVES
2 Drawers,
2 Doors w/o center mullion,
2 Adj. pull-out shelves

	Semi-Custom
ES	●
RD	
ID	
PW/PB	●
FHD	●
FO	●
IF	●
FDBS	

BASE CABINETS WITH PULL-OUT SHELVES

BASE CABINET WITH PULL-OUT SHELVES
1 Drawer,
2 Doors w/o center mullion,
2 Adj. pull-out shelves

	Semi-Custom
ES	●
RD	
ID	
PW/PB	●
FHD	●
FO	●
IF	●
FDBS	

DRAWER BASE CABINETS

DRAWER BASE CABINETS
4 Drawers

	Semi-Custom
ES	●
RD	●
ID	
PW/PB	
FHD	
FO	●
IF	
FDBS	

THREE DRAWER BASE CABINETS
3 Drawers

	Semi-Custom
ES	●
RD	●
ID	
PW/PB	
FHD	
FO	●
IF	
FDBS	

Base cabinets, sliding shelves, drawers. (Wellborn Cabinet, Inc.)

extra appliance. That means checking to see if the kitchen and its jobs are real interests of the client, or simply something done in passing. You need lifestyle information for this kind of work, and the only way to get it is to ask questions. If the kitchen is primarily for appearance, you'll possibly want to slip in a standard design. If it is for real use, you want to find out exactly what use and how that use goes. (These days, even pizza ovens in residential kitchens aren't too unusual, and the trend to large, commercial-style stainless-steel appliances seems set to continue, so lots of judgments must be made before you draw the first line, or take the first measurement.) For the simple job, cabinet replacement is in order, and that simply means getting the best fit with the best quality that your client will pay for. Design in such cases tends to focus more on the style of cabinet than on any work surface arrangement, because the kitchen is in place, the client likes the working setup, and the cabinets and appliances are ratty-looking, dirty, battered, out of style, or otherwise less than satisfactory.

As in factories, you'll want to make your cabinets as close to modular in size as possible. This approach saves material (which usually comes in 2-, 4-, and 8-foot modules, with stock countertops and laminates running as much as 12 feet). Because you're already set up to work with certain size materials, the modular approach saves time. So your first decision on any job will be the size of the modules, after which you can begin to lay things out.

Start with a floor plan. Floor plans are simple—nothing but scale drawings done to show the kitchen viewed from above. The amount of detail you need is going to

> **Start with a floor plan.**

vary, but draw the outside room walls first. Next, carefully measure and site all windows and doors, showing the swing of the doors (into the room doesn't allow as much freedom of placement as does out-of-the-room swing). Next, locate electrical outlets. Finally, locate plumbing stub-outs. These measurements need to be correct within $1/4$ inch, so use care in taking them. Always measure each distance twice. If the two measurements don't agree, take a third, and then check that one.

Scaling Drawings

Scale sizes on the drawing can be important: Too small, and needed detail may be lost; too large, and the drawing is unwieldy. At the least, $1/2$ inch = 1 foot is a good bottom-end size, while 1 inch = 1 foot is a good top-end size. The $1/2$ inch = 1 foot scale is nicely set up for most

DRAWER BASE CABINETS

DRAWER BASE CABINETS 5 Drawers

	Semi-Custom
ES	●
RD	●
ID	
PW/PB	
FHD	
FO	●
IF	
FDBS	

34 1/2   24   30

SINK AND RANGE BASE CABINETS

SINK AND RANGE BASE CABINETS 2 Drawer blanks, 2 Doors

	Semi-Custom
ES	●
RD	●
ID	
PW/PB	●
FHD	●
FO	●
IF	●
FDBS	●

34 1/2   24   30, 33, 36, 39, 42, 48

SPICE DRAWER BASE CABINET

24

SPICE DRAWER BASE CABINET

34 1/2

DRAWER FRONT SIDE PROFILE

6

SINK AND RANGE BASE CABINET 2 Drawers, 2 Drawer blanks, 4 Doors

34 1/2   24   60

	Semi-Custom
ES	●
RD	●
ID	
PW/PB	●
FHD	●
FO	●
IF	●
FDBS	●

BASE CHEF PANTRY

BASE CHEF PANTRY CABINET 2 Full height doors

	Semi-Custom
ES	●
RD	
ID	
PW/PB	
FHD	▲
FO	
IF	
FDBS	

34 1/2   24   36

TILT OUT TRAY

2 1/4   3 3/4   11 1/4

SINK AND RANGE BASE CABINETS

SINK AND RANGE BASE CABINETS 1 Drawer blank, 2 Doors w/o center mullion

	Semi-Custom
ES	●
RD	●
ID	
PW/PB	●
FHD	●
FO	●
IF	●
FDBS	●

34 1/2   24   24, 27

Base cabinets: drawer, sink and range. *(Wellborn Cabinet, Inc.)*

SINK AND RANGE BASE CABINETS

TILT OUT TRAYS

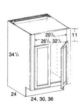

JAPANESE RANGE BASE CABINETS
2 Doors w/o center mullion

Semi-Custom	
ES	●
RD	●
ID	
PW/PB	●
FHD	
FO	●
IF	
FDBS	●

DRAWER RANGE BASE CABINETS
3 Drawers

Semi-Custom	
ES	●
RD	●
ID	
PW/PB	
FHD	
FO	●
IF	
FDBS	

BREAD BOX DRAWER BASE CABINET

BREAD BOX DRAWER BASE CABINET
1 Deep top drawer with factory installed bread box, 3 Small drawers

Semi-Custom	
ES	●
RD	
ID	
PW/PB	
FHD	
FO	
IF	
FDBS	

SINK AND RANGE BASE CABINETS

GLASS FRONT TWO DRAWER BASE CABINET
2 Drawers

Semi-Custom	
ES	●
RD	
ID	
PW/PB	
FHD	
FO	
IF	
FDBS	

GLASS FRONT FOUR DRAWER BASE CABINET
4 Drawers

Semi-Custom	
ES	●
RD	
ID	
PW/PB	
FHD	
FO	
IF	
FDBS	

SINK FRONT FLOOR

SINK MAT

RANGE FRONT

Semi-Custom	
ES	●
RD	
ID	
PW/PB	
FHD	
FO	▲
IF	
FDBS	

A variety of base cabinets. *(Wellborn Cabinet, Inc.)*

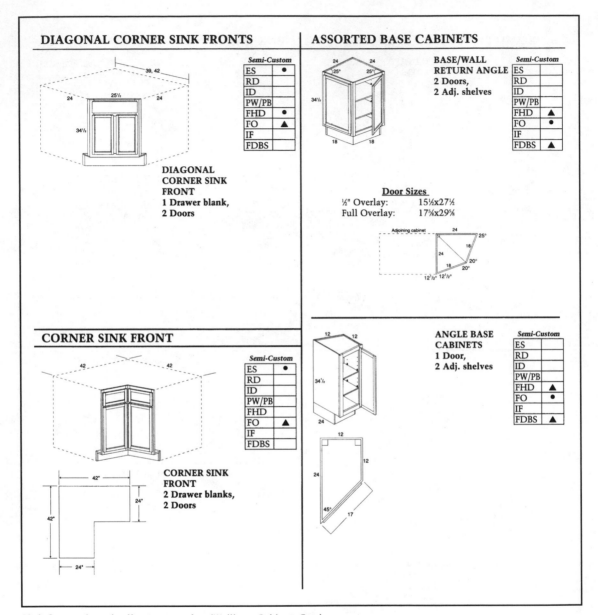

DIAGONAL CORNER SINK FRONTS

	Semi-Custom
ES	●
RD	
ID	
PW/PB	
FHD	●
FO	▲
IF	
FDBS	

DIAGONAL CORNER SINK FRONT
1 Drawer blank,
2 Doors

ASSORTED BASE CABINETS

BASE/WALL RETURN ANGLE
2 Doors,
2 Adj. shelves

	Semi-Custom
ES	
RD	
ID	
PW/PB	
FHD	▲
FO	●
IF	
FDBS	▲

Door Sizes

½" Overlay:	15½x27½
Full Overlay:	17⅝x29⅝

CORNER SINK FRONT

	Semi-Custom
ES	●
RD	
ID	
PW/PB	
FHD	
FO	▲
IF	
FDBS	

CORNER SINK FRONT
2 Drawer blanks,
2 Doors

ANGLE BASE CABINETS
1 Door,
2 Adj. shelves

	Semi-Custom
ES	
RD	
ID	
PW/PB	
FHD	▲
FO	●
IF	
FDBS	▲

Sink fronts, base/wall return angle. *(Wellborn Cabinet, Inc.)*

ASSORTED BASE CABINETS

ASSORTED BASE CABINETS

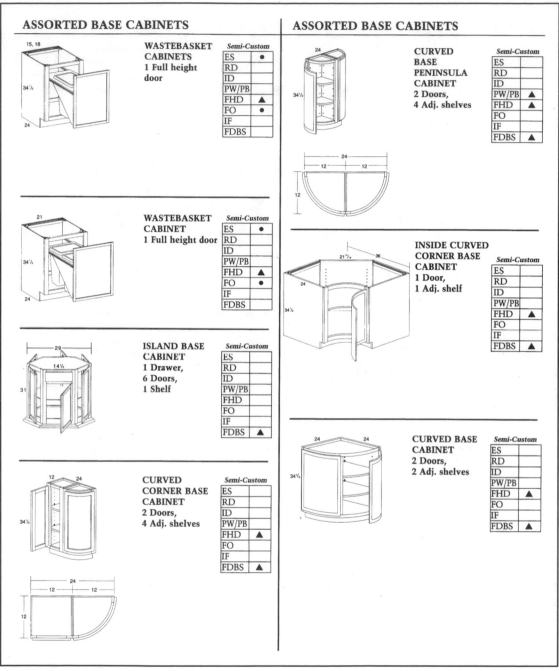

WASTEBASKET CABINETS
1 Full height door

Semi-Custom	
ES	●
RD	
ID	
PW/PB	
FHD	▲
FO	●
IF	
FDBS	

WASTEBASKET CABINET
1 Full height door

Semi-Custom	
ES	●
RD	
ID	
PW/PB	
FHD	▲
FO	●
IF	
FDBS	

ISLAND BASE CABINET
1 Drawer,
6 Doors,
1 Shelf

Semi-Custom	
ES	
RD	
ID	
PW/PB	
FHD	
FO	
IF	
FDBS	▲

CURVED CORNER BASE CABINET
2 Doors,
4 Adj. shelves

Semi-Custom	
ES	
RD	
ID	
PW/PB	
FHD	▲
FO	
IF	
FDBS	▲

CURVED BASE PENINSULA CABINET
2 Doors,
4 Adj. shelves

Semi-Custom	
ES	
RD	
ID	
PW/PB	▲
FHD	▲
FO	
IF	
FDBS	▲

INSIDE CURVED CORNER BASE CABINET
1 Door,
1 Adj. shelf

Semi-Custom	
ES	
RD	
ID	
PW/PB	
FHD	▲
FO	
IF	
FDBS	▲

CURVED BASE CABINET
2 Doors,
2 Adj. shelves

Semi-Custom	
ES	
RD	
ID	
PW/PB	
FHD	▲
FO	
IF	
FDBS	▲

Wastebasket, island base, curved-corner, and inside-curved-corner base cabinets. *(Wellborn Cabinet, Inc.)*

CORNER BASE CABINETS

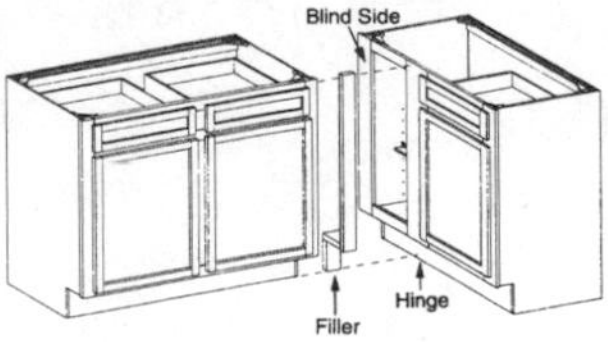

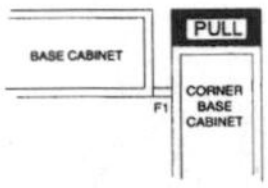

Maximum/Minimum Pull Measurements

	½" Overlay	Full Overlay
maximum pull	40¾	40¾
minimum pull w/hardware	37¼	40
minimum pull w/o hardware	36¼	37
maximum pull	43¾	43¾
minimum pull w/hardware	40¼	43
minimum pull w/o hardware	39¼	40
maximum pull	46¾	46¾
minimum pull w/ hardware	43¼	46
minimum pull w/o hardware	42¼	43
maximum pull	49¾	49¾
minimum pull w/ hardware	46¼	49
minimum pull w/o hardware	45¼	46
maximum pull	52¾	52¾
minimum pull w/ hardware	48	48
minimum pull w/o hardware	48	48

ON ALL FULL OVERLAY DOOR STYLES USE AT LEAST A 1" BASE FILLER.

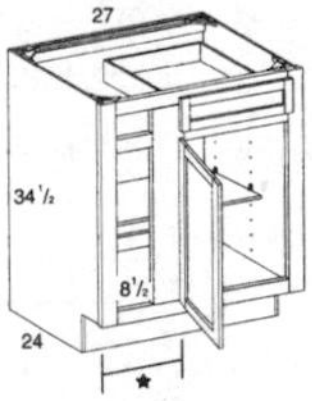

36" CORNER BASE CABINET
1 Drawer,
1 Door,
1 Adj. shelf

Semi-Custom	
ES	•
RD	•
ID	
PW/PB	•
FHD	•
FO	•
IF	
FDBS	•

*½" Overlay = 16¼"
Full Overlay = 15³⁄₁₆"

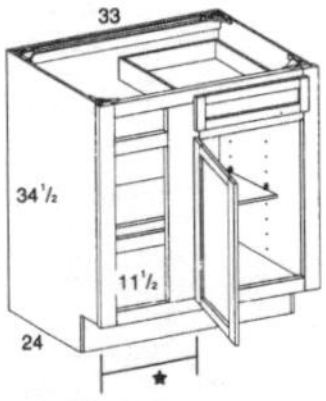

39" CORNER BASE CABINET
1 Drawer,
1 Door,
1 Adj. shelf

Semi-Custom	
ES	•
RD	•
ID	
PW/PB	•
FHD	•
FO	•
IF	
FDBS	•

*½" Overlay = 19¼"
Full Overlay = 18³⁄₁₆"

CORNER BASE CABINETS

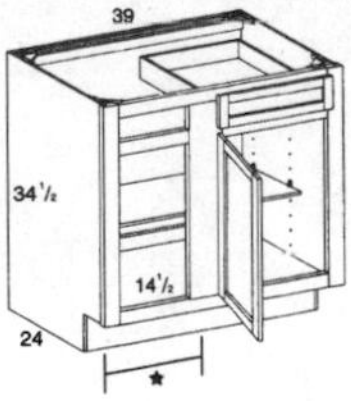

42" CORNER BASE CABINET
1 Drawer,
1 Door,
1 Adj. shelf

Semi-Custom	
ES	•
RD	•
ID	
PW/PB	•
FHD	•
FO	•
IF	
FDBS	•

* ½" Overlay = 22¼"
Full Overlay = 21³⁄₁₆"

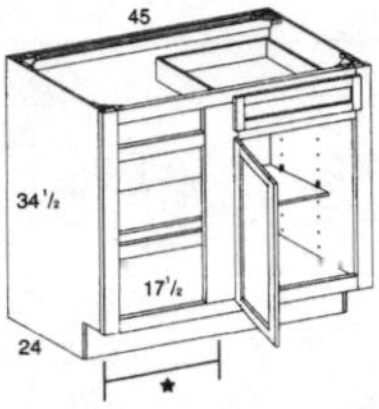

45" CORNER BASE CABINET
1 Drawer,
1 Door,
1 Adj. shelf

Semi-Custom	
ES	•
RD	•
ID	
PW/PB	•
FHD	•
FO	•
IF	
FDBS	•

*½" Overlay = 25¼"
Full Overlay = 24³⁄₁₆"

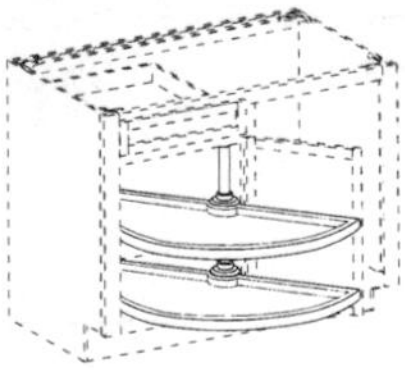

SWING-OUT SHELVES FOR CORNER BASES

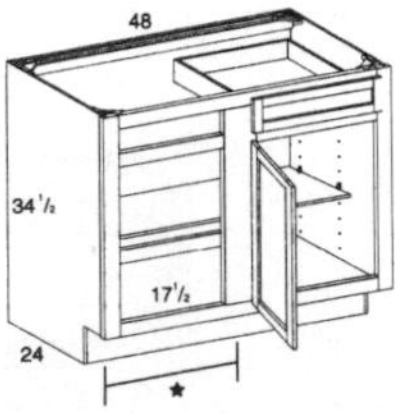

48" CORNER BASE CABINET
1 Drawer
1 Door,
1 Adj. shelf

Semi-Custom	
ES	•
RD	•
ID	
PW/PB	•
FHD	•
FO	•
IF	
FDBS	•

*½" Overlay = 18½"
Full Overlay = 20⅝"

Corner base cabinets. *(Wellborn Cabinet, Inc.)*

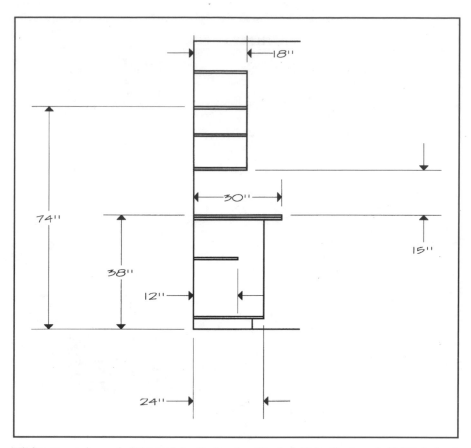

Minimum-maximum height and depth drawing. The height to the top shelf (74 inches) is a maximum, as is the height to the countertop. The minimum height to the counter-top is normally 30 inches, but may be modified for those who are disabled. The minimum inside base cabinet shelf width is 12 inches. The maximum countertop width is 30 inches, and the minimum is 15 inches. The maximum base cabinet depth is 24 inches. The minimum countertop height is 30 inches, and the maximum (shown) is 38 inches.

work, and allows you to draw exterior stud walls as 6-inch-thick entities in $1/4$-inch spacing. Of course, 1 inch provides a $1/2$-inch space on such walls, and allows easier detailing of studs, etc. You do need to know exactly where the studs are in order to hang the wall cabinets. Interior walls are thinner, usually not much more than $4^1/2$ inches, though that can vary tremendously. Old houses with plaster walls and full-thickness 2 × 4s can easily have walls 8 inches thick. Cheap new construction will have walls $4^1/2$ inches thick. Better new construction has double-

UTILITY CABINETS

UTILITY CABINETS

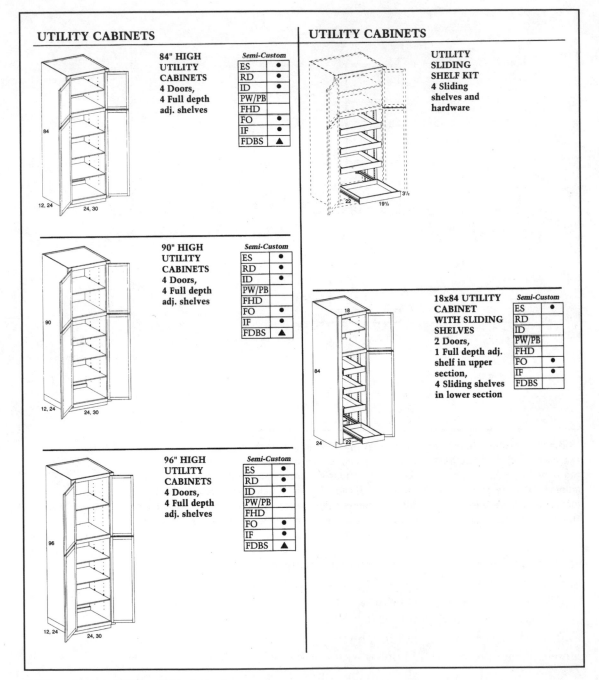

84" HIGH UTILITY CABINETS
4 Doors,
4 Full depth adj. shelves

Semi-Custom	
ES	•
RD	•
ID	•
PW/PB	
FHD	
FO	•
IF	•
FDBS	▲

90" HIGH UTILITY CABINETS
4 Doors,
4 Full depth adj. shelves

Semi-Custom	
ES	•
RD	•
ID	•
PW/PB	
FHD	
FO	•
IF	•
FDBS	▲

96" HIGH UTILITY CABINETS
4 Doors,
4 Full depth adj. shelves

Semi-Custom	
ES	•
RD	•
ID	•
PW/PB	
FHD	
FO	•
IF	•
FDBS	▲

UTILITY SLIDING SHELF KIT
4 Sliding shelves and hardware

18x84 UTILITY CABINET WITH SLIDING SHELVES
2 Doors,
1 Full depth adj. shelf in upper section,
4 Sliding shelves in lower section

Semi-Custom	
ES	•
RD	
ID	
PW/PB	
FHD	
FO	•
IF	•
FDBS	

Utility cabinets. *(Wellborn Cabinet, Inc.)*

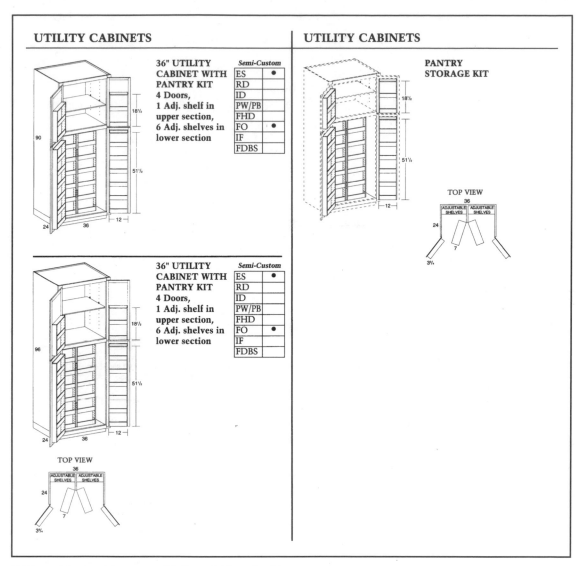

UTILITY CABINETS

36" UTILITY CABINET WITH PANTRY KIT 4 Doors, 1 Adj. shelf in upper section, 6 Adj. shelves in lower section	Semi-Custom
ES	●
RD	
ID	
PW/PB	
FHD	
FO	●
IF	
FDBS	

UTILITY CABINETS

PANTRY STORAGE KIT

TOP VIEW

36" UTILITY CABINET WITH PANTRY KIT 4 Doors, 1 Adj. shelf in upper section, 6 Adj. shelves in lower section	Semi-Custom
ES	●
RD	
ID	
PW/PB	
FHD	
FO	●
IF	
FDBS	

TOP VIEW

Utility cabinets with pantry. *(Wellborn Cabinet, Inc.)*

layer $1/2$-inch gypsum wallboard, so on $3^1/2$-inch-thick modern 2×4s, walls are about $5^1/2$ inches thick. For exterior walls with stucco, brick, and other facings, you may have to take a depth measurement at a window to get a correct thickness, but in most cases such accuracy is not needed for the exterior wall thickness. It is useful on interior walls, especially when you must build pass-throughs and similar items.

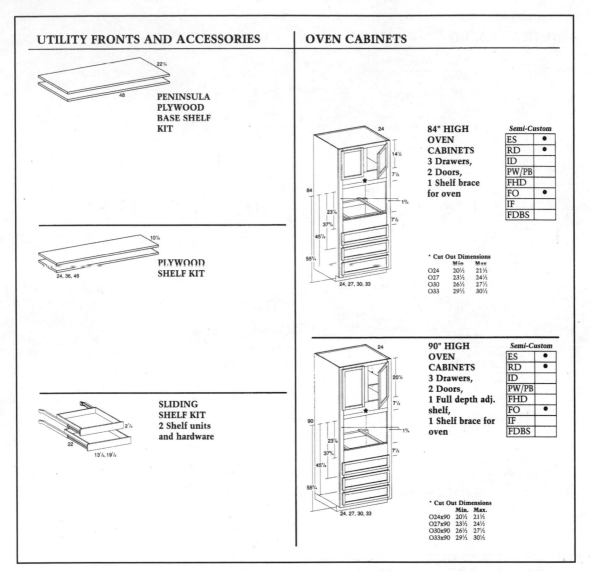

UTILITY FRONTS AND ACCESSORIES

PENINSULA PLYWOOD BASE SHELF KIT

PLYWOOD SHELF KIT

SLIDING SHELF KIT 2 Shelf units and hardware

OVEN CABINETS

84" HIGH OVEN CABINETS 3 Drawers, 2 Doors, 1 Shelf brace for oven

Semi-Custom	
ES	●
RD	●
ID	
PW/PB	
FHD	
FO	●
IF	
FDBS	

* Cut Out Dimensions

	Min	Max
O24	20½	21½
O27	23½	24½
O30	26½	27½
O33	29½	30½

90" HIGH OVEN CABINETS 3 Drawers, 2 Doors, 1 Full depth adj. shelf, 1 Shelf brace for oven

Semi-Custom	
ES	●
RD	●
ID	
PW/PB	
FHD	
FO	●
IF	
FDBS	

* Cut Out Dimensions

	Min.	Max.
O24x90	20½	21½
O27x90	23½	24½
O30x90	26½	27½
O33x90	29½	30½

Ninety-inch-tall oven cabinets. *(Wellborn Cabinet, Inc.)*

When making wall measurements, measure from corner to corner (diagonal) on all four corners to check for room squareness. The room almost certainly will be out of square, and this needs to be figured into your planning, or else the resulting cabinets are not going to fit. There is always a possibility of walls being out of plumb, too. Check this with a 4-foot-long spirit level. Note such defaults from the straight and level on your floor plan sheet. It is not necessary at this stage to check the floor for level in the areas where the cabinets will run, unless the house is very old or the floor appears way out of level. This can make a difference in adjusting toeboard height from one side to another. In an area where there is a severe dip (which can exceed $1^1/_2$ inches in a 14-foot-wide kitchen in my experience, so probably can be much more in other people's experience), you might want to place a tapered toeboard to assist in leveling the countertops.

When all the needed measurements are written down on paper, the best thing to do is to make one more check of those measurements. Make sure that the measurements are accurate and that nothing is left out. Next comes the design part of the job. You've got one drawing, which you worked on for a couple hours. Do you draw designs on that, erase and scribble and mess up your only decent sheet, at which point you can spend more time putting together another base? I don't. Head for your nearest office supply store, and have the thing copied. For a few dollars you can get all the copies you'll ever need to experiment with several designs. Three or four copies will be plenty, unless you've got no ideas at all, and will be doing lots of erasing and changing.

Start Drawing

You're not going to draw all the complexities, all the flecks in the countertop material, or anything like that. Start with a line drawn to scale 24 to 28 inches or so inside the inside wall line. Use whatever figure you plan to use for countertop width. Then draw the 24-inch line for the actual base cabinet fronts behind that line. Do this for the entire room, and come back and erase where the doors or other openings are. Or you can do your breaks as you do the lines and save the erasures.

Fill in appliance placements. Draw lines 12 inches in from the interior wall to show wall-hung cabinets. Break those at windows, doors, and pass-throughs.

Erase and scribble. Try different layouts, especially if you're not locked in by plumbing runs. Add kitchen islands. Try peninsulas. Check with your client's needs

> When all the needed measurements are down on paper, that is the time to make one more check of those measurements. Make sure that the measurements are accurate and nothing is left out.

for work and storage space. Make sure the work triangles you produce are within reasonable limits. Keep trying until you start getting rough drawings that look like something your client might like. Unless it has already been discussed, don't start considering moving walls or changing the locations of doors and windows. This kind of change usually is not within the work provisions of the cabinetmaker. It's your job to present the kitchen design as close to what your client needs as possible, within established limits of space, plumbing, wiring, and shape (including walls and windows that may already be in place)—not to recommend expensive and massive changes to the home.

Of course, if you're lucky, not a bit of the above is necessary. Take in a photo book of your executed designs (or those of a factory designer), and go over what the client wants as cabinet designs to fit within the architect's layout.

Once the overall design is completed, you need to get into the installation basics. Because cabinets are always built square, and houses don't stay that way for very long after they are built (if they even got built that way, something of an iffy proposition in some areas and eras), you need to make allowances for out-of-plumb walls, out-of-level floors, and similar problems. First, of course, you must know they exist. If you design and build to fit a space tightly and then discover that the wall is out of plumb in a direction that causes the upper or lower part of a cabinet case to be too tight, adjustments are in order. For that reason, always plan for filler strips. If you screw up a 3- to 4-inch-wide filler strip, its replacement costs a great deal less than replacement of the cabinet case.

Out-of-plumb and uneven walls also create difficulties with countertops where they meet the wall, whether flat or at a backsplash. This is a simple problem to deal with by using a compass to scribe a line on the back of the countertop and a belt sander to sand to that line. The belt sander gives better control, and a cleaner edge, than a jigsaw. A similar method can be used on toeboards to bring cabinets level, though in most cases the toeboard is removed, cut individually on a slant—as needed—and then replaced, with quarter-round molding over the finish flooring covering up any differences and gaps.

Starting and ending fillers are a good idea in longer cabinet runs. They make the overall spacing easier, and the sizes of the fillers can be shrunk as needed, so they are less obvious fillers and seem more like a part of the overall design.

Designing Cabinets to Suit: Base Cabinets

The basic cabinet with door is a simple thing; it is in multiples that complexities ensue. Kitchen and bath cabinets are normally nothing but boxes with face frames that hold doors. (In the instance of Eurostyle 32-mm cabinets, there is no face frame, which changes the

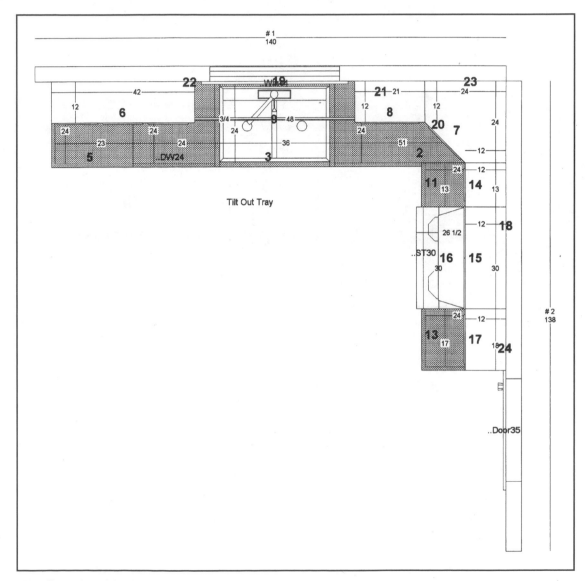

KCDw floor plan: raised-panel cherry, not to scale.

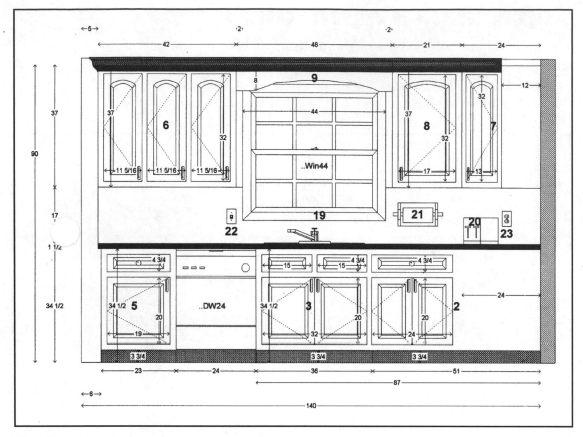

KCDw floor plan: raised-panel cherry, not to scale.

design features a lot but the planning features little, if at all.) The boxes are generally sized to fit particular spaces or to hold particular appliances or fixtures (sink bases are seldom less than 48 inches wide, but may be as wide as 96 inches, for example). The size of the box tends to determine the number and size of the doors. Drawers, or drawer blanks, are added as you, and your customer, desire in the storage planning needs, or to suit your overall feeling of how the design looks in place. Many cabinetmakers design a drawer to top each base cabinet. Others do not. I like the drawer on the top for each cabinet, but it is up to you to form your own idea of what works best, in line with your clients' desires.

Cabinet width, then, determines door numbers in that particular unit. Check out the pages of stock and semicustom cabinet drawings to

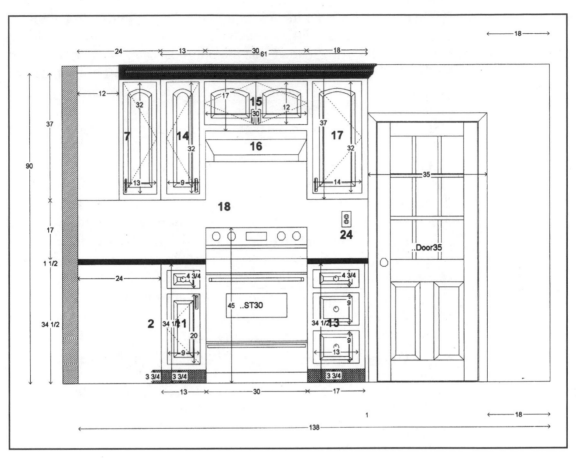

KCDw floor plan: raised-panel cherry, not to scale.

see some ideas. In general, cabinets up to 24 inches wide may have one door, although with Eurostyle, that's really pushing the limits. At 24 inches, for Eurostyle, two narrow doors (12 inches) work best, whether one drawer or two are above. With face frame construction, I'd tend to go with a single drawer up to about 26 or 28 inches in case width, because the frames narrow both the real and apparent door widths.

Drawer widths can create similar challenges: drawers up to 24 inches wide make sense. Drawers more than 24 inches wide look strange, unless you're storing architectural drawings, and require expensive, extremely heavy-duty slides. As a personal preference, I work with drawers no wider than 30 inches whenever possible (and it should always be possible, unless the client demands wider drawers).

Drawers wider than 24 inches require heavier bottoms, stronger over-all construction, as well as heavier-duty slides because there is the chance of their being loaded past the point of acceptance (for example, $1/4$-inch plywood drawer bottoms in a dado $1/4$ inch off the bottom of a poplar side). For wider cabinets, thicker bottoms and wider dados are obviously necessary, but it is also sensible to raise the bottom of the dado $1/8$ inch farther from the bottom of the side to give the overloaded drawer more "meat" to hang on to.

Base cabinets may also contain nothing but drawers. It is common for such cabinets to use different-height drawers, with the tallest on the bottom. It is also common for all the drawers to be the same height. Again, width comes into play here, and drawers over 32 inches are not recommended; my personal recommendation is to treat drawers over 24 or 26 inches as above. It is amazing how much weight can be shoved into a kitchen drawer, even the minimal top drawers provided in many stock cabinet sets. Add a full-extension slide, and you've got a lot of weight depending from the case, and from the sides of the drawer, sometimes as much as 75 pounds. Design to take care of the weight, if you can design to reduce drawer size.

In most instances, smaller drawers are going to be handier, although a few larger ones make for handy storage. Smaller drawers, especially where small items such as flatware and small kitchen accessories are stored, are easier to divide and have less of a tendency to hide the contents from the user.

Wall Cabinets: Spacing to Match the Base Cabinets

Wall cabinets are similar in construction to base cabinets, but have no countertop hangers and are shallower—normal depth is 12 inches. Of course, there also is no toeboard, and there are almost never drawers. Recently, drawers have become popular as add-ons under the cabinets, still maintaining the basic 15 inches above countertop minimum. Whether or not that popularity will continue, no one can tell, but it's based on good sense: It gives eye-level access to small items such as spices, thus reducing the need for add-on wall-hung racks of indeterminate style and construction after the kitchen is in use. In fact, you might want to build special spice drawers as a feature of your cabinets, designing them to seal tightly, or to hold already sealed containers (so the spices retain flavor).

Wall cabinets also offer a lot of design features that are, to me, of iffy utility. I number among these plate dowels and wine racks. Kitchen air carries dust and grease in about equal quantities. Plates standing exposed in racks are going to have to be washed after use and washed again before use, unless use is almost daily. A similar objection holds for wine bottles stored in open kitchen racks; those bottles will need to be cleaned with a soapy rag before being opened at the table.

I feel the jury is still out on appliance garages, but a lot of people love them, and they have been around for a good number of years. They do, generally, give you the chance to incorporate small-scale tambour doors into your design. They do protect small appliances that are seldom used from grease and dust.

Spacer use is similar to that for base cabinets, but spacing of wall cabinets requires some more thought, and often an extra unit or two, if you wish to match the cabinets to the base cabinets. This happens because base cabinets are usually 23 inches deep, while wall cabinets are 12 inches deep, leaving an 11-inch gap at the end of the run, if you start everything from the corner. With wall cabinets, start the run even with the end of the base cabinet run, and run it in toward the wall, instead of out. Keep cabinets matched for size along that run (unless your design decides otherwise, though matching breaks tend to look neater). At the end of the wall cabinet run, where you meet either a wall or another cabinet, add an 11-inch cabinet to fill the space. This is less obvious and less intrusive from a neat design standpoint than is adding the 1-inch cabinet at the end of the run, although that, too, can be done if you choose.

With this part of the design done, you've roughly sited all your cabinets and gotten rough sizes. The drawing is a bit messy, but can be easily transferred to a clean sheet, with the accurate base measurements in place. At this point, do everything in scale and carefully, because you want final measurements as close to perfect as possible. To further your aim of saving time, I'd waste a little and go back and measure a third time.

It is at this point that an age-old device for taking and transferring measurements comes into its own.

Story Poles

A *story pole* in its simplest form is laid against a wall and marked to give a height—or width. For cabinetry, both directions work well, and some jobs require a number of poles. Most cabinetmakers I know store the poles, with job names and dates written on them. I wondered about this, because it is very, very seldom that one client wants two kitchens in the same room, but recently, while talking to J. R. Burnette, of Burnette's Cabinets, in Huddleston, Virginia, I discovered the real utility of such storage. J. R. took over the business from his father, and has now been working in the same place for well over two decades. His father was also there about two decades. In that time, several customers have worn out two or three kitchens, but have not moved. The story poles come in handy the second and third time round, too, and unless major remodeling has been done, there are sometimes almost no changes needed, reducing preparation time considerably.

So what should you do to get a story pole stack working for you? You start by drawing some marks across the width of the stick, which is probably best made $1^1/_2$ to 2 inches wide of $^3/_4$-inch stock—plywood works nicely—for durability and ease of use. These are your witness marks, and they work to help you build exactly what is needed.

Story sticks are also cheap. You have to measure anyway, so making a mark on the stick as you measure is easy to do. You don't need a laser level, a computer, or anything else that costs much money. A good scribe or pencil, a good measuring tape, and an accurate square, and you're ready to go. If cabinets are already in place, you may not even need a tape measure; just mark the new against the old. Accuracy is guaranteed.

The same style of story pole works to design framed or frameless cabinets and entire rooms of cabinets such as are in kitchens. Visualization is also easy, since the story pole is on a 1-to-1 scale. The stick can be used later as a fixture or guide when you assemble cabinets into the kitchen. And, as noted, the sticks store easily (stuff them up over a rafter, or slip them on an openwork dowel shelf) for future reference.

> **A story pole in its simplest form is laid against a wall and marked to give a height—or width.**

Stick Construction

Your story stick is best made from a light-colored wood. Close grain is handy, too, so a hardwood such as poplar or birch or a softwood such as pine works fine. Length is going to vary, but with kitchen cabinets it will almost never be more than 8 feet, regardless of whether a horizontal or vertical stick is needed. As noted, make the stock $1^{1}/2$ inches wide, of $^{3}/4$-inch-thick stock. If you use the two wide surfaces for height and width representations, you've still got a $^{3}/4$-inch edge for the depth. Today, we can use colored pencils to distinguish specifics, maybe using red for fronts, green for frames, blue for drawer boxes, and so on, as needed.

Start with horizontals, since that, as we've seen, calls for the greatest number of design decisions. Double-check your points of reference. Although the story stick, once finished, is infallible, you are not. Take it slowly, measure carefully, two or three times mark the stick with great precision, and realize you've saved a ton of time.

Modern Methods

Naturally, some readers will want to try more modern layout methods, ones that do not involve graphed or otherwise scaled

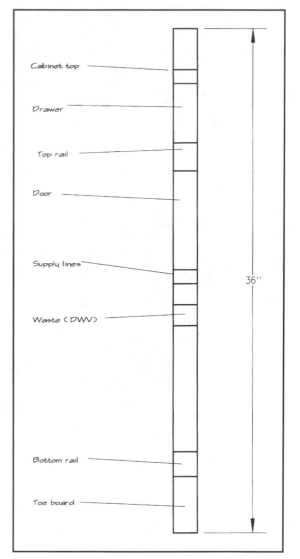

Story pole, simple vertical.

paper, erasures, story stocks, or anything old-fashioned. Such things don't really impress customers, either.

To check out new things, I got a copy of a KCDw program in a trial version. After autoloading, I discovered an easy-to-use program that has a manual and on-line help, and even offers a shot at getting hold of KCDw Cabinetmakers Software technical support to walk you through the software. The premise is, of course, that you'll buy after you try, and you well might. I've been using drafting software for years, and this was like a breath of fresh air—no more complex commands unless you want them. It is incredibly simple to add sinks, appliances, walls, islands, etc. For those having little time or patience for learning new things, but with good experience of cabinet layout needs, I recommend this one.

I also used AutoCAD Lite recently, and found that a delight. Auto-Sketch, their simplest (and cheapest) program, is easy for even me to use (I have the drafting talents of a…well, let's just say computer-aided drawings are a great help to my neatness). These two programs take more effort, but are more flexible in what they can produce, and they offer the chance to use the files in full-blown AutoCAD, in whatever version. This means that if your customer decides to work with an architect or professional planner, the odds of your plan being visible and fitting are very, very good.

Contact information for both companies can be found in App. G.

You must remember for both of these units, though, that you have to take accurate measurements and decide on placement ahead of time, because it is difficult to bring the computer to the job site. (Although a top-grade laptop will work, the screens are all still small enough to give me a major headache when working with detailed graphics like these.)

So you escape only the hard work of erasure, though that may well be enough. Sometimes being able to skip some dirty chores frees up creativity, and allows you to better use your design sense, before you get into the real work, and the real creativity, of cutting the parts and assembling and installing your cabinetry.

(*Text continues on p. 79.*)

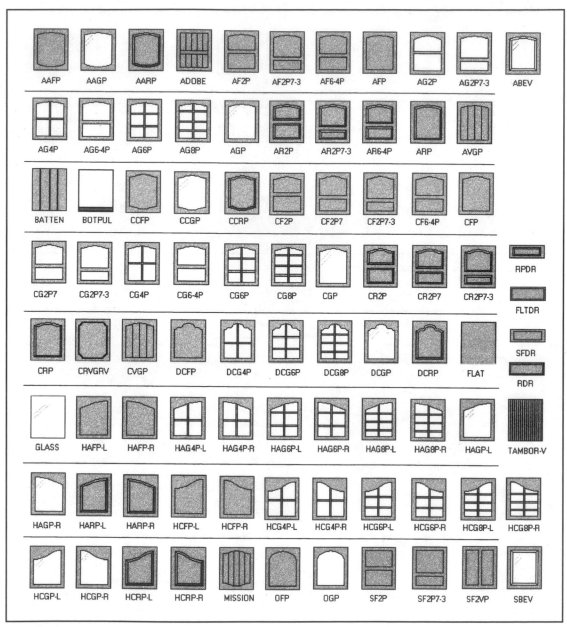

KCDw Cabinetmaker door and drawer styles.

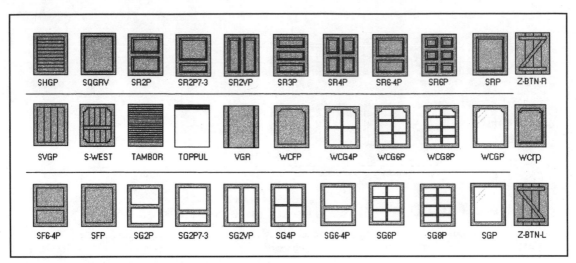

KCDw elevation no. 2: raised-panel cherry, not to scale.

Solid Cherry Wood Cabinets
Laminate Tops

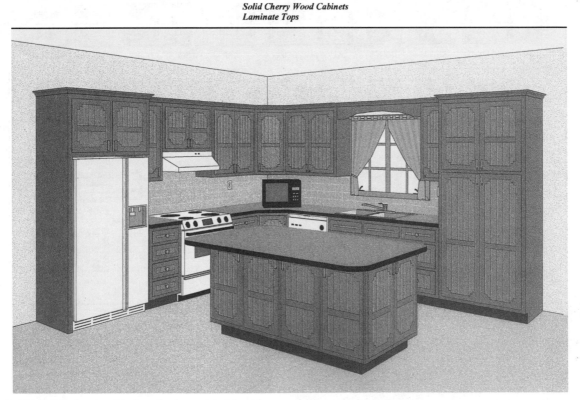

KCDw Cabinetmaker kitchen design, with island.

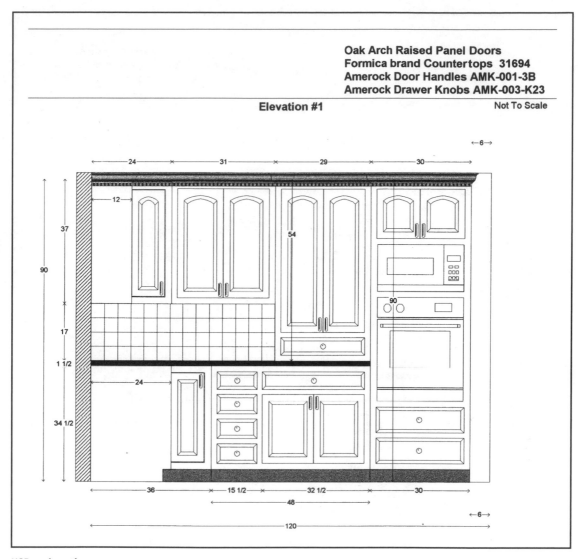

Oak Arch Raised Panel Doors
Formica brand Countertops 31694
Amerock Door Handles AMK-001-3B
Amerock Drawer Knobs AMK-003-K23

Elevation #1   Not To Scale

KCDw elevation.

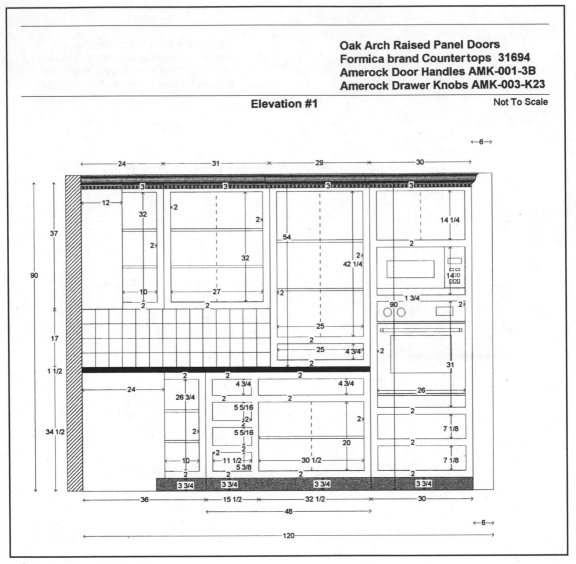

KCDw elevation, with detail.

Oceanside - Albert (Final) ARP/SRP 1/4" Overlay Ogee
Kitchen - Very White Pickled Oak - White Mico Hinges
Kitchen - Magna Sahara Corian 4" Loose BS
3 Vanities - Alm. Pickled Oak - White Mico Hinges
3 Vanities - Heather Onyx Cult. Marble

KCDw peninsula.

Master Bathroom Vanities with Paneling for Hot Tub
Oak Cabinets
White Porcelin Knobs
Wilsonart Brand Counter Tops D455-60 Montpellier
Counter Top Edges D421-60 Midori

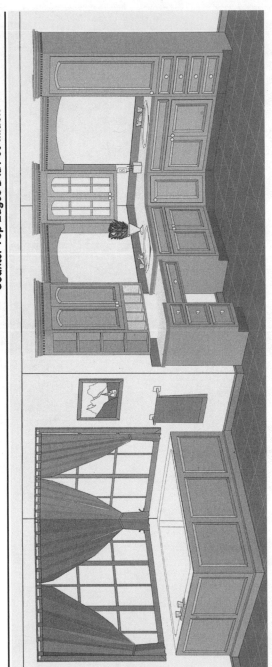

KCDw bathroom with hot tub.

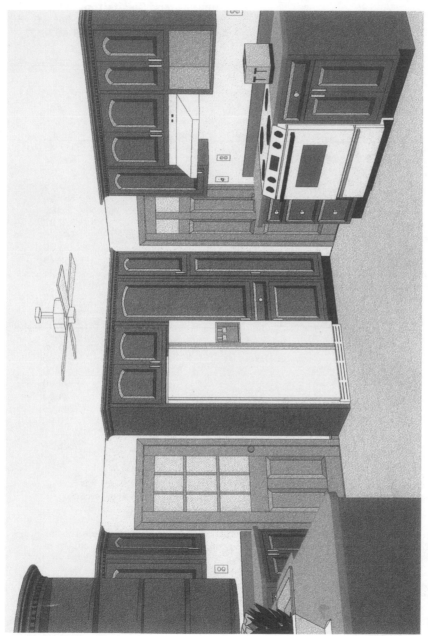

KCDw kitchen with cherry cabinets.

Cherry Arch Raised Panel Doors
Formica brand Countertops 31694
Amerock Door Handles AMK-001-3B
Amerock Drawer Knobs AMK-003-K23

CONTRACT

Cabinets   -   Cherry
Top doors   -   Arch Raised Panel
Base doors   -   Square Raised Panel
Drawer Fronts -   Raised Panel Drawer

Unit #	Name	Width	Length	Depth	Cost
2	Sink Base	30	34 1/2	24	317.41
3	Base Lazy Susan	36	34 1/2	36	480.34
5	Wall Cabinet	38	90	24	379.87
6	Wall Cabinet	16	37	12	173.89
8	Wall Lazy Susan	24	37	24	351.54
9	Valance	52	8	3/4	36.00
10	Base Cabinet	32 1/2	34 1/2	24	372.97
11	Wall Oven	30	90	25	447.76
12	Filler (Base)	2	34 1/2	24	55.00
13	Drawer Base	15 1/2	34 1/2	24	258.48
14	Wall Cabinet	31	37	12	345.25
15	Wall Cabinet (Drawer Below)	29	54	21	376.85
16	Base Cabinet	48	34 1/2	24	583.83

Subtotal	4179.19
Formica Countertops	400.00
Tile Backsplash	245.00
Handles 17	17.00
Knobs 8	8.00
Hinges 28	28.00
3 1/4" Crown Molding (417-Inch)	121.63
Tax 5%	249.94
Installation	420.00
Total Cost	**5668.76**

   All details left unspecified on this contract will be left to the discretion of the cabinet shop. It is the customer's responsibility to verify all dimensions and to have all walls, floors, and ceilings, straight, level, and square.

TERMS: We agree with all specifications. We also understand that any changes may incur added costs. Payment due 10 days from final installation date. A finance charge of 1 1/2% monthly, which is an Annual Percentage Rate of 18%, will be added to overdue balance. In the event this Cabinet Shop institutes any action for the enforcement of collection of this account, there shall be immediately due from the under-signed, in addition to the unpaid balance and interest, all costs and expenses including a reasonable attorney's fee.

CUSTOMER'S SIGNATURE_____DATE_____

CUSTOMER'S SIGNATURE_____DATE_____

KCDw Cabinetmakers Software.

Cherry Arch Raised Panel Doors
Formica brand Countertops 31694
Amerock Door Handles AMK-001-3B
Amerock Drawer Knobs AMK-003-K23

ESTIMATE

The following Estimate below labeled "Total Cost" will be valid for 30 days beyond date posted above. This Estimate was determined through information provided by you the home owner and is subject to change based on the accuracy of that information.

Cabinets - Cherry	4179.19
Formica Countertops	400.00
Tile Backsplash	245.00
Handles 17	17.00
Knobs 8	8.00
Hinges 28	28.00
3 1/4" Crown Molding (417-Inch)	121.63
Tax 5%	249.94
Installation	420.00
Total Cost	5668.76

KCDw Cabinetmakers Software.

Cherry Arch Raised Panel Doors
Formica brand Countertops 31694
Amerock Door Handles AMK-001-3B
Amerock Drawer Knobs AMK-003-K23

Square Foot Report

Material Type	Square Foot Amount	# of Sheets 4'x 8'	Material Cost
3/4" Solid Wood	99.33	N/A	178.79
5/8" Poplar	26.50	N/A	25.44
3/4" Finish Ply	72.76	2.27	123.69
1/2" Interior Ply	121.19	3.79	103.01
1/4" Interior Ply	144.74	4.52	104.21
			Total Material Cost
			535.14

Hardware Report

Hardware Type	Hardware Cost
Hinges 28	42.00
Knobs 8	4.00
Vertical Handles 17	21.25
Drawer Guides 11	55.00
	Total HardWare Cost
	122.25

Molding Report

	Feet of Molding	Cost of Molding
Molding	34.75	86.88

Total Cost
744.27

Square Foot of Doors Report

Material Type	Square Foot Amount
Doors and Drawer Fronts	54.72

Lineal Foot of Cabinets Report

Material Type	Lineal Foot Amount
Top Cabinets	11.50
Base Cabinets	13.67
Tall Cabinets	2.50

KCDw Cabinetmakers Software.

Cherry Arch Raised Panel Doors
Formica brand Countertops 31694
Amerock Door Handles AMK-001-3B
Amerock Drawer Knobs AMK-003-K23

DOORS

Door Species   cherry

Qty	Name	Width	Height	Unit Cost	Total
2	Drawer (Raised Panel Drawer)	26 1/2	7 5/8	14.05	28.10
1	Drawer (Raised Panel Drawer)	12	5 7/8	4.90	4.90
2	Drawer (Raised Panel Drawer)	12	5 13/16	4.85	9.70
1	Drawer (Raised Panel Drawer)	31	5 1/4	11.31	11.31
1	Drawer (Raised Panel Drawer)	26 1/2	5 1/4	9.67	9.67
1	Drawer (Raised Panel Drawer)	25 1/2	5 1/4	9.30	9.30
1	Drawer (Raised Panel Drawer)	12	5 1/4	4.38	4.38
3	Drawer (Raised Panel Drawer)	14 3/16	5	4.93	14.79
2	Door (Square Raised Panel)	9 7/8	26 1/2	18.17	36.34
3	Door (Square Raised Panel)	14 3/16	21	20.68	62.04
2	Door (Square Raised Panel)	15 7/16	20 1/2	22.08	44.16
2	Door (Square Raised Panel)	13 3/16	20 1/2	18.87	37.74
2	Door (Arch Raised Panel)	12 11/16	42 3/4	52.86	105.72
2	Door (Arch Raised Panel)	13 11/16	32 1/2	46.04	92.08
1	Door (Arch Raised Panel)	12 1/2	32 1/2	43.22	43.22
2	Door (Arch Raised Panel)	9 7/8	31 3/4	36.77	73.54
2	Door (Arch Raised Panel)	17 3/16	16 1/2	34.77	69.54
2	Door (Arch Raised Panel)	13 3/16	14 3/4	28.58	57.16

Subtotal		713.69
Tax 5%		35.68
Total Cost		749.37

KCDw Cabinetmakers Software.

Raised Panel Cherry

Frame List

#)Unit	Qty	Width x Length	Name	Material	Note
1)14	1	3 x 57	Rail	3/4" Solid Wood	
1)6	1	3 x 38	Rail	3/4" Solid Wood	
1)8	1	3 x 17	Rail	3/4" Solid Wood	
1)7	1	3 x 13	Rail	3/4" Solid Wood	
3)3	3	2 x 58	Rail	3/4" Solid Wood	
1)6	1	2 x 38	Rail	3/4" Solid Wood	
1)15	1	2 x 30	Rail	3/4" Solid Wood	
3)5	3	2 x 19	Rail	3/4" Solid Wood	
1)8	1	2 x 17	Rail	3/4" Solid Wood	
1)17	1	2 x 14	Rail	3/4" Solid Wood	
1)7 4)13	5	2 x 13	Rail	3/4" Solid Wood	
3)11 1)14	4	2 x 9	Rail	3/4" Solid Wood	
1)2	1	3 x 30 3/4	Stile	3/4" Solid Wood	
2)6 2)8 1)14 1)17	6	2 x 37	Stile	3/4" Solid Wood	.
2)7	2	2 x 37	Stile	3/4" Solid Wood	
1)14 1)17	2	2 x 34	Stile	3/4" Solid Wood	.
2)6	2	2 x 32	Stile	3/4" Solid Wood	.
1)3 2)5 2)11 2)13	7	2 x 30 3/4	Stile	3/4" Solid Wood	.
1)3	1	2 x 20	Stile	3/4" Solid Wood	.
2)3	2	2 x 4 3/4	Stile	3/4" Solid Wood	

KCDw Cabinetmakers Software.

Raised Panel Cherry

Boxes & Kick List

#)Unit	Qty	Width x Length	Name	Material	Note
1)5	1	23 1/4 x 34 1/2	Side (Finished Left)	3/4" Finish Ply	
1)13	1	23 1/4 x 34 1/2	Side (Finished Right)	3/4" Finish Ply	
1)6 1)8	2	11 1/4 x 37	Side (Finished Left)	3/4" Finish Ply	Dado for shelf
1)6 1)17	2	11 1/4 x 37	Side (Finished Right)	3/4" Finish Ply	Dado for shelf
1)17	1	11 1/4 x 36 1/4	Side (Exposed Left)	3/4" Finish Ply	Dado for shelf
1)14	1	11 1/4 x 36 1/4	Side (Exposed Right)	3/4" Finish Ply	Dado for shelf
1)2 1)5 1)11	3	23 x 29 3/4	Side (Unfinished Right)	1/2" Interior Ply	
1)3 1)11 1)13	3	23 x 29 3/4	Side (Unfinished Left)	1/2" Interior Ply	
1)7	1	11 x 37	Side (Unfinished Left)	1/2" Interior Ply	
1)7	1	11 x 37	Side (Unfinished Right)	1/2" Interior Ply	
1)8	1	11 x 37	Side (Unfinished Right)	1/2" Interior Ply	Dado for shelf
1)14	1	11 x 37	Side (Unfinished Left)	1/2" Interior Ply	Dado for shelf
1)3	1	23 x 28 3/4	Partition	1/2" Interior Ply	
1)7	1	23 1/4 x 23 1/4	Top	1/4" Interior Ply	
1)14	1	11 x 59 5/8	Top	1/4" Interior Ply	
1)6	1	11 x 40 3/4	Top	1/4" Interior Ply	
1)8	1	11 x 19 5/8	Top	1/4" Interior Ply	
1)7	1	23 1/4 x 23 1/4	Bottom	1/2" Interior Ply	
1)3	1	23 x 85 3/4	Bottom	1/2" Interior Ply	
1)5	1	23 x 21 3/4	Bottom	1/2" Interior Ply	
1)13	1	23 x 15 3/4	Bottom	1/2" Interior Ply	
1)11	1	23 x 11 1/2	Bottom	1/2" Interior Ply	
1)6	1	11 x 41	Bottom	1/2" Interior Ply	
1)15	1	11 x 30 1/2	Bottom	1/2" Interior Ply	
1)8	1	11 x 19 3/4	Bottom	1/2" Interior Ply	
1)17	1	11 x 17	Bottom	1/2" Interior Ply	
1)14	1	11 x 11 3/4	Bottom	1/2" Interior Ply	
2)7	2	37 x 23 1/4	Back	1/4" Interior Ply	
1)6	1	36 x 41	Back	1/4" Interior Ply	
1)8	1	36 x 19 3/4	Back	1/4" Interior Ply	
1)17	1	36 x 17	Back	1/4" Interior Ply	
1)14	1	36 x 11 3/4	Back	1/4" Interior Ply	
1)3	1	30 3/4 x 85 3/4	Back	1/4" Interior Ply	
1)5	1	30 3/4 x 21 3/4	Back	1/4" Interior Ply	
1)13	1	30 3/4 x 15 3/4	Back	1/4" Interior Ply	
1)11	1	30 3/4 x 11 1/2	Back	1/4" Interior Ply	
1)15	1	16 x 30 1/2	Back	1/4" Interior Ply	
1)3	1	5 1/4 x 87	Kick	3/4" Solid Wood	
1)5	1	5 1/4 x 23	Kick	3/4" Solid Wood	
1)13	1	5 1/4 x 17	Kick	3/4" Solid Wood	
1)11	1	5 1/4 x 15 1/2	Kick	3/4" Solid Wood	

KCDw Cabinetmakers Software.

Unit # 1

Num	Width	x	Length	Name	Material	Note
1	3 3/4	x	64 5/8	Rail	3/4" Solid Wood	
1	1 3/4	x	30	Rail	3/4" Solid Wood	
1	1 3/4	x	18	Rail	3/4" Solid Wood	
1	1 3/4	x	13 1/8	Rail	3/4" Solid Wood	
2	1 3/4	x	37	Stile	3/4" Solid Wood	.
2	1 3/4	x	33 1/4	Stile	3/4" Solid Wood	.
1	11 1/4	x	37	Side (Finished Right)	3/4" Finish Ply	Dado for shelf
1	11 1/4	x	36 1/4	Side (Exposed Right)	3/4" Finish Ply	Dado for shelf
1	11 1/4	x	36 1/4	Side (Exposed Left)	3/4" Finish Ply	Dado for shelf
1	11	x	37	Side (Unfinished Left)	1/2" Interior Ply	Dado for shelf
1	11	x	66 3/4	Top	1/4" Interior Ply	
1	11	x	30 1/2	Bottom	1/2" Interior Ply	
1	11	x	20 1/2	Bottom	1/2" Interior Ply	
1	11	x	15 3/8	Bottom	1/2" Interior Ply	
1	36	x	20 1/2	Back	1/4" Interior Ply	
1	36	x	15 3/8	Back	1/4" Interior Ply	
1	23	x	30 1/2	Back	1/4" Interior Ply	
1	11	x	30 1/2	Shelf	3/4" Interior Ply	
2	11	x	20 1/4	Shelf	3/4" Interior Ply	
2	11	x	15 1/8	Shelf	3/4" Interior Ply	

Depth = 12

KCDw Cabinetmakers Software.

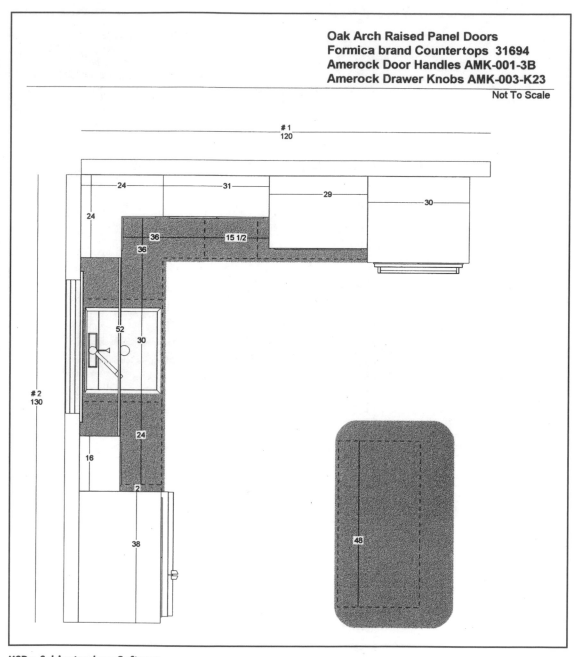

Oak Arch Raised Panel Doors
Formica brand Countertops 31694
Amerock Door Handles AMK-001-3B
Amerock Drawer Knobs AMK-003-K23

Not To Scale

KCDw Cabinetmakers Software.

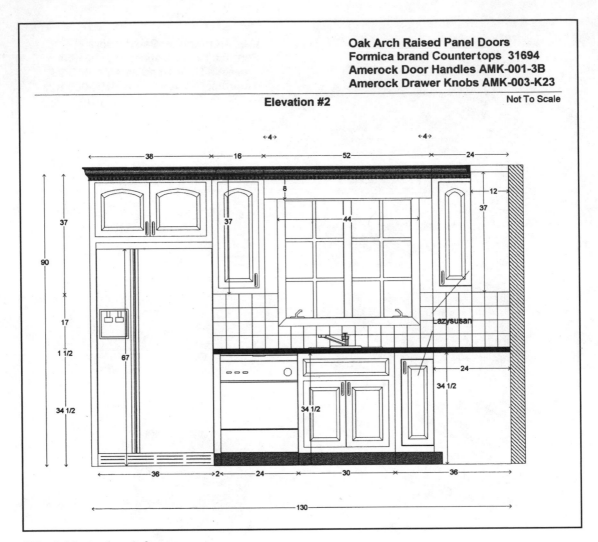

Oak Arch Raised Panel Doors
Formica brand Countertops 31694
Amerock Door Handles AMK-001-3B
Amerock Drawer Knobs AMK-003-K23

Elevation #2

Not To Scale

KCDw Cabinetmakers Software.

Oak Arch Raised Panel Doors
Formica brand Countertops 31694
Amerock Door Handles AMK-001-3B
Amerock Drawer Knobs AMK-003-K23

Elevation #2

Not To Scale

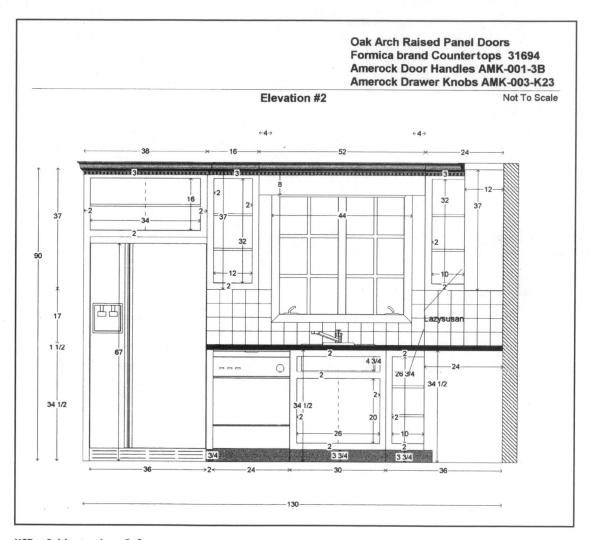

KCDw Cabinetmakers Software.

Woods for Cabinetry

We can begin by saying we cannot even conceive of covering all the wood types that are available today for the manufacture of cabinetry. Given that, we'll cover a few of the most useful and popular domestic and foreign woods. The custom cabinetmaker has a distinct advantage over factories in this area. It is difficult to produce cabinets on a production line if the woods keep changing, because of different working characteristics (and the possibilities of misassemblies), so being able to work with unusual hardwoods or softwoods is a great boon to the small custom shop. Plywoods and veneers of most of the listed woods are available from various suppliers around the country and into Canada, so these should pose no problems. Veneered medium-density fiberboard (MDF) in facings other than oak, maple, ash, cherry, and a very few other species may be hard to find, but veneers can be bought and added when MDF is determined to be the material of choice.

We start with softwoods and progress to hardwoods.

First, we discuss cedars. Incense cedar is *Libocedrus decurrens,* a California wood that works its way into southwestern Oregon and on to extreme western Nevada. Sapwood is a white to creamy color, and the heartwood a light brown, going toward red for a resulting pink overall tinge. The wood has a fine, uniform texture and a spicy aroma when cut or sanded. Incense cedar is lightweight and low in strength,

Merillat's oak illustrates its lightness.

with low shock resistance. Incense cedar is locally inexpensive, and a good choice *if* you can accept its peckiness (it has localized pockets of decay that do not spread after the wood is dried). It makes superb small-box material as well as lining material for larger boxes, cabinets, and closets. The wood works easily with conventional tools.

Port Orford cedar is *Chamaecyparis lawsoniana,* called Lawson cypress, Oregon cedar, and white cedar. It grows in a narrow belt (never more than 40 miles inland) along the coast of Oregon, from Coos Bay south into California. Heartwood of Port Orford cedar is a light yellow to a pale brown, and sapwood is thin and hard to tell from the heartwood. The wood is nonresinous, slow-growing, and fine-textured, with generally straight grain and a pleasant spicy odor. Port Orford cedar is resistant to decay, moderately lightweight, stiff, moderately strong, and hard. It is useful for lining blanket chests, making large and small boxes, and making interior moldings. Port Orford cedar works easily with conventional hand and power tools.

Eastern red cedar is another aromatic cedar, and it grows about everywhere in the eastern United States, except Maine and Florida. *Juniperus virginiana* is the major species of eastern red cedar, but in some south Atlantic and Gulf coast plains areas, there is *Juniperus silicicola*. The heartwood of the red cedar is bright red, sometimes almost purple, tending to a duller red. The thin sapwood is almost white and clearly defined. The wood is moderately heavy, moderately low in strength, and hard. It seasons well and has a tight, fine texture. Grain is straight. The wood is easily worked with conventional tools of all kinds, but will deposit lots of pitch. It is stable when seasoned and shrinks only a little while it is drying. Selected pieces make fine small boats, and as with the other aromatic cedars, it is great for cabinetry that involves clothing, and for work for people who simply like the look of cedar.

Western red cedar grows in the Pacific northwest, and along the Pacific, or left, coast to Alaska. *Thuja plicata* is also called canoe cedar, giant arborvitae, shinglewood, and Pacific red cedar. Principal production is in Washington, but much comes from Oregon, Idaho, and Montana. The aroma is pungently attractive. Heart-

Canac's oak has a darker look in keeping with its design.

wood of the western red cedar lumber is a reddish brown to a dull brown, and the sapwood is nearly white, and narrow, seldom forming a band more than 1 inch thick. The wood is generally straight, with a coarse texture. Shrinkage is slight, and the resulting wood is light-weight, moderately soft, and low in strength. Primary uses are for shingles, boxes, boat building, greenhouse construction, millwork, and similar items. You may wish to use it to line some cabinets, especially those containing clothing. Western red cedar works easily—most red cedar shakes are hand-split—with all conventional tools, both hand and power.

Northern and Atlantic white cedar are two distinct species that are clumped together. Northern white cedar is *Thuja occidentalis* (also

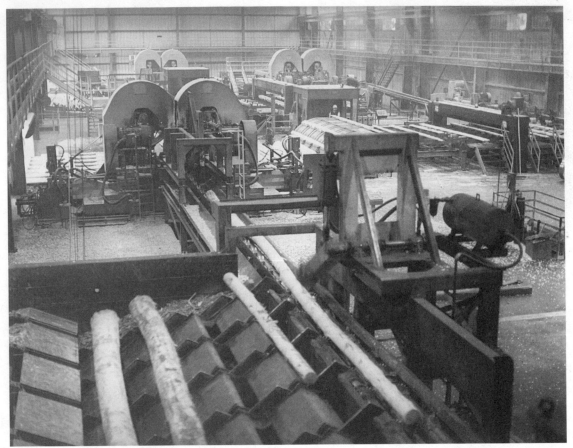

Debarking wood. *(Georgia-Pacific Corp.)*

called simply cedar and arborvitae). Southern white cedar is *Chamae-cyparis thyiodes* (southern white cedar, swamp cedar, and boat cedar). Northern white cedar grows from Maine along the Appalachians, into the northern part of the Great Lakes states. Atlantic white cedar grows along the Atlantic coast from Maine down to northern Florida. The latter is strictly a swamp tree. Production of northern white cedar is heaviest in Maine and the Great Lakes states, while swamp cedar is produced primarily in North Carolina and along the Gulf coast. White cedar heartwood is light brown, and the sapwood is nearly white and forms only a thin band. The wood is lightweight, soft, and not strong. Shrinkage during seasoning is low, and the wood is stable afterward. This cedar works easily with conventional tools, both hand and power.

Next is cypress, sometimes known as bald cypress. Cypress is called southern cypress, red cypress, yellow cypress, or white cypress. The wood is less available than in years gone by, and forced-growth and current early-growth heartwood cypress is not as durable as old-growth heartwood. Cypress is another tree with a small sap-wood band, which is nearly white, contrasting strongly with the light yellowish brown to dark reddish brown, brown, or even chocolate color of the heartwood. *Taxodium distichum* is moderately heavy, moderately strong, and moderately hard. It's easy to work with conventional hand and power tools, and shrinks a bit more than cedars. Durability (decay resistance) ranges from truly great for old-growth cypress to moderate for new-growth materials. Fungus attacks the living trees and creates a pecky condition that stabilizes on seasoning, so that pecky cypress is a very attractive wood for cabinetry. Cypress is great for general millwork, boxes of most sizes, shingles, molding, and cabinetry.

Firs are next on our list. There are Douglas firs and true firs, in eastern and western divisions, to choose from.

Douglas fir is called red fir, or yellow fir, and Douglas spruce. Most production comes out of coastal states, Oregon, Washington, and California, and from the Rocky Mountain states. Douglas fir is classified as an outstanding softwood throughout the world. *Pseudotsuga menziesii* has narrow sapwood in old-growth trees, though it may be as much as 3 inches wide in newer growth, even trees of commercial size. On older trees, the bark is often 1 foot thick. Fairly young trees have a reddish heartwood and are called red fir. Narrow-ringed wood on old trees is yellow-brown, and the wood is sometimes marketed as yellow fir. Weight and strength vary widely. Douglas fir is used for building and construction, and in plywood, as well as in sashes, doors, general millwork, boat construction, and similar items, with its plywood used in all types of construction, furniture, cabinets, and elsewhere. The wood works well with common hand and power tools. Very sharp blades on power tools help reduce tear-out and splintering caused by the coarse texture. Staining is a problem, although paint and clear finishes over unstained wood are fine. Use a sealer before staining. The wood grips nails and screws well and works with all types of adhesives. It is very slightly resinous and so gums up saw blades and other tools.

While the Douglas fir is a marvelous tree, it isn't a true fir. True firs are of the species *Abies,* found throughout much of North America. The wood is nonresinous and works easily with sharp tools.

True eastern firs consist of two species: balsam fir, or *Abies balsamea,* which is primarily a tree of New York, New England, and Pennsylvania, plus the Great Lakes states; and Fraser fir, or *A. fraseri,* which grows in the southern Appalachians. Wood from true firs, both western and eastern species, is creamy white, shading to a pale brown, with heartwood and sapwood nearly impossible to tell apart. Balsam fir is lightweight, low in bending and compressive strength, and soft.

True western firs include five species: subalpine fir, or *Abies lasiocarpa*; California red fir, or *A. grandis*; noble fir, or *A. procera*; Pacific silver fir, or *A. amabilis*; and white fir, or *A. concolor.* Western firs are lightweight. Shrinkage is small to moderate. Lumber comes from Washington, Oregon, California, western Montana, and northern Idaho and is marketed as white fir throughout the United States. High-grade noble fir is used for interior finish and moldings, but other firs go into general construction lumber, boxes, crates, sashes, doors, and general millwork. Ladder rails are often made from noble fir. Like eastern true firs, the wood is readily worked with standard hand and power tools.

The United States and Canada share two species of hemlocks. North America is the only source of commercial hemlock timber. Western hemlock is commercially the more important of the two species, and the wood is more attractive.

Eastern hemlock grows from New England to northern Alabama and Georgia and into the Lake States, as well as into eastern Quebec and southern Maritime Provinces. *Tsuga canadensis* is known as Canadian hemlock and hemlock spruce. Production is split about evenly among New England, the middle Atlantic, and the Great Lake states. Heartwood of Eastern hemlock is pale brown, with a reddish tint. The sapwood doesn't form an easily identifiable separation from the heartwood. Hemlock is coarse textured, and uneven, moderately lightweight, moderately hard, moderately low in strength. The wood is used for lumber and pulpwood, with the lumber used primarily for building framing, sheathing, subflooring, and in the making of pallets and boxes. It works easily with conventional tools. Not of great interest to the cabinetmaker, Eastern hemlock may serve for small projects, and as part of larger projects. It works well with regular tools, but is subject to tear-out in crosscutting. Screws and nails and glues all hold well. Eastern hemlock doesn't take finishes as well as the western variety.

Western hemlock is west coast hemlock, Pacific hemlock, British Columbia hemlock, hemlock spruce, and western hemlock-fir. Like eastern hemlock and many of the firs, it is often sold as hemlock-fir. It grows along the Pacific coast of Oregon and Washington, and in the northern Rockies, north to Canada and Alaska. *Tsuga heterophylla* has a relative—mountain hemlock (*T. mertensiana*)—that inhabits mountainous country from central California to Alaska. Heartwood and sapwood are almost white, with a slight purplish tint, and the sapwood is somewhat lighter in color and generally not more than 1 inch thick. The wood frequently has small, sound black knots that are tight and stay in place. Western hemlock is moderately lightweight and has moderate strength. Shrinkage is about the same as that for Douglas fir, but green hemlock contains more water than Douglas fir and so requires more kiln drying time to reach the same percentage of dryness. Western hemlock's principal uses are as lumber and in plywood, and end uses include sheathing, siding, subflooring, joists, studs, planks, rafters, boxes, pallets, crates, and small amounts for furniture and ladders. For the average cabinetmaker, it's not a very attractive wood, but it does a nice job for larger projects that need hidden frames. It works well with conventional hand and power tools and glues readily, while taking finishes well, after sanding to a silky smooth surface.

Western larch grows in western Montana, northern Idaho, northeastern Oregon, and the eastern slope of Washington's Cascade Mountains. The largest percentage of *Larix occidentalisis* is produced in Idaho and Montana, with the rest coming from Oregon and Washington. Sapwood is generally 1 inch or less thick, and the wood is stiff, moderately strong, hard, moderately high in shock resistance, and moderately heavy. Shrinkage is moderately large in this straight-grained wood, and distortion during drying can be a problem. The wood splits easily. Knots are common, but are usually small and tight. Western larch is useful primarily for rough dimension building lumber, small timbers, planks, and boards; and the lumber is sometimes sold mixed with Douglas fir, which has similar properties. Some of the best material is made into interior finish products, sashes, and doors. For cabinetmaking purposes, it is most commonly used in the framing of projects.

Pines provide some good, basic cabinetry woods with attractive figure and grain, as can spruces. Pines are more prolific than any other wood, and they are probably used for more things than most other woods. Pines come in species that are very soft and low in strength, but still strong enough for building construction. They accept shaping amazingly well, so that a great many items from toys to doors are produced from pine today, as has been true for centuries and will be for centuries to come. And pines come in species that are almost as hard as a midrange hardwood (many of the yellow pines), that do not take nails easily, and that are not as easy to shape, but that are wonderfully strong and useful in many ways.

Eastern white pine grows all the way from Maine to northern Georgia, and over into the Great Lakes states. You can also find *Pinus strobus* under the names *white pine, northern white pine, Weymouth pine,* and *soft pine.* One-half of the production of eastern white pine lumber comes out of the New England states, with the other half spread between the Great Lakes states and the middle Atlantic and south Atlantic states. Heartwood is light brown, often tinged red, but turns considerably darker when exposed to light and air. The wood has uniform texture and is straight-grained. It kiln-dries easily, with little shrinkage, and ranks quite high in stability. It is also easy to work

Mark Martin moves logs in his sawmill.

with all sharp tools, and it is easily glued with almost all cabinetmaking adhesives. Eastern white pine is lightweight, moderately soft, and moderately low in strength and has a low resistance to shock. Second-growth knotty lumber is used for containers and packaging, while high-grade lumber goes into casting patterns, sashes, doors, furniture, trim, shade and map rollers, toys, and dairy and poultry supplies. Caskets and burial boxes also get their share, and the cabinetmaker finds it works easily with conventional tools. It takes finish well, especially if you seal—with shellac—all the knots. Although it is relatively low in pitch, cleaning the blades and bits frequently helps to prevent burning. When aged, eastern white pine is a pumpkin color, and it often derives the name pumpkin pine from that color. White pine is one of the lower-cost woods, and it comes in many grades.

Pitch pine, *Pinus rigida,* is found along the mountain spine from Maine through northern Georgia, and it is a resinous wood with a brownish-red heartwood. The sapwood is thick and a light yellow, while the wood of the pitch pine is moderately heavy, fringing toward heavy, moderately strong, stiff and hard, and with moderately high shock resistance. Shrinkage is moderately small to moderately large, and use is mostly as lumber and pulpwood. The wood works easily with conventional hand and power tools, but the buildup of resin is quick and must be cleared from blades frequently.

Ponderosa pine is known as western soft pine, western yellow pine, bull pine, and blackjack pine. *Pinus ponderosa* has a companion tree that grows in close association, the Jeffrey pine (*P. jeffreyi*), and is marketed under the same name. Major growth areas are California, Oregon, and Washington, with smaller amounts coming from the southern Rockies and the Black Hills of South Dakota and Wyoming. This is a true yellow pine, not a white pine, but the wood is similar to the wood of white pine in appearance and properties. The heartwood is a light, reddish brown, and the wide sapwood band is nearly white, varying to pale yellow. Wood of the outer portions of the tree is moderately lightweight, moderately low in strength, moderately soft, and moderately stiff. It is straight-grained and has modest shrinkage during drying. The wood is uniform in texture, with little tendency to warp or twist. It is used mainly for lumber, and to a small extent for veneers, poles, posts, and so forth; but clearwood makes sashes, paneling, mantels, moldings, and cabinets. Much now goes into particleboard and paper making. It works well with conventional hand and power tools. Wood can be painted or varnished, but is best finished after a sealer is used. It finishes out exceptionally well for a softwood.

Red pine is Norway pine, hard pine, and pitch pine. *Pinus resinosa* grows in New England, New York, Pennsylvania, and the Great Lakes states. Heartwood is pale red varying to reddish brown, while the nearly white sapwood may have a yellowish tinge. Sapwood is 2 to 4 inches wide. The wood resembles lighter-weight southern pine. Latewood is distinct in the growth rings. Red pine is moderately heavy, moderately strong and stiff, and moderately soft. It is straight-grained, but not as uniform in texture as eastern white pine. It is resinous, but not extremely so. Shrinkage is moderately large, but stable when seasoned. Red pine goes into sashes, doors, general millwork, and crates. The wood is easy to work with conventional hand and power tools. Resin buildup is moderate.

Sugar pine is the world's largest species of pine, and it is sometimes called California sugar pine. Most lumber is produced in California. *Pinus lambertiana* has a buff to light-brown heartwood, sometimes tending toward red. Sapwood is a creamy white, and the wood is straight-grained, fairly even in texture, and very easy to work with sharp hand or power tools. Shrinkage is small, and the wood seasons well without checking or warping. It is also stable when seasoned, lightweight, moderately low in strength, and moderately soft. Sugar pine is used primarily for lumber products, with most going to boxes, crates, sashes, doors, frames, general millwork, building construction, and foundry patterns. Like eastern white pine, sugar pine is suitable for use in nearly every part of a house because of its working ease, its ability to remain in place, and its easy nailing properties. It is also superb for the cabinetmaker, as a moderate-cost, high-return wood for many projects. It glues well and finishes nicely, too.

Western white pine is known as Idaho white pine or simply white pine. About 80 percent of the cut comes from Idaho, with the rest coming from Washington, Montana, and Oregon. The heartwood of *Pinus monticola* is a cream color to light reddish brown, and it darkens on exposure to light and air. (A number of woods do this. Cherry is a classic example; darkening extends over a year or so.) The sapwood is 1 to 3 inches wide and yellow-white. The wood is straight-grained, easy to work, easily kiln-dried, and stable after drying. Western white pine is moderately low in strength, moderately soft, moderately stiff, and lightweight. It glues easily and finishes well. Shrinkage is moderately large, but the wood is stable after drying. Most becomes lumber. Some is made into siding and exterior and interior trim and finish. Overall, a marvelous wood for boxes, furniture, toys, and many cabinetmaking projects, western white pine also is relatively low-cost and needs only conventional tools.

There is a lot to say for redwood in cabinetmaking: It doesn't put cherry or walnut to shame, but gives a different look to a room. Very sharp tools and a little extra care reduce possible problems. Make sure you crosscut with a very fine toothed, sharp blade to prevent tear-out. Ripping requires little or no special technique or blade. Allow no resin buildup.

Redwood is a very large tree, growing along the coast of California. *Sequoia sempervirens* has a closely related species, giant sequoia (*Sequoiadendron giganteum*), that grows in a very limited area of the Sierra Nevada of California. Use of the giant sequoia is very limited. Redwood is also called coast redwood, California redwood, and sequoia. The heartwood of redwood varies from a light cherry red to a dark mahogany color, and the narrow sapwood is almost white. Old-growth redwood is moderately lightweight, moderately strong and stiff, and moderately hard. The wood is easy to work, is generally very straight-grained, and shrinks and swells very little. Heartwood from old-growth trees has extremely high decay resistance, but heartwood from second-growth trees ranges from resistant to moderately resistant. Most redwood lumber is used for building, and it is made into siding, sashes, doors, finish, and containers. It is used for all sorts of tanks, silos, cooling towers, and outdoor furniture, where durability is important. At one time, plywood was manufactured, but my sources tell me no more is being made, without specifying a reason. The wood works easily with conventional power or hand tools, and resin buildup is slight. The wood splits, but doesn't splinter easily, so you can often work with less damage to your hands.

Redwood leads us to the spruces. Spruces provide light-colored, reasonably strong wood that can go into building construction or pulpwood, or into cabinets. The Sitka spruce is really the only one of the group suitable for cabinetry.

Sitka spruce grows along the northwestern coast of North America, from California to Alaska. *Picea sitchensis* is also known as yellow spruce, tideland spruce, western spruce, silver spruce, and west coast spruce. Sitka spruce heartwood is a light, pinkish brown, and the sapwood is creamy white and shades gradually into the heartwood. The sapwood may be 6 inches wide. The wood has a fine, uniform texture and is generally straight-grained, with no distinct taste or odor. It is moderately lightweight and moderately low in bending and compressive strength. Shrinkage is fairly small, and straight, clear-grained pieces are easily obtained. Principal uses are in lumber, furniture, millwork, and boats. It works very nicely indeed with sharp conventional hand and power tools, and it makes superb drawer boxes of almost any size, while the lack of taste and odor makes it excellent for canisters and similar projects. It is great for paneled projects, too, and glues up nicely. It takes stain fairly well and takes good finishes with both paint and varnishes. Sitka spruce is on the costly side for softwoods.

Hardwoods shine next and really dominate the cabinetmaking field.

Domestic Hardwoods

Hardwoods are the furniture woods of the world, and as such they are thought to be greater beauties than softwoods, while also being more costly and more difficult to work than softwoods. The woods listed here are domestic hardwoods of species you'd almost have to go way out of your way to avoid using if you're at all serious about custom cabinetmaking. Some are useful only as detail woods (most of the fruitwoods, such as apple and pear, are too twisted in grain most of the time to use otherwise, and holly is too small and hard to find).

Red alder is a Pacific coast species, used commercially in Washington and Oregon, where it is the most abundant commercial hardwood species. *Alnus rubra* has no visible boundary between heartwood and sapwood, with wood that varies from almost white to a pale, pinkish brown. Growth is in the lowlands and usually within 50 miles of salt water. Tree growth is rapid, and tree life is

> Domestic hardwoods are the primary woods of cabinetry; they are considered more beautiful than softwoods and are a great deal more durable as well. The array of such woods is wide, but in any given period, three or four are most useful.

short, at 60 to 80 years. The wood is moderately lightweight and intermediate in most strength characteristics. Alder rotary peels well for veneer. Shrinkage during drying is relatively low. Red alder is principally used for furniture, doors, panel stock, and millwork. Use sharp tools, whether hand or power. Unlike some hardwoods, alder doesn't insist that the cabinetmaker show up with carbide-tipped tools, although they help. It is exceptionally good for many projects and has been used to make artificial limbs.

Apple often brings a puzzled look to a commercial wood dealer's face. *Malus sylvestris* is available on a local and infrequent basis, and it is not a commercial wood; but it is grown, or is found, in most of the world today. The tree is often not shaped well, producing distorted wood. Apple is a pale to medium-pinkish brown and has a fine, even texture. The wood weighs about the same as beech and is hard to dry. It is stable once dried. The wood is hard and strong. It saws and machines well with sharp tools, but to get a good finish requires a lot of work. Irregular grain can tear out. Apple accepts fine detail and holds it, making it good for carving; carved or turned details in custom kitchens may profit from its use. It is often used for wooden screws and saw handles, and as a decorative inlay.

Ash comes in several important species, including white ash (*Fraxinus americana*), green ash (*F. pennsylvanica*), blue ash (*F. quadrangulata*), black ash (*F. nigra*), pumpkin ash (*F. profunda*), and Oregon ash (*F. latifolia*). The Oregon ash is the only one native to the west. The wood is straight-grained, sometimes with a wavy grain that improves its appearance. It dries readily with moderate shrinkage and is relatively stable after seasoning. It steam-bends well. Nail and screw holding power are high, and gluing presents no difficulties. The wood takes a nice finish and sands easily. Ash works well with conventional hand and power tools. Darken it slightly, and ash resembles oak. Black ash has the most interesting grain and figuring. (Black ash may be marketed as brown ash.) Heartwood is brown, and the sapwood is very light, almost white. Second-growth trees have a large proportion of sapwood. Second-growth white ash is heavy, strong, hard, and stiff. Because of these qualities, ash is typically used for tool handles, oars, and baseball bats. Ash has no taste or odor to impart to contents, so is superb for parts that will contain or contact food, such as cutting boards. Lighter-weight ash wood is often sold as cabinet ash and is great for most cabinetmaking projects. Some is peeled into veneer.

Aspen consists of two species, bigtooth aspen (*Populus grandidentata*) and quaking aspen (*P. tremuloides*), that are mainly cut in the northeastern and Great Lakes states, with some production in the Rocky Mountain states. Sometimes called popple and poplar, aspen is also confused with tulip poplar (yellow poplar). The wood resists splitting during nailing and works easily with hand tools. Sharp tools are necessary because the wood fibers tend to fuzz when worked. Apply a sealer before staining, or else the stain will blotch. Aspen finishes well. Glues work well on aspen. It wears well indoors without splintering. Heartwood is gray-white to light gray-brown, with lighter-color sapwood that merges gradually into the heartwood without a clear defining line. The wood is straight-grained, with a fine uniform texture, and is easily worked with conventional tools. If properly seasoned, aspen doesn't impart flavor to foodstuffs. Aspen wood is lightweight and soft, low in strength, and moderately stiff. Uses include toothpicks, boxes, crates, paneling and chipboard, pulpwood, and veneer, as well as a range of turned articles. For cabinetmaking purposes, it is great for many parts, but seldom is used as a finish wood. It almost never splinters.

American basswood is the most important of the native basswood species. *Tilia americana* is followed in importance by white basswood (*T. heterophylla*). They are sold with no attempt to tell the difference. Basswood grows in the eastern half of the United States and Canada, southward. Most basswood lumber comes from the Great Lakes, middle Atlantic, and central states. Heartwood is pale yellow-brown, with some darker streaks. Basswood has wide, cream-colored or pale brown sapwood that merges gradually into the heartwood. When seasoned, the wood has neither odor nor taste. It is soft, lightweight, and of a fine, even texture. The wood is straight-grained and easy to work with hand tools. It holds detail well. Shrinkage during drying is great, but basswood seldom warps after seasoning. The wood takes finishes well, although stains blotch, and holds nails and screws well. It is easy to glue. Basswood is often used in apiary supplies (bee hives, etc.), door frames, molding, woodenware, cooperage, and for veneer and pulpwood.

American beech is native to the United States and Canada. It grows in the eastern third of the United States and up into abutting Canadian provinces. The greatest production of beech lumber is in the central and middle Atlantic states. Wood color varies from white sapwood to a reddish-brown heartwood, with little or no demarcation between the types visible. Sapwood may be 5 inches thick. Beech has little figure and is of close, uniform texture, with no characteristic taste or odor. Grain is straight. Beech is classified as a heavy, strong hardwood that is great for steam bending. *Fagus grandifolia* shrinks substantially during seasoning. It also tends to move a lot under changing humidity conditions. Beech works well with hand and power tools. Nails and screws hold well; beech glues easily and well and stains and polishes very well; and it is readily rotary-peeled for veneers. Most beech goes into flooring, brush blocks, handles, veneer, woodenware, cooperage, containers, and similar areas; but because it is difficult to dry, a great deal also goes into pulpwood and clothes pins. Beech heartwood and sapwood have markedly different expansion and contraction rates, so cannot be mixed in a project. It is unusual in cabinetry and difficult to use, and it can suit demanding buyers.

Birch for the cabinetmaker's purpose means yellow birch (*Betula alleghaniensis*), paper birch (*B. papyrifera*), and sweet birch (*B. lenta*). River birch (*B. nigra*), gray birch (*B. populifolia*), and western paper birch (*B. papyrifera* var. *commutata*) make up the remaining mix. The first three grow mainly in the northeastern and Great Lakes states and into the adjoining Canadian provinces. Yellow birch and sweet birch also grow along the Appalachians down to northern Georgia. These are the source of most birch lumber and veneer. Yellow birch has white sapwood and light, red-brown heartwood, while sweet birch's light-colored sapwood meets a dark-brown heartwood tinged in red. The wood from sweet and yellow birch trees is hard, heavy, and strong. It has a fine, uniform texture. Birch holds nails and screws well, glues well, and is fairly easy to machine with power tools. It can be worked with hand tools, but with difficulty, and machine working is far easier. Birch stains nicely if a sealer is used, but blotches without a sealer. It polishes out to a fine finish. Yellow and sweet birch lumber and veneer go mainly into furniture, boxes, woodenware, interior-finish products, paneling, and doors.

Butternut is called white walnut, partly because, when stained, it so closely resembles black walnut. *Juglans cinerea* grows from southern New Brunswick, into Maine and west to Minnesota, with a

Void-free birch plywood.

southern range down into northeastern Arkansas and east to western North Carolina. Sapwood is narrow, and almost white, with heartwood a light brown showing some pink tones or dark-brown streaks. Butternut wood is lightweight—not much heavier than eastern white pine—and coarse-textured, as well as moderately weak in bending and endwise compression. It is also relatively low in stiffness and fairly soft. Butternut machines easily, with both power and hand tools, and finishes well. It glues well. Sharp tools help prevent tear-out of the soft fibers. It holds nails and screws securely. Butternut is a fine furniture, cabinet, paneling, trim, and veneer wood, but may be hard to find commercially because most cabinetmakers don't ask for it—though carvers do.

Black cherry started its cabinetmaking life, or a large part of it, as a cheaper substitute for mahogany, which is an insult to a superb working wood that is exceptionally beautiful. *Prunus serotina* is scattered from southeastern Canada throughout the eastern half of the United States. Cherry heartwood is a medium red-brown with its own characteristic luster. Sapwood is narrow and nearly white. Grain is straight, finely textured, and close, with usually a gentle waving figure. Ripple and quilt patterns sometimes crop up, too. Cherry has a uniform texture and excellent machining properties, except for a tendency to burn if dull tools or a too-slow feed is used (with power tools). High-speed steel blades create less of a burning problem than do carbide-tipped blades. Use slow drill press speeds (250 rpm, or even 200), and do not pause or you'll again get some burning. Routing demands light passes, because even a split-second pause creates burn marks. Cherry needs a fairly fast, absolutely steady feed with saws and routers, and a fairly slow, but steady feed with drills. Twist drills always create burn marks. It is a medium-heavy, strong, stiff, moderately hard wood. Cherry is lighter than hard maple and about two-thirds as hard, but just as strong. Cherry shrinks a lot when drying, but is stable after it is seasoned. Black cherry glues well, but is affected strongly by excess glue squeeze-out—it mars the finished work badly when clear finishes are applied. There is little reason to use cherry if you're not also going to use a clear finish, so extra care is needed. Cherry finishes beautifully. Black cherry is used for fine furniture, fine plywood veneers, architectural woodwork, and woodenware novelties, and in a wide range of other projects.

American chestnut once grew in vast commercial quantities from New England to northern Georgia. Almost all *Castanea dentata* has been killed by a blight introduced early in this century, and current supplies come from standing dead timber. It is not a cabinetmaking wood. This was once the dominant tree in the deciduous forest of eastern North America.

Cottonwood consists of several species of the genus *Populus,* with the most important being eastern cottonwood (*P. deltoides*), known as Carolina poplar and whitewood; swamp cottonwood (*P. heterophylla*), also called cottonwood, river cottonwood, and swamp poplar; black poplar (*P. trichocarpa*); and balsam poplar (*P. balsamifera*). Eastern cottonwood and swamp cottonwood grow throughout the eastern half of the United States. Black cottonwood grows in the west coast states

Cherry raised-panel kitchen. *(Kraftmaid)*

and in western Montana, northern Idaho, and western Nevada, while balsam poplar grows from Alaska across Canada and in the northern Great Lakes states. Heartwood of the three cottonwoods is gray-white to light brown, with a whitish sapwood that merges gradually with the heartwood. The wood is fairly uniform in texture, straight-grained, and odorless when seasoned well. Eastern cottonwood is moderately low in bending and compressive strength, moderately stiff, and fairly soft. Black cottonwood is slightly below eastern cottonwood in most strength properties, and cottonwoods generally shrink a fairly large amount during seasoning. After curing, the wood is stable, and nail and screw holding ability is very good. The wood works easily and accepts all glues nicely. A sanding sealer helps reduce fuzzing. Cottonwood generally is okay to stain, but not great. Principal uses are for veneer, pulpwood, lumber, and fuel. Cottonwood cabinets would be exceptionally rare, and it might be worth some experimentation to see what sort of finishes work best and how difficult building is. I don't recall ever seeing a veneer, so most of your experimenting would be with solid wood, even for doors and drawer fronts.

American elm is another species of wood that will bring tears to your eyes. It is seriously threatened and in some areas nearly extinct because of two diseases, so most of what's sold as elm today is from other elms. Other useful elms found in the United States and Canada include slippery elm (*Ulmus rubra*), rock elm (*U. thomasii*), winged elm (*U. alata*), cedar elm (*U. crassifolia*), and September elm (*U. serotina*). Each of these is known by a number of names, depending on the area of the country in which it is located. American elm is called white elm, water elm, and gray elm; slippery elm converts to red elm; rock elm is cork elm or hickory elm; winged elm is called wahoo; and cedar elm becomes basket or red elm, while September elm is red elm. Elm sapwood is nearly white, and the heartwood is a light brown, often with a reddish tint. Hard elms are rock elm, winged elm, September elm, and cedar elm. American elm and slippery elm are the soft elms. Most elms are best worked with power tools, as the wood is strong, tough, and coarse-textured as well as moderately heavy. Grain is sometimes straight, but also interlocks. Elm glues well and stains and polishes nicely. It holds nails and screws solidly, too. Elms have superb steam bending qualities. Most elm comes from the Great Lakes, central, and southern states, and its main uses are in boxes, baskets, furniture, and veneer, and for fruit, vegetable, and cheese boxes. Elm is another wood that has no characteristic odor or taste when seasoned, making it ideal for food contact uses such as cutting boards. Elm burls are used in fine veneers. This is an underused wood for cabinetmakers, and as such it may prove to be a bargain in some areas.

Pecan hickory includes bitternut hickory (*Carya cordiformis*), pecan (*C. illinoensis*), water hickory (*C. aquatica*), and nutmeg hickory (*C. myristiciformis*). Bitternut hickory is found through the eastern half of the United States, and into Canada, while pecan hickory grows from central Texas and Louisiana to Missouri and Indiana. Water hickory resides from Texas to South Carolina, and nutmeg hickory is usually found in Texas and Louisiana. Wood of the pecan hickory differs little from true hickory, with a fairly wide band of nearly white sapwood and a darker, whitish heartwood. The wood is heavy and hard, with large shrinkage during drying. Pecan hickory is used in tool and implement handles and in flooring. High-grade logs go for decorative veneer and paneling.

True hickory is included with pecan hickories because of similarities in the wood, and because you'll almost never find it differentiated

Hickory kitchen. *(Kraftmaid)*

at any retail level. True hickories cover most of the eastern half of North America, once you get below the absolute total frost areas. Most important species include the shagbark, or *Carya ovata,* pignut (*C. glabra*), shellbark (*C. laciniosa*), and mockernut (*C. tomentosa*). Hickory splits and checks a lot when drying, and it is rough on tools. It holds screws and nails well—but you absolutely have to drill first, for both. Hickory also glues reasonably well. Wood texture is coarse, and grain is usually straight, though wavy, and other irregular graining does occur. Hickory looks somewhat like ash, but is much heavier and harder. It is dense and tough. It dulls tools quickly and almost demands carbide-tipped blades. It can be stained nicely and finishes up nicely, usually needing a filler. For all commercial uses, the greatest production of true hickory comes from the central and middle Atlantic states. The southern and south Atlantic states produce nearly one-half of all hickory lumber. Sapwood is nearly white, and the heartwood tinges to red, with both sapwood and heartwood weighing the same. True hickory is exceptionally hard, heavy, tough, and strong, and it shrinks a lot in seasoning. Tool handles use a lot of hickory, as do ladder rungs, poles, and furniture. It bends reasonably nicely with steam.

Holly is often considered a domestic exotic, because it tends to be both hard to find and costly. It is found in most of the eastern United States (from eastern Pennsylvania south and east to east Texas). Even after drying it is not stable, having a lot of movement as humidity changes, and therefore reinforcing the need to work it in small sizes. Holly is the whitest wood known, with pure-white sapwood and a creamy heartwood and with little to distinguish between the two. It is moderate in weight, but hard enough to make working it difficult with hand tools. It glues up nicely and resists splitting, so it holds screws and nails well. It is fine-grained and sands to an extreme smoothness. Holly also accepts stain beautifully. Holly turns well and is a good choice for cabinetmakers who want to do some luxury inlay work for their clients.

California laurel is a local wood, also called Pacific myrtle and pepperwood (the leaves have a sharp, spicy aroma when bruised), and has as a distant cousin the more widely available sassafras. *Umbellularia californica* is at home on the range between the mountains and the Pacific, reaching down from southern Oregon almost to Mexico's Baja peninsula. Laurel's sapwood is tan, and the heartwood is close-grained and a light brown. There is a pronounced figure in almost all cuts, and the wood produces many bird's-eye figures, mottles, and swirls. The wood is hard, heavy, and strong, comparing favorably to red oak in these characteristics, but is much tighter-grained and less apt to splinter than any oak. Laurel finishes nicely but doesn't work well with hand tools. Laurel glues nicely, takes screws and nails well, and is so tight-grained that sanding sealers aren't needed. It takes stain and all finishes well.

Magnolias—southern, sweetbay, and cucumbertree—are a group of trees that are classified as magnolia for lumber purposes. *Magnolia grandiflora* (southern), *M. virginiana* (sweetbay or swamp magnolia), and *M. acuminata* (cucumbertree) have an extended southern United States range, all the way up into Ohio and southern New York, but Louisiana leads in the production of magnolia lumber. Sapwood of the southern magnolia is yellowish white, and heartwood a light to dark brown, tinged with yellow or green. The wood has a close, uniform texture and is straight-grained, closely resembling tulip poplar (in appearance). It turns nicely and steam-bends well, too. Magnolia has no characteristic odor or taste. Magnolia takes finishes nicely and stains well. It glues up well and doesn't split easily when nails or screws are used. Wood is moderately heavy, moderately low in compressive and bending strength, and moderately hard. Principal uses are in furniture, boxes, sashes, doors, veneers, and general millwork. Magnolia is an often ignored wood that costs about the same as poplar and is at least as useful. It is readily available throughout its range, and it may sometimes be mistaken for soft maple.

It has an appearance that is distinctive enough to make it a great conversational piece for upscale homeowners, if you can find a source.

Sugar maple and soft maple are the major divisions one should expect when buying maple in the United States and Canada. *Acer saccharum* and black maple (*A. nigrum*) are hard maple, and all else is soft maple. Silver maple (*A. sacchariunum*), red maple (*A. rubrum*), boxelder (*A. negundo*), and bigleaf maple (*A. macrophyllum*) provide woods excellent within their own limits. Sugar maple, though, known also as hard or rock maple, is the maple of bowling pins, bowling alley lanes, and other uses where extreme shock resistance, resistance to abrasion, good looks, and easy upkeep are essential.

Maple cabinets. *(Kraftmaid)*

Maple lumber production is primarily from the middle Atlantic states and the Great Lakes states. The hard maples are found from Indiana and Wisconsin east, dipping down into Tennessee, but swinging up along the Appalachians, hanging around mostly on the western mountain slopes. Softer maples grow almost everywhere, in one variety or another. Maple sapwood is commonly white and as much as 5 inches thick, with heartwood a light, reddish brown. Hard maple's texture is fine, very uniform, and the wood is medium heavy, strong, stiff, and has large shrinkage, though it's stable after seasoning. The grain is straight, with little figure, but different patterns add value to the wood in some applications, such as curly, fiddleback, and bird's-eye figures. Maple can be hard to work and stain, which may blotch. Rock maple also has a severe tendency to burn when sawed; extremely sharp tools and carbide-edged tools help. It holds screws and nails well, without splintering, and takes glue well. It finishes and polishes very well. Soft maple is lighter and weaker than hard maple, and easier to work. Maple is especially useful for lumber, veneer, and pulpwood.

Soft maple.

Mesquite brings to mind tumbleweeds and a charcoal grill or smoker, but is also an excellent choice for cabinetmaking of several kinds. Three varieties of mesquite (*Prosopis glandulosa, P. juliflora,* and *P. pubsecens*) provide wood for carvings, turnings, and furniture. There won't be a lot of straight wood in the trunk, as it twists and turns its way upward, but the mesquite wood has a lovely, tightly interlocking grain with a narrow tan sapwood bordering deep-brown heartwood. Mesquite has a limited range, but covers a lot of ground within that range, 75,000,000 acres in Texas alone, extending down into Mexico, Oklahoma, and Arizona. During seasoning, shrinkage occurs evenly, so mesquite remains stable. You need power tools for this heavier-than-oak wood that also has a high silica content—which means you also need carbide-edged tools to retain cutting edges. The wood is brittle, so with the silica content you need to use screws and glue for joinery. Wipe down surfaces with alcohol or lacquer thinner before gluing. You can get a superb, fine finish, and it takes clear finishes well. Stains are not necessary. Mesquite is great as an accent wood, but I'd hate to be using it for drawer fronts and doors.

Red oaks are generally found in the southern states, the southern mountain regions, the Atlantic coastal plains, and the central states, and well up into southeastern Canada. Primary species are the northern red oak (*Quercus rubra*), scarlet oak (*Q. coccinea*); Shumard oak (*Q. shumardii*), pin oak (*Q. palustris*), Nuttal oak (*Q. muttallii*), black oak (*Q. velutina*), southern red oak (*Q. falcata*), water oak (*Q. nigra*), laurel oak (*Q. laurifolia*), and willow oak (*Q. phellos*). Red oak works well with hand and power tools; although it doesn't require carbide-edged blades, power tool cabinetmaking is faster and easier with them. Red oak sapwood is nearly white, 1 or 2 inches thick, while heartwood is reddish brown and coarse-grained. Sawn lumber of the various species cannot be readily separated on the basis of wood alone. Red oak lumber can be separated from white oak by the size and arrangement of pores in latewood, and because it lacks tyloses in the pores (see the discussion of white oaks below for information on tyloses). Red oak splinters, warps, and cups if not perfectly dried, and it is generally hard to handle. Quartersawn oak lumber is easily distinguished by the broad rays, which add to its good looks, and the rays are more evident in white oak. It glues well, holds screws and nails well if pilot holes are drilled—oak has a tendency to split, a tendency that is worse close to edges—and takes stains and finishes nicely. Red oak is heavy, and second-growth oak is usually both harder and tougher than old-growth. Red oaks shrink a lot in drying. Much red oak goes into cabinets, flooring, furniture, general millwork, boxes, woodenware, and handles.

White oaks come mainly from the south, south Atlantic, and central states, including the southern Appalachians. Principal species are white oak (*Quercus alba*), chestnut oak (*Q. prinus*), post oak (*Q. stellata*), overcup oak (*Q. lyrata*), swamp chestnut oak (*Q. michauxii*), bur oak (*Q. macrocarpa*), chinkapin oak (*Q. muehlenbergii*), swamp white oak (*Q. bicolor*), and live oak (*Q. virginiana*). Heartwood is grayish brown, and the sapwood, which may be 2 or more inches thick, is nearly white. The pores of the heartwood of white oaks are plugged with tyloses. *Tyloses* are bubblelike structures that appear in the transition grain which forms from sapwood as it changes to heartwood, and they remain, of course, in the heartwood. It takes a good hand magnifying glass to see them, but remember that tyloses plug the wood so no water, or other liquids, can flow through the heartwood. White oak is a hard, tough, straight-grained wood, with even more of a tendency to splinter than red oak. It holds screws and nails well, if pilot holes are drilled, resists shock well, and generally looks great. White oak figures are more attractive than red oak figures. White oak takes a good finish, with no filler required, and glues easily with all cabinetmaking adhesives *except* epoxy. Green white oak bends easily, and rip sawing is a simple matter with a sharp blade and a slow feed. Use carbide-edged tools whenever you can. Old-growth white oak is not as tough as newer-growth material.

The live oak is a separate tree, but is listed among the white oaks for convenience. Live oak is an evergreen. It is the toughest of all the oaks to work, and it is suitable for use only with power tools with carbide-edged blades. Generally, live oaks are tropical or near tropical trees, and a lot of research remains to be done on their value, and use, as cabinet woods. Do your share if you can find a few board feet. It's hard, but fun, to work. With white oaks in general, take shallow, multiple passes with planers, jointers, and routers, and always use backing boards on cross-grain cuts to reduce splintering.

Pear is another fruitwood that is widely available, mostly from wood from old orchard trees, as is applewood. *Pyrus communis* has wood that is a pale pink-brown, with a very fine texture—finer than apple's texture. The grain is straight if the stem bearing the wood is straight, irregular otherwise. It weighs about the same as beech and dries slowly, with a tendency to distort. If the grain is irregular, it distorts even more. Pearwood is strong, tough, and difficult to split. It is fairly hard to saw and is rough on tool edges, so carbide blades are best. Pear takes a fine finish and reacts well to most glues. Small sizes are available. The wood is used for drawing instruments, T-squares, and cabinetwork.

Persimmon is another backyard tree that provides cabinetmakers with some unusual possibilities. *Diospyros virginiana* is of the same family as ebony, and it is sometimes called white ebony. The tree is usually found in the south, but is seen up into New York and Connecticut. The wood is off-white with a gray-brown tint, heavy, and of little commercial use because the lumber it provides is not very good. Wood texture is fine, and grain is straight. The wood is dense, about 15 percent heavier than beech, but not nearly as dense as black ebony. Persimmon dries with ease, but shrinks a great deal. Even when dry, it moves a lot under changing humidity. It works well with power tools and can be worked with hand tools. Carbide-edged tools are recommended for power working with persimmon. Strength qualities exceed those of beech, and the wood takes an excellent finish and wears well. In the textile industry, shuttles may be made from persimmon, and it is traditionally used in golf club heads because of its exceptional hardness. It is a good cabinetmaker's accent wood.

Sassafras is probably better known for its qualities as a beverage base than as a furniture-quality wood, but *Sassafras albidum* covers much of the eastern half of the United States. It resembles black ash in color, grain, and texture. Sassafras works easily with hand tools, but planing has a tendency to lift the grain if edges aren't very sharp. It glues well, sands easily, and takes a fine finish. Drill pilot holes for screws and nails. Sassafras sapwood is light yellow, while heartwood ranges from dull gray to brown and dark brown, occasionally with a red tinge. The wood is moderately light for a hardwood, weak in bending and endwise compression, soft, and brittle. It is not useful where weight must be borne, but has frequent unusual grain patterns. The roots, with their oil of sassafras, are used to make tea, perfume, and soaps, and were once used as a patent medicine. The small amount of oil in the wood gives it a characteristic odor; it is aromatic, with a medicinal odor that is mild enough to be quite pleasant.

Sweetgum ranges from southern Connecticut west to Missouri and south to the Gulf, but lumber production is almost entirely from southern and south Atlantic states. *Liquidambar styraciflua* has sapwood and heartwood that are solidly differentiated, to the point where each is classified as a different wood: Sapgum is the light-colored sapwood, and redgum is the red-brown heartwood. You don't want the sapgum. The following features are for redgum, or heartwood sweetgum, only. Sweetgum, like tupelo (black gum), has interlocking grain, which makes it very useful for some cabinetmaking applications and awful for anything that requires splitting. Tools must be sharp to do well. Wipe the wood with alcohol before gluing. It holds screws and nails very well, and is not subject to excessive splitting once dry. Quartersawn sweetgum with interlocked grain produces a lovely ribbon stripe figure that is superb for many cabinetry uses. Sweetgum takes finishes well and comes up to a truly lovely luster. The wood is moderately heavy and hard. It is fairly strong and stiff. Mainly it is used for lumber, veneer, plywood, and pulpwood; sweetgum goes into boxes, cabinets, furniture, interior trim, and millwork.

American sycamore is sometimes called the buttonwood tree and the planetree. It grows from Maine west to Nebraska, and south to Texas, then on east to Florida. The central states produce the most *Platanus occidentalis* lumber. Sycamore heartwood is reddish brown, with lighter-colored sapwood, from $1^1/_2$ to 3 inches thick. The wood is finely textured with even, interlocking grain. Sharp tools reduce tear-out. Sycamore glues well and finishes well. It shrinks moderately in drying and is moderately heavy, moderately stiff, and moderately strong. Quartersawn stock seasons easily, but plain-sawn sycamore warps and cups a lot. Quartersawn sycamore shows rays, as does white oak, but is seldom seen. The wood has no characteristic odor or taste. Sycamore handles a lot like cherry—fast feed, light cuts, several passes, sharp tools—but it has even more of a tendency to burn, as if it were maple. Sycamore sands without effort and takes a glasslike finish. All glues perform well, and the wood resists splitting, so pilot holes are only essential when you're hand-driving screws. Plain-sawed sycamore takes stain well, but quarter-sawn stock requires testing to see if a particular stain works properly. Apply a sealer. Sycamore primarily makes furniture, flooring, handles, and butcher blocks; and the veneer is used for fruit and vegetable baskets.

Tupelo has a bunch of names for several species: water tupelo (*Nyssa aquatica*), also called swamp tupelo and tupelo gum; black tupelo (*N. sylvatica*), known also as black gum; and sourgum and several other variants. It is resistant to splitting, so nails and screws hold well, even without pilot holes. All, except for black gum, grow primarily in the southeastern United States, while black gum grows from Maine down to Texas and Missouri. About two-thirds of tupelo lumber comes from the southern states. Wood of the different tupelos is similar in appearance and properties. Heartwood is a light brown, casting to gray, and merges gradually into the lighter-colored sapwood. The sapwood is several inches wide. Tupelo wood has a fine, uniform texture and an interlocked grain. It can be hard to work because of the interlocked grain, and it may also be difficult to finish. Use carbide tools whenever possible. Feed boards at a slight angle during planing, and use a slow feed when ripping. Too fast a feed causes burning of edges with tupelo. Take several light passes when jointing or routing, and use slow drilling speeds. Tupelo glues up well, if surfaces are smooth, and it takes all stains and finishes nicely. It rates as moderately heavy, moderately strong, moderately hard, and stiff. Main uses are in lumber, veneer, and pulpwood.

Black walnut was once North America's premier furniture wood, and its most costly. Today's trends to lighter colors have made it a secondary wood. Cabinetry of walnut is gorgeous, but does tend to darken a room. The natural range of American black walnut is from southern Canada on down into Texas, touching Louisiana, and over into the coastal Carolinas and up the coast to southern New York. *Juglans nigra* has a nearly white sapwood up to 3 inches thick in open grown trees. The heartwood varies considerably, from purplish brown with thin, darker veins to a grayish brown, and an orange-brown. The grain is fairly open and straight, without figure; but in some walnut figures such as fiddleback, burl, stump, and crotch, the grain is coarse-textured. It is heavier than cherry. Walnut is heavy, hard and stiff, straight-grained and easily worked with almost all tools, power or hand, and stable when seasoned. Take shallow cuts and multiple passes when jointing or routing to avoid tear-out. Keep glue squeeze-out to a minimum. Like cherry, walnut suffers badly from sloppiness on the part of the cabinetmaker. I don't fill straight-grained walnut when finishing, though I thin the first coat about 2-to-1 as a sort of primer or filler coat, with all clear finishes.

Custom-made country-look solid walnut cabinets are detailed with turned walnut knobs, raised panels, and brass hinges.

Black willow is the only willow commercially handled under its own name. *Salix nigra* is produced mainly in the Mississippi Valley, from Louisiana to southern Missouri and Illinois. Black willow heartwood is gray-brown, or a light red-brown, with darker streaks often found. Sapwood is a whitish, creamy yellow. Grain is interlocked, and the wood is finely textured. Willow is lightweight, low in strength, and moderately soft. Willow works well with all tools, power and hand, but favors those who bring sharp tools to the work. It accepts glues readily and takes finishes well. Willow principally makes lumber, but small amounts go for veneer, cooperage, pulpwood, artificial limbs, and fenceposts.

Yellow poplar is a widely used, lightweight hardwood that is known as poplar, tulip poplar (here, that's the local name), and tulipwood. Sapwood from yellow poplar is sometimes called whitewood or white poplar. The *Lirodendron tulipfera* range runs from New York and Connecticut down to Florida, and west to Missouri.

> Yellow poplar is a widely used, lightweight hardwood that is usually known as poplar or tulip poplar. The *Lirodendron tulipfera* range runs from New York and Connecticut down to Florida, and west to Missouri. Its primary use is for drawer sides. It works easily, is relatively low-cost, but is not often thought strong enough, or attractive enough, for cabinet fronts and frames.

The greatest commercial production is in the south and southeast. Yellow poplar sapwood is often several inches thick, while the heartwood is a yellowish brown, sometimes streaked with green, purple, black, blue, or red. The wood is straight-grained and uniform in texture. Tulip poplar works easily with hand and power tools, especially those that are kept sharp, and it glues up easily with all cabinetmaking adhe-

Poplar is usually glazed. It may be any of several woods, including aspen, cottonwood, and yellow poplar

sives. It doesn't split easily, so holds nails and screws well, but pilot holes are a help when hand-driving screws, or when within 1 inch of edges. Poplar takes stains, varnishes, and paints well. Old-growth timber is lighter, weaker, lower in bending strength, and softer than newer-growth. Lumber goes primarily into furniture, boxes, pallets, and interior finish. I particularly like it for drawer sides and backs, even bottoms in reproduction furniture. It dries readily and is stable once dried.

Those are many of our domestic high-end woods, middle-range woods, and low-end woods, covering almost all the needs a cabinet-maker is likely to find. It is in the exotics that we find the stretch into luxury cabinetry, with wood grains and colors much different, if only in degree, from the ones found in North America.

Exotics

African blackwood is a relative of the rosewoods, and it is a lovely wood, if very costly. *Dalbergia melanoxylon* comes from a small, usually misshapen tree from Mozambique and Tanzania. Characteristically black, it is shipped only in short billets. Price is on a par with rare Brazilian rosewood, and this wood is also rare; it is suitable only for accents in cabinetry.

Andiroba is widespread in tropical America, and it also is known as cedro macho, carapa, crabwood, and tangare. *Carapa guianensis* is the primary source of andiroba wood, but *C. nicaraguensis* is another, a slightly inferior wood. Heartwood color varies from red-brown to dark red-brown, and the texture is like that of American mahogany (*Swietenia*). The wood is sometimes used in place of mahogany. Grain is interlocked, but it rates as easy to work, paint, and glue. The wood is very durable, heavier than mahogany, and is superior in bending properties, compression strength, hardness, and toughness. For gluing, no special tricks are required. Andiroba makes excellent flooring, furniture and cabinetwork, millwork, veneer, and plywood where durability and beauty must be combined.

Angelique comes from French Guiana and Surinam. *Dicorynia guianensis* has heartwood that is russet-colored when fresh cut, turning to a dull brown, with a purple cast. Heartwood stays more distinctly red, often showing wide bands of purplish color. Texture is coarser than that of black walnut, and the grain may be straight or mildly interlocked. Angelique is stronger than teak and white oak, whether green or air-dried. It is highly resistant to decay and marine borers, working qualities differ according to seasoning and silica content. Once completely dried, angelique is only workable with carbide-tipped power tools. Mating surfaces must be carefully cleaned with alcohol or lacquer thinner before gluing. This is another accent wood for lighter-colored cabinets.

Avodire ranges from Sierra Leone west to the Congo, and south to Angola. It is most common on what was the Ivory Coast, and it is scattered elsewhere. *Turreanthus africanus* forms a medium-sized tree of the rain forest. Wood is cream to a pale yellow in color, with a high natural luster. It eventually darkens to a golden yellow. Grain is straight on occasion, but is more often interlocking in a wavy or irregular pattern. The figure produced when avodire is quartersawn is attractively mottled. Avodire is identical in strength to oak, though considerably lighter (about 85 percent as heavy) and lower in shock resistance. It works reasonably easily with hand and power tools and finishes well. It accepts most glues well, too. Most imported avodire is in the form of highly figured veneer, and it is used in fine joinery, cabinetry, paneling, and furniture.

Bubinga is often also called African rosewood (it is not a rosewood, which requires membership in the *Dalbergia* genus) and comes from west Africa. Heartwood from *Guibourtia tessmanni, G. demeusei,* and *G. pellegriniana* is pink, vivid red, or red-brown with purple streaks. The latter becomes yellow or medium brown with a red tint when exposed to light and air. Sapwood is a creamy vanilla in appearance. Great density and fine texture are features of bubinga. Moderately hard and very heavy, bubinga works well with machine or hand tools. Bubinga is useful in furniture, cabinets, and decorative veneers. It takes clear finishes with no problems, and anyone who stains it probably should be run out of town. Toxicity appears to be low.

Cocobolo is very similar to rosewood: It's the Central American variant, from the same genus, *Dalbergia retusa.* While the other parts of the genus prefer the rain forests of South America, cocobolo thrives in drier upland country of Central America's Pacific coast. Cocobolo is about twice as heavy, air-dried, as cherry, and it is a wood too dense to float. Heartwood has a wide range of shades, including red, yellow, pink, and black, with the occasional streaks of green, purple, and blue. The sapwood is a creamy white. Cocobolo is best worked with power tools or exceptionally sharp hand tools. Sanding and polishing bring out an exceptional luster in the wood. Nail and screw gripping power is great, and splitting isn't a problem, but you must drill pilot holes. Gluing is difficult, because the wood, like rosewood, contains a lot of oils, plus silica. Wipe all mating surfaces with lacquer thinner or alcohol. Use slow-set epoxy, as with teak. Cocobolo needs a surface protector because the wood darkens with age; without some form of finish, whether wax over a penetrating oil or wax alone (other finishes do not work well), the wood turns black as it ages. Cocobolo should never be used for food-handling projects, or in any way that comes up in a project that requires touching of the lips. Cocobolo is readily available, gorgeous, and really expensive.

Ebony is an incredibly expensive wood, and it is rare as well as being unique in color. It is the only truly black wood I know of. Sapwood is yellowish white, but heartwood of *Diosyros* species is jet-black in its most desirable form, but may also be a medium brown to dark brown, with black stripes. It is fine and even in texture, and extremely heavy. Ebony does not work or handle easily. It is hard and brittle after drying. Ebony does not glue well. It can, with care and sharp tools, be brought to a truly excellent finish. Ebony is also a superb turning wood. The wood is used in turnery, musical instruments, cutlery handles, and inlays.

If you get clients who can afford ebony inlays, make sure they're on nonfood surfaces, and consider yourself very fortunate.

Goncalvo alves first came from Brazil, but the species range from southern Mexico through Central America and into the Amazon Basin. Both *Astronium graveolons* and *A. fraxinifolium* are found throughout the area. The wood is also known as tigerwood, zebrawood, and kingwood. Fresh heartwood is russet, orange brown, or red-brown to red, with narrow to wide irregular stripes of medium to very dark brown. The wood is hard, medium-textured, and very dense. It is very strong. In most cases, it is imported for its beauty. Grain varies from straight to wavy and interlocked. Goncalvo alves is difficult to work, but has a natural luster and polishes to a superb finish. Carbide tool edges are almost essential, but it glues up well. It holds screws and nails well, with pilot holes essential to ease of working. It is usually seen as a veneer.

Jarrah is one of the Australian offerings to the world. *Eucalyptus marginata* has a small range, along the coast south of Perth, but it is very common there. The heartwood is medium to dark brown after exposure, but when freshly cut it's a uniform pink to dark red, a rich mahogany color. Sapwood is pale in color and very narrow in older trees, wider in newer growth. Grain is often interlocked and wavy, and texture is even but moderately coarse. The wood glues satisfactorily, but drill pilot holes if you are nailing or driving screws—it tends to be splintery in some situations. Gum veins create some problems in this hard, heavy wood. The wood is difficult to work by hand because of the high density and interlocking grain. Jarrah is often used for heavy construction, exterior and interior millwork, furniture, turnery, and decorative veneers.

Its popularity, thus availability, is increasing in North America, especially for outdoor uses, although for decorative cabinetry jarrah is also hard to beat.

Kapur is found in Malaya, Sumatra, and Borneo. *Dryobalanops* species give us a reddish-brown heartwood, clearly marked from the pale sapwood. The wood is uniform in texture, but fairly coarse, and is straight-grained. A little heavier than oak, it is stable and durable once dried. It is moderately heavy, and the heartwood is resistant to decay. It works with moderate ease with hand and power tools, but is best machined with carbide-tipped tools because it has a high silica content which blunts edges quickly. Dull cutters also tend to raise the grain rather badly. Kapur takes nails and screws well, after pilot holes are drilled, but it is not easily bonded with adhesives. Use urea formaldehyde adhesives for best results.

Kingwood is another of the *Dalbergia* species, this time *D. cearensis*; the wood is often called violetwood or violetta. Kingwood is finely textured and even, heavy, and lustrous. Heartwood has a variegated striped figure of violet and black on brownish purple. Sapwood is white. The wood is hard and heavy, but works well with sharp tools and takes a fine polish. It glues well. Most uses are for inlays and marquetry.

Koa is Hawaii's contribution to the exotics. It is a U.S. wood, but *Acacia koa* grows in quantity only in the Hawaiian Islands, where it seems to grow just about everywhere. Koa weighs about 50 percent more than black walnut and has similar high shock-absorbing qualities. Koa, though, has interlocking grain, so that a superb fiddleback figure shows up often. Sapwood is light-colored, and heartwood is primarily red-brown, going to dark brown, sometimes with tones of gold, black, and deep purple. Bending strength is great, and the wood works well with conventional hand and power tools, though it may burn during routing, or cross-cutting. It is best planed at a slight angle, with a fast feed, to keep the grain from tearing. Kao is best joined with screws as well as glue because there are resin pockets that sometimes prevent good gluing. Koa sands to a truly superb, silky look. Uses include gunstocks, veneers, paneling, furniture, sculpture, and general turnings.

Lacewood covers a grouping of species. The version I've worked with, *Cardwellia sublimis,* is from Queensland, Australia, while another version comes from Great Britain, *Platanus acerafolia.* The others are *Grevillea robusta* (from southern Australia, Africa, India, and Sri Lanka; this can cause a rash similar to poison ivy), and *Panopsis rubescens* (Brazil). Lacewood in its Australian guise is also called silky oak, and in the British Isles version, it goes as planewood and harewood, as well as London plane, while *Panopsis* is also called leopard wood. Silky oak has a light-to-medium cherry background color that looks a lot like the rays exposed when white oak is quartersawn. The figure is consistent, with flaked silvery grain and large, regular medullary rays. The wood is medium-weight, with straight grain except for the rays. Boards are unusually free of defects and knots. Texture is medium to coarse, and flat-sawn planks are apt to warp and cup. The cell walls surrounding the rays are prone to chipping out and so must be cut carefully—use very, very sharp tools, and use backer boards wherever possible. With routers, planers, and similar tools, take several light passes instead of hogging all the material out in a single run. Hand-applied stains blotch, but alcohol-based spray stains work okay. The wood finishes well, holds nails and screws well, and glues up nicely. I understand that very little of the *Cardwellia* is now exported, as it is considered an endangered species in Australia.

Lauan-Meranti grouping is the heading for Philippine, or lauan, mahoganies in three genera: *Shorea, Parashorea,* and *Pentacme.* I'm not going into the fine differences of each type. As a whole, the species have a coarser texture than American mahogany (*Swietenia*) or the African mahoganies (*Khaya*), and do not have dark-colored deposits in the pores. All lauan species have axial resin ducts aligned in long tangential lines in the end surfaces of the wood, and sometimes the ducts contain white deposits visible to the naked eye. The wood is not resinous. Species are very dense. Strength and shrinkage properties compare favorably with those of oak, and all four groups machine easily, except white meranti which has a high silica content. In the United States, most lauan shows up as inexpensive plywood, and is suitable for backing boards for cabinets and shelving.

Limba is also called ofram and afara. It is widely distributed from Sierra Leone to Angola and Zaire, and it is favored as a plantation tree in west Africa. Heartwood in *Terminalia superba* varies from gray-white to creamy or yellowish brown, with frequent almost-black streaks. The resulting figure is highly prized for decorative veneers. The wood is straight-grained and uniform, but coarse, in texture. Strength is moderate, and some timber may be brittle. Limba seasons easily, with slight shrinkage, and is easy to work with hand and power tools. It peels for veneering without problems and is used in furniture, plywood, veneers, and interior joinery. It is sold in the United States as plywood under the copyrighted name Korina.

Mahogany is the name currently given to several different kinds of commercial woods. The original, *Swietenia,* came from the West Indies and was the premier wood for fine furniture in Europe in the 1600s. American mahogany is sometimes thus referred to as true mahogany, while a related African wood, *Khaya,* is marketed as African mahogany. Stability is excellent, and so are workability, glueability, and holding of nails and screws, though pilot holes, as always in hardwoods, work best with the latter. Mahogany has set the standard for other furniture hardwoods, and for hardwoods in general in many ways, for several centuries, as it doesn't warp, cup, check, or otherwise give problems, and it works easily with any kind of sharp tool. And it is gorgeous when polished to a superb sheen. *Khaya* is less stable than *Swietenia,* but only by a single percentage point.

African mahogany is shipped from west central Africa and is a widely distributed and plentiful species, found in the coastal belt of the so-called high forest. *Khaya ivorensis* is the most favored and plentiful, while *K. antotheca* is found farther inland. Heartwood is pale pink to a dark red-brown. The grain is often interlocked, and the texture is medium-coarse, comparable to that of American mahogany. The wood seasons easily, but machining properties may vary. Nailing and gluing properties are good, and it takes an excellent finish. It also rotary-peels easily for veneers and is claimed to be moderately durable. Most uses are for furniture, boat work, cabinetry, interior finish, and veneer.

American mahogany, also called Honduras mahogany, ranges from southern Mexico through Central America and into South America as far south as Bolivia. Plantations have been established throughout its natural range and elsewhere. Heartwood varies from a pale pink to a dark red-brown. Grain is generally straighter than that of African mahogany, but with a wide variety of figures. Texture is fine to rather coarse. *Swietenia macrophylla* air-dries easily, without warping and checking difficulties. It rates as durable, and both heartwood and sapwood resist impregnation with preservatives. The wood works easily by hand, and with power tools, and it slices into a rotary-cut veneer with no problems. It also takes a fine finish quite easily. Air-dried strength of mahogany is similar to that of American elm, and principal uses are for fine furniture and cabinetmaking, interior trim, pattern making, boat construction, fancy veneers, carving, and almost anywhere else an attractive wood with exceptionally good dimensional stability is needed.

Merbau comes from the Philippines, New Guinea, and Malaya, and it is also called ipil and kwila. *Intsia bijuga* is found throughout the Indo-Malay region, and into Australia, as well as on many of the western Pacific islands. Fresh-cut heartwood is a yellow color ranging to orange-brown, and it turns brown to dark red-brown on exposure. Texture is fairly coarse, and the grain is apt to be wavy and interlocked. Quartersawn figures are striped with the interlocked grain, but otherwise the figure is plain. Air-dried strength compares to that of hickory, but the density is a little lower. Merbau seasons well, with little loss, but stains black in the presence of iron and moisture. It gums sawteeth and dulls cutting edges, so is difficult to work, but dresses smoothly and takes a nice finish. It is inclined to brittleness. Merbau glues up well, if mating faces are cleaned with alcohol. Merbau is used in furniture, turnery, cabinetry, musical instruments, and specialty items.

Obeche trees in west central Africa have wood that is creamy-white to pale yellow, with no clear difference between sapwood and heartwood. *Triplochiton scleroxylon* is a fairly soft wood with a medium-coarse texture. The wood is strong. It glues and takes stain well. The grain is often interlocked. Quartersawing produces a stripe figure.

Obeche works easily with hand and power tools, and it takes nails and screws without problems with splitting. The wood is useful for veneers.

Padauk is often called barwood, or camwood, and is made up of seven species belonging to the genus *Pterocarpus.* African padauk, or *P. soyauxii,* is sometimes called vermillion. The wood is hard and heavy, with interlocking grain, and a moderately coarse texture. As with most interlocking-grain woods, quartersawing produces an attractive figure, a strong ribbon stripe. Heartwood is red-purple-brown with red streaks, and the sapwood is a pale beige (sapwood may also be as much as 8 inches thick). As it ages, the wood turns to a deep maroon. The wood works well with all sharp tools, takes glue well, and takes a fine finish. It holds nails and screws well, after pilot holes are drilled, and the heartwood is very durable. The wood must be dried slowly, but dries well, with moderate to low losses from distortion. The wood is used in joinery, for tool handles, and even as flooring in areas where wear would otherwise be extreme. It is also a fine boat-building wood, and has been popular for top-grade furniture. It makes superb cutting board stock, too. Another padauk, muninga (*P. angolensis*), may be available again shortly, as most production has always been from South Africa, making it unavailable in the United States and in many other areas in the immediate past. This is from a much smaller tree—an open-forest tree described as squatty in comparison to the African padauk— and is golden brown to deep brown, with irregular grain. Texture is medium, and it dries slowly, but well. It is also durable, makes a good floor, and has been used in veneer and top-grade furniture in the past, and probably will be again in a few years.

Pau marfim, sometimes called moroti, grows in a limited range from southern Brazil and Paraguay into northern Argentina. *Balfourodendron riedelianum* has heartwood similar in appearance to the sapwood of maple. The pale yellow color points up a fine, even texture. The grain is usually straight, but on occasion may be wavy. The wood is heavy, about like hickory, and is strong and tough, more so than ash, but not up to hickory. It dries fairly easily and takes a smooth finish. The wood is used for striking tool handles and flooring, and the fine texture and light color suggest it as a substitute for boxwood for rules. It also turns well and holds nails and screws strongly, though pilot holes are almost essential to easy working. It works readily with all sharp tools.

Purpleheart is also called amarantha, and it comes from the north central areas of the Brazilian Amazon region, though the species ranges from Mexico through Central America, and on down to southern Brazil. *Peltogyne* species comes from about 20 species of trees and is a dull brown when first cut. On exposure to air and light, the heartwood turns a deep purple, fading on prolonged exposure to light. Final wood tone is a rich red-brown. Textures are fine to moderately coarse, while the grain is often interlocked, though it may also be straight. Quartersawing produces an attractive figure in logs with interlocked grain. Purpleheart is a heavy wood. It is a very strong wood and is stable in use, but difficult to saw. It is best cut and machined with carbide tools when power is used, as it dulls tools quickly because of its hardness. It finishes to a fine-looking gleam. It holds screws well, but demands pilot holes for easy working. The wood is resilient, so splitting isn't a problem. It is also high in resin. It turns nicely and glues up well. Purpleheart stains nicely, too, though I can't imagine why anyone would stain it. The wood is used in fine furniture, cabinetry, and carvings, but the largest use is seen in heavy construction because of its strength and durability.

Ramin is one of a few moderately heavy woods that are classified as blond woods. *Gonystylus bancanus* is found throughout much of southeast Asia, and it was first introduced to the import market in about 1950. The heartwood and sapwood are both a pale, creamy color. Grain is usually straight, but on occasion may interlock. The wood is plain-looking, with no wild or fancy figures in normal logs, but the texture is middling fine and even, much like true mahogany. Ramin is hard and heavy, but easy to work with all sharp tools, finishes nicely, and glues up with no extra preparation. Ramin takes stains, paints, and varnishes very well. The wood is used for furniture, veneer, turnery, toys, carving, and flooring.

Rosewood encompasses a variety of woods of the *Dalbergia* genus. *Dalbergia* also includes African blackwood, tulipwood, and cocobolo (all covered separately). Rosewoods are rare, highly decorative, and highly prized, so are costly. The Indian rosewood tree is *D. latifolia,* while the Brazilian version is *D. nigra.* Texture is uniform and moderately coarse, and the wood dries slowly, but well. Indian rosewood has a subtle ribbon grain figure because of interlocked narrow bands, and the color is a golden brown to a deep purple-brown, with streaks of dark purple or black. The wood is hard to work and glues up best if mating surfaces are cleaned with a solvent such as alcohol. It dulls tools quickly and only works easily if tools are superbly sharp. Uses include all those of Brazilian rosewood, plus musical instruments and veneers. Indian rosewood is more readily available, more stable in use, and less costly—though it is far from cheap. Brazilian rosewood varies from rich brown to a dark violet-brown, with black streaks. The wood is oily and strong and bends under steam well. The wood is heavy, hard, and strong, and moderately hard to work with hand tools, and it keeps on resisting during power tool operations. The lumber contains calcareous (mineral) deposits, which dull tools quickly. Resin levels are also high enough to quickly gum sawblades and sanding belts. Brazilian rosewood turns nicely and has good screw-holding strength, although pilot holes are essential. Pores need filling to get the smoothest surface. Wiping mating surfaces with acetone, lacquer thinner, or alcohol improves gluing. Rosewood is beautiful and is used for furniture, joinery, turning, carving, and decorative veneers. Rosewood is very costly.

Sapele occurs widely in tropical Africa. *Entandrophragma cylindricum* is a wood similar to mahogany in appearance, with a notable striped figure when quartersawn. It sometimes has a fiddleback or mottled figure and is darker in color, and finer in texture, than African mahogany. Sapwood may be 4 inches thick, and grain is interlocked, producing the narrow, uniform striped pattern mentioned. It has greater strength than white oak and is a heavy wood that works easily with sharp tools, although the interlocked grain creates problems in planing and molding operations. Sapele takes finishes well and glues up nicely. It is used for a decorative veneer and is rotary-peeled for plywood; also it is used for window frames, staircases, and flooring.

Satinwood comes from central and southern India and Sri Lanka, and is sometimes called East Indian satinwood. *Chloroxylon swietenia* wood is distinctive and good-looking, ranging in color from a pale yellow to a golden yellow, with a fine, even texture. Grain is wavy, sometimes interlocking, and produces decorative stripe or mottled figure when quartersawn. Resin veins may mar the looks. The wood is strong and durable, difficult to work and season. Satinwood tends to surface-check and distort. It is hard on tools, dulling edges quickly, thus requiring carbide cutting edges. Satinwood can be brought to a truly fine finish. It is hard to glue. Clean mating surfaces with alcohol, and make sure joints are tightly cut. Satinwood is used as an inlay and as a quarter-cut veneer.

Snakewood has a lot of names, including letterwood, leopard wood, and tortoiseshell wood, indicating a range of patterns that may be found in the wood. *Piratinera guianensis* gives only the heartwood of commercial interest; it is truly distinctive-looking, with irregularly shaped, dark markings somewhat like the markings on some snakes, or a little like Egyptian hieroglyphics (thus the name letterwood). The color is a deep mahogany-red, with markings darker, almost black. Snakewood is strong, but brittle, and almost twice as heavy as oak. The density makes snakewood a problem to season, and it checks and splits while drying. It splinters readily and splits along the grain, meaning screws must have pilot holes drilled and great care taken during fastening. It is hard to cut, but takes a high natural finish. The wood is oily, like teak and cocobolo, so is hard to glue. Use acetone, alcohol, or lacquer thinner to swab off mating surfaces for best results. Snakewood is usually sold by weight and may be difficult to locate. It is very expensive. It is used for umbrella handles, inlays, marquetry, and bows for viols and other stringed instruments. It is a great accent wood.

Teak is found throughout much of southeast Asia, and it is now heavily grown in plantations throughout its natural range, and into Latin America and Africa. *Tectona grandis* has coarse and uneven texture, and the wood has an oily feel. Grain may be straight or wavy. Burma teak tends to be a uniform golden brown, while teak from other areas is a darker brown, with more black markings. Teak is strong, with good steam bending qualities, and is very durable. There is a marked growth-ring figure on flat-sawn surfaces. The wood is heavier than mahogany, but not as heavy as oak, and is stable. It is about on a par with the oaks for strength. Teak works reasonably well, but the silica in the wood dulls tools quickly, so carbide-edged tools are helpful. Teak glues and finishes well, if pretreatment is carried out in both cases. Clean all mating surfaces with alcohol or lacquer thinner, and give the entire piece a good wipe-down with thinner before applying a finish. The wood holds nails and screws well, but pilot holes are needed. Teak does not cause rust or corrosion in contact with metals, so it has a field of use not open to other woods—for tanks and vats that must resist acids. Teak is a costly wood currently used in the construction of expensive boats, lawn furniture, flooring, decorative objects, and decorative plywood.

To add to the list, seven newly imported woods are now available in the United States. In fact, I'm working with the importer, Global Resources [2713 Piedmont Street, Kenner, La. 70062 (877-467-7500)], to help work up some industry familiarity with these Peruvian woods. Three of the seven are softwoods, and the other four are hardwoods that may supplant, or complement, other woods. None are superdramatic woods in appearance, but all are easily worked with standard tools, fastened with standard fasteners, and glued with standard glues. Finishing is easily done without special preparation, and the results are superb. Generally lower in cost than similar woods, these Incan treasures are definitely worth considering for any number of applications.

Incan lacewood [*Brosimum alicastrun (Congona)*]. Cross-grained, regular rays, wheat color to tan, even figure. Very attractive, similar to lacewood or quartersawn sycamore, white oak. Silky on edge and face after jointing or planing. Nails easily; nails and screws hold very well. Face and edges joint easily; some chip-out on face jointing. Hand-planes easily, cleanly; needs lower angle plane for face planing (standard angle chatters on cross-grain) and exceptionally sharp tools. Cross-cuts remarkably cleanly. Carbide blade recommended. Sands easily, but sanding obliterates rays. Does not fuzz. Drills easily and cleanly. Stains blotch on rays. Work with a shellac first coat before staining. Glues well with standard woodworking glues. Accepts most finishes. Little or no aroma.

Peruvian white cedar [*Cedralinga catenaeformis ducke (Tornillo)*]. Straight grain, darker streaks, light tan, with a pale-pink cast when freshly cut. Smooth, soft. Light cedar aroma. Smooth, something of a matte feel. Nails easily; nail and screws hold well. Drill $3/4$ inch from edges and ends to prevent splitting. Face and edges joint easily. Hand-planes easily; clean finish, edge and face. Cuts easily and well. Rips cleanly. Sands well; doesn't fuzz. Drills easily, cleanly. Stains well. Glues well with standard woodworking glues. Accepts most finishes.

Incan red cedar [*Cedrella montana (Cedro virgen)*]. Straight grain, mildly varying figure, open pores. Similar to red cedar but more regular grain, very smooth, soft. Characteristic cedar aroma, fairly strong. Smooth, almost silky. Nails easily; nails and screws hold well. Faces and edge joint easily. Hand-planes well; face and edge (very well on edge planing) clean finish. Cross-cuts easily and well. Rips cleanly. Sands well, doesn't fuzz. Drills very easily, cleanly. Stains well; can give appearance of mahogany, though it is much lighter, softer. Glues well with standard woodworking glues. Accepts most finishes.

Incan cherry [*Cordonia fruticosa (Huamanchilca)*]. Straight grain with almost no figure. Color and feel resemble those of American black cherry. Red color, more pinkish (salmon) than black cherry. Little or no aroma. Smooth, not silky, but very nice. Nails easily; predrill $1/2$ inch from ends, edges. Nails and screws hold well. Edges and face joint well, some chip-out on face jointing. Hand-planes exceptionally well, face and edge. Cross-cuts nicely, very smooth, but has a tendency to burn with slow feeds. Rips cleanly, but has a tendency to burn with slow feeds. Sands well; slight tendency to fuzz. Drills well and cleanly. Accepts stain well. Glues well with standard woodworking glues. Accepts most finishes.

Peruvian walnut [*Juglans netropica (Nogal negro)*]. Very similar to North American black walnut, but no fresh-cut purplish cast. Very dark brown. Little or no aroma. Smooth with the grain, cross-grain feeling is coarse. Nails easily; predrill 1 inch from ends and edges. Nails and screws hold well. Edges and face joint well. Hand-planes nicely, face and edge. Cross-cuts nicely, with a slight tendency to burn on fast feeds. Rips cleanly; burns slightly with overfast feeds. Lighter areas of figure fuzz badly. Drills nicely, but needs backer board fitted tightly to prevent chip-out on exit. Soaks up stains and darkens. Glues well with normal woodworking glues. Accepts most finishes.

Incan white mahogany [*Nectrana S. P. (Moena)*]. Regular grain, open pores, dark-brown striations, whitish tan. Substitute in appearance for mahogany, with stain. Little or no aroma. Smooth feel, more of a matte, with slight roughness cross-grain. Nails easily; predrill $1/2$ inch from sides, $1/4$ inch from ends. Nails and screws hold well. Edges and face joint well. Hand-planes well, but face is easier to plane with a low-angle plane. Cross-cuts nicely. Rips cleanly. Power sanding obscures cross-grain markings. Drills nicely; clean except for light chip-out at exit. Use backer board. Stains okay; can be stained to very strongly resemble Cuban mahogany. Glues well with standard woodworking glues. Accepts most finishes.

Peruvian pine [*Podocarpus rospigliossi (Ulcumano)*]. Looks much like pine. Pale tan-yellow, moderate figure, slightly uneven grain. Slight resinous aroma. Slightly silky feel when freshly planed. Nails easily, when predrilled $3/4$ inch from edges and ends. Nails and screws hold well. Edges and face joint well. Hand-planes extremely easily and well. Cross-cuts very cleanly. Rips very cleanly. Sands well, with no fuzzing. Drills very nicely and quickly. Takes stain well. Glues well with standard woodworking glues. Accepts most finishes.

This synopsis does not come close to covering all the available woods today, but does give you an idea of what to look for when designing and building custom cabinetry for discerning clients.. The search for great woods to use goes on, and I recommend that you get some other research material to add to your files: *World Woods in Color* by William A. Lincoln (Linden Publishing, Fresno, Calif., 1997, 800-345-4477) is a fine book and deserves a place of honor on any cabinetmaking shop's shelves. More detailed information on woods in general can be found in R. Bruce Hoadley's classic *Understanding Wood* (Taunton Press, 52 Church Hill Road, Newtown, Conn. 06470, 1980). My own *Woodworker's Guide to Selecting and Milling Wood* (Betterway Books, 1994, 800-289-0963) contains much of the above information, plus a lot more. There is also the *Encyclopedia of Wood,* in its various editions, from the *Wood Handbook,* issued by the U.S. Government Printing Office, and reprinted in 1989 by Sterling Publishing (387 Park Avenue South, New York, N.Y. 10016).

Engineered Woods: Plywoods and Medium-Density Fiberboard

Plywoods, and engineered woods in general, are my idea of products that serve a multiplicity of purposes, some very well, others well, and still others not so well. Plywood is a stack of veneer plies. The substrate doesn't have to be fancy or expensive, but the final layer can be gorgeous, making the entire piece lovely to look at—and to use. Marquetry isn't the only use for veneers, and never has been. There is no other way to apply burl patterns to a project, at least that is in any way practical, because of the conflicting wood movements within solid burls.

Into the wide world of softwood plywoods we go, with an emphasis on those of greatest use to the widest number of cabinetmakers.

Plywood

Plywood improved in slow steps. By the time World War II PT boats were made of the engineered material, it had reached a stage of high quality, reliability, and durability. Whether hardwood or softwood, plywood is *usually* made up of an odd number of plies (some types of hardwood plywood have an even number of plies). Face plies have the grain running the long way of the panel, and intermediate plies have the grain oriented at 90° (the reason for the odd number of plies). The fact that the plies run in different directions makes plywood both stronger, and more dimensionally stable, than solid wood of the same species and thickness.

Dimensional stability is what makes plywood such a great wood for cabinetmakers. You can make very wide panel faces serve as cabinet paneling, countertops, and other jobs without having to worry about excessive movement, warping, and cupping.

Plywood also makes marvelous backs, sides, and bottoms for cabinets and other projects where those backs, sides, and bottoms are not seen. It may also be used as a substrate for laminates (underlayment, essentially) and for veneers.

Softwood Plywood

Softwood plywood is both a supporting and a leading material for some cabinetmaking projects, and as such it is a good component to start checking for savings in time and effort. Wherever softwood plywood can be reasonably used, the cabinetmaker sees a savings in time and energy, and may actually see some cash savings. There are enough vari-

Sheets ready to become plywood. *(Georgia-Pacific Corp.)*

ations on the theme to keep you reading for hours, but I'm going to explain only those of importance to cabinetmakers—that also includes MDF, which I'll discuss in detail, in most of its different uses. Softwood plywood face grades begin, at the top, with N, for natural finishes, and go on down to D, used as a backup grade only, for CDX and other panels, and for interior plies.

These are American Plywood Association (APA) standards. You can often tell grades by a simple glance at the faces: CDX is a rough grade, meant totally for sheathing and similar uses, and the clean face is backed by the not-so-clean (in terms of gaps, etc.), while exterior glue is used (the X). It is obvious at a glance if you have any experience with plywood at all; but, like all APA-graded plywood, it will be stamped with grade uses and span allowances for the particular thickness of the panel. Span allowances are important to the cabinetmaker only when the uses involve shelving.

Softwood Plywood Grades

Grade N softwood plywood panels are intended for natural finishes, so are smoothly cut of 100 percent heartwood, or 100 percent sapwood,

Plywood being assembled on the production line. *(Georgia-Pacific Corp.)*

and must be free of knotholes, pitch pockets, open splits, other open defects, and stains. The panels, in 48-inch widths, cannot be made up of more than two pieces; wider panels can go to three. Synthetic fillers may be used on cracks and checks $1/32$ inch and narrower, and small splits no more than $1/16$ inch wide and 2 inches long. Faces are limited to a total of six, and they must be well matched for grain and color. This is the top-grade softwood plywood panel, and it is aimed at people wanting a clear finish over a stain—or without a stain.

More frequently, plywood projects call for a coat of paint, and that's where grade A comes in. The veneer must be smooth, firm, and free of knots, pitch pockets, open splits, and open defects. Synthetic fillers may be used to fill cracks and checks to $1/32$ inch wide, and small splits to $1/16$ inch wide, and 2 inches long. On interior panels, small cracks or checks may be as much as $3/16$ inch wide when filled, and depressions to $1/2$ inch by 2 inches may also be filled. Patches are limited to 18 in number, parallel to the grain, and each shall not be more than $2^1/4$ inches in width.

Another preferred paint face ply is grade B. It is more roughly sanded than grade A, but is also lower in cost. This is something you have to work out on an individual basis. Saving a few bucks on a panel that then requires an hour or two of work to make it usable may not be economical in lots of shops. The veneer is to be solid and free of open defects and broken grain, except as listed below. Slightly rough grain is allowed. Minor sanding and patching defects, including sander skips, must not exceed 5 percent of the panel area. Small splits and checks, and chipped areas as in grade A can be filled, in exterior panels. In interior panels, cracks and checks may be larger, as in A grade. Knots are allowed but can't be larger than 1 inch, and they must be sound and tight. Pitch streaks over 1 inch in width are not allowed. Splits to $1/32$ inch may be left open, as may vertical holes up to $1/16$ inch in diameter. The holes shall not exceed one per square foot in number. All in all, this panel takes more work to ready for the paint can or spray gun, but the savings over A grade can be considerable. If A is affordable, go with it; but if the project completion hinges on a matter of dollars, grab B grade, a random orbit sander (one of the brands with more than 3 amperes of power for this work), and sanding disks in 80 and 120 grit, plus a pint or so of top-quality wood filler. Random orbit sanders are the current marvel of the cabinetmaking world, and deservedly so, as they do a fantastic job of removing excess wood, while leaving a very smooth, swirl-free surface for finish. The more powerful brands take off almost as much wood as a belt sander, but are far more easily controlled and leave a far smoother surface at the end. They're not cheaper, when you start talking about the heavy-duty models.

Grade C plywood is of nearly no use to the cabinetmaker, because it's probable that even backer boards and floors of projects are best made of B-B or B-C grade plywood to save excessive filling and labor. Grade C is a rough grade mostly, as already noted, useful for sheathing and, in C-plugged, for underlayment for wood floors. Grade D is a backer veneer in CDX and a filler ply in other plywoods.

Softwood plywoods are available in differing combinations of face veneers. You may buy N-N for clear finishing on both sides, or N-A for clear and paint, and there is also N-B where the back side is not visible. In fact, you can find N-D where the back can't be seen or handled. Grade A comes in A-A and A-B at most suppliers, and B comes in B-B and B-C, with lesser variants available on special order (by lesser, I mean a lower-grade back veneer).

In general, you'll find softwood plywood available in 4- by 8-foot sheets, with 4- by 9-foot and 4- by 10-foot also easily found. For wider panels, which cabinetmaking almost never requires (in many, many years, I've used 5-foot-wide plywood panel very seldom), special order is almost a certainty.

Study your needs carefully and specify the size and type of panel you want. There is no point in working with N-N when one face will be turned to the wall. Too, cabinets seldom take more than one good side: Use a top grade for the outside and a lower grade for the inside.

Overlays

Overlaid plywoods can serve as countertops and provide an especially smooth finish for different paints, including enamels where finishing is much easier if the underlying surface is very smooth.

There are two types of overlay, with medium-density overlay (MDO) probably of greatest use to most cabinetmakers. The basic plywood is an exterior-grade type (with waterproof glue), and it is available in marine as well, in B-B, with inner plies at least C grade. The overlay is a resin treatment on fiber to provide a smooth and uniform textured surface aimed at taking paint extremely well. For the MDO panels, overlay thickness has to be at least 0.012 inch.

High-density overlay (HDO) differs in that hardness is added to the specifications for the surface, so that the phenolic resin overlay is the same thickness, but the resin contains more solids than that used in MDO.

Special overlays are available, but small cabinetmakers often aren't going to be able to afford any such thing, as all that is specially engineered for a purpose, and usually is manufactured in short runs, something that raises costs dramatically.

Hardboards

Tempered hardboard is mostly found these days in the form of pegboard and in various sizes as backer boards for cabinets and low-end bookshelves. It is available in most panel sizes, from 2 by 2 on up to 4 by 8 and larger.

Hardwood Plywoods

I've already looked over the reasons for using hardwood plywoods, but it may be worth recovering some ground to get an idea that paying a premium for solid wood can be worthwhile.

First, and foremost, as with any plywood, is the dimensional stability the laid-up plies add to any job. Layers are cross-banded (laid at 90° to one another), in laid-up construction. Particleboard between veneers is also stable, and is cheaper, but is very heavy and doesn't hold screws well. Lumber core hardwood plywoods have a solid wood core with face and back veneers, and two more cross-band veneers. All veneer has face and back veneers, with cross-band veneers and a veneer core. Strength is close to uniform—it is not exactly uniform—across and with the face grain because of the crossbanding. Plywoods come in a wide variety of thicknesses with no need to crank up the planer. You can almost always get $1/8$, $1/4$, $3/8$, $1/2$, and $3/4$ inch, with $7/16$, $5/8$, $7/8$, and 1 inch often available, especially from professional cabinetshop suppliers. When you are working with these sizes, though, it pays to do a minimum-maximum size check on thicknesses. If you're dadoing lots of pieces to accept $3/4$-inch shelving, the job is much neater-looking if you discover beforehand that your $3/4$-inch plywood is actually $1/32$ inch less than that in thickness.

Color and grain matching is available in premium grades.

Hardwood Plywood Cost

In most cases, hardwood plywood may cost a little more per square foot than solid wood of the same species. In other cases, it costs less. Actually, buying through a supplier, you're probably going to spend considerably less on the hardwood plywood than you would on similar board-foot amounts of the actual hardwood. The difference can be appreciable, except when you're dealing with exotics, which can knock your client's socks off fiscally, no matter the form.

American hardwood plywoods are the least costly. Baltic birch is stronger than American, but it's also more costly. The Taiwanese, Japanese, and Indonesian versions of various American hardwoods are not accurately machined to thickness. Being off $1/32$ inch doesn't seem like a lot, but the day you run a 4- or 8-foot dado to accept a $3/4$-inch thick piece and find it off that much is the day you swear off using junk. I'm not writing here of a consistent $1/32$-inch drop in size because the sheet was sized metrically, but of an inconsistent variable that can waste time and money. Too, the veneers on face and back may be too thin to work properly. Most American veneers are about $1/30$ inch, while walnut may slip in at $1/32$, but foreign veneer is often thinner than that, which creates problems with both sawing and sanding.

The voluntary grades contained here establish the minimum characteristics for the face and back grades of hardwood and decorative plywood produced in Canada. These are defined as follows:

Face Grades

Face means the better side of a plywood panel in which the outer plies are of a different grade, or either side of the panel in which the outer ply grades are identical. The face grades available under this program include AA, A, B, C, D, and E.

Back Grades

Back means the side of a panel with the lower grade when the outer veneer plies are of different grades. The back grades available under this program include 1, 2, 3, and 4. Veneer cuts are rotary cut, rift cut, or flat-sliced maple veneers, which is used throughout North America. Hardwood plywood is produced with a variety of different core materials. The most common are plywood, particleboard, and medium-density fiberboard. The grading of a hardwood plywood panel features a letter designation for the face grade and a number designation for the back grade. For example, a typical sheet of rotary-cut white oak hardwood plywood is A1. This panel has an A face grade and a number 1 back grade.

Grade Marking

Hardwood plywood being graded in this system is identified by a grade mark on each panel, stamped on the edge, indicating the cut, species, and grade of the hardwood plywood sheet (face and back grade). The manufacturer's name or recognized identification appears either on the panel or on the accompanying invoice and/or shipping label.

Cutting Veneers

Many of the hardwood plywoods work very much as their solid wood counterparts do—and the lumber-core types work even more closely as their parent woods. But there are considerable differences, in both grading and working, starting with the way the wood is cut to make face veneers.

Rotary-peeled veneer cuts are probably the most common in both hardwoods and softwoods when plywood is to be the end result. The log is mounted on a lathe and turned against an exceptionally sharp blade, peeling off a very thin veneer. The cut follows the annual growth rings, producing a very bold, variegated figure, one that is fairly characteristic of softwood plywoods.

Plain or flat-sliced veneer cuts find a flitch, made of half a log (cut longways) mounted with the heart side flat against the guide plate. The knife slices parallel to a line through the center of the log and produces a figure similar to that seen with a flat-sawn log and solid wood.

Half-round slicing uses an off-center mount for log segments (trimmed halves), giving a cut slightly across the annual growth rings, and it creates a pattern that shows modified characteristics of both rotary and flat-sliced veneers. This is called quarter-slicing by some, and the method is often used on red and white oak to show grain figures.

Rift cuts are used in oaks, and they show the medullary rays to best advantage. Also called a comb cut, the cut is made perpendicular to the medullary rays, which radiate from the center of the log outward.

Veneer Matching

Veneer matching is a feature of hardwood plywood production (and one that takes place, to a slight degree, in grade N softwood faces). There are two primary types of matching and four other methods of veneering hardwood plywood:

Book match is made like the opened pages of a book, with identical opposite patterns. The book match is made by turning over every other piece of the veneer peeled from the *same* log.

Slip match is made by joining progressive pieces of veneer side by side to give a more uniform figure overall than is possible with book matching.

Whole-piece veneering does what it says: A single piece of veneer exposes a figure across the entire panel.

Pleasing match means the face veneer is matched for color at the veneer joint, but not necessarily matched for figure characteristics.

Mismatch or random match occurs when veneer pieces are joined to create a casual unmatched effect. You'll see a lot of this in real veneer wall paneling.

Unmatched veneer is assembled with no attention paid to figure, color, or uniformity; such panels are usually used for project backs and are considerably cheaper than either book or slip matches.

The last two are the cheapest.

Further lowering of the cost of hardwood plywoods is possible if you design your project to use plywood, getting the greatest use out of a standard-sized (4- by 8-foot) panel as possible. There are several computer programs to help with this, but none seem perfect at this moment. If you do your own layouts, don't forget to leave room for the saw kerfs, about $1/8$ inch minimum at each cut.

True hardwood grades seem to depend at least in part on whom you buy the stuff from and on who makes it. Georgia-Pacific likes A, B, C, and D. (Cabinet grade allows color streaks, color variations, mineral streaks, pin knots, small burls, some sound or filled larger knots, repaired joins, slight shakes, mixed sapwood and heartwood, and no limit on maximum number of components. In other words, it ain't much, and you should pay accordingly.) Craftsman, or C, grade is bet-

ter, but only by a little, so should be correspondingly more costly—
about a dollar per panel with most woods. Select, or B, is another
story. In some woods (maple, for one) color mismatches and streaks
are allowed, but otherwise you're looking at a reasonably good wood.
Grade A is a step up from there, with only slight color streaks, variations, and pin knots allowed, with no shake or pin knots allowed, and
is generally a superb-looking wood.

But, too, we need to consider uses when thinking of superb and not-
so-superb woods. Some of the uses will find C a much more sensible
choice than A or B; the same goes for the ratings used by others.

For other makers, hardwood plywood breaks down into the following choices:

A, premium.   Pieces are slip or book matched unless rotary-cut one-
piece is used. Small burls and pin knots, but not many, are allowed.

1, good.   Unmatched slices are allowed, but there may be no sharp
contrasts in color, grain, or figure. More variations are allowed,
including burls, color streaks, pin knots, and small patches in limited numbers.

2, sound.   No figure, color, or grain match is made. Smooth patches,
sound knots, discoloration, and variable color are allowed.

3, utility.   Rejected material for the first three grades. Open knots, splits,
wormholes to 1 inch long, and major discoloration are all to be found.

Note that A, 1, 2, and 3 pretty much correspond to Georgia-Pacific's
standards of A, B, C, and D. Note, too, that hardwood plywood is
pretty variable stuff, and different species will be allowed different
numbers and types of faults before a panel is rejected. Maple, birch,
and ash, for example, are allowed a lot more color variation than are
lauan, cherry, and mahogany.

Buying hardwood plywood is more difficult than buying softwood
plywood. Most lumberyards don't stock much of a selection, but can
special-order it. You may have to try several times to get what you
want: I wanted something more than cabinet-grade oak a while ago,
and I had to try several times to get it ordered, because the distributor
kept claiming to be out.

Always inspect the lumber before loading, if you're hauling, or
before and during unloading, if you paid for delivery.

Edging

Covered plywood edges are a sign of a good job, whether you're using hardwood, softwood plywood, or MDF. There are many ways to cover edges, some simple, others less easy.

Edge banding of hardwood plywoods is the "hot" method, where real-wood edge banding is applied by a machine to glue it in place. A 250-foot coil of real wood is used, with a heat-glue backing. This is unfurled as the wood is moved by, placed, and finally cut against the edge to be covered. There are many types of banding for use with such edge-banding systems, including a number of woods—I keep at least birch and white oak on hand and find walnut and cherry are also handy. You can also buy plastic edge banding for a contrast in finishes—or to match or contrast with plastic laminates. If you don't wish to buy the edge-banding machine, use a home flat iron. Run the iron over the front of the band, melting the glue and providing a good edge cover. The banding stores well, too; recently, I found a white oak roll that disappeared in the shop at least 5 years ago. I tested it on some scrap plywood and found the adhesion superb.

Other types of edge banding use straight strips, with or without glue, installed with any type of adhesive, including contact cement which requires no clamping. If you use regular adhesives, clamping is needed; you may use 2-inch-wide masking tape to clamp the edge band to the structure, at 12-inch intervals. This procedure is very common and can be done with uncommon combinations, both as simple plywood edging and as edging for countertops, using single strips of wood, double strips of wood, shaped wood strips, double wood strips shaped and with an insert of the countertop laminate, and on to almost anything your imagination and your customer's desires can encompass.

Cutting a miter on corners automatically covers raw edges. It's a simple process if your table saw is equipped with an aftermarket fence and you have an accurate blade angle set up. Otherwise, long miters (really, bevels) can drive you nuts.

Solid wood edging is often really nice. Cut to just a fraction oversize for the plywood; it is glued in place. Contact cement is not suitable for thicker materials such as this, so yellow glue should be used. That means clamping, best done with edge clamps.

The smooth and stable surfaces of MDF provide an excellent substrate for painting or the application of decorative lamination or wood veneers. There are water-resistant varieties for use on or around sink cabinets. MDF also serves well as shelving and is often used, with veneers, as doors and other exterior cabinetry parts.

You may use solid wood edging in matching or contrasting colors. Walnut looks great on maple, and on cherry, while maple looks fine on walnut, and so on. Properly concealed plywood edges appear as attractive as solid wood.

MDF

The smooth and stable surfaces of MDF provide an excellent substrate for painting or the application of decorative lamination or wood veneers. It is just about the best overall material for most kinds of countertops now available; there are water-resistant varieties which should always be used on or around sink cabinets. MDF also serves well as shelving, and it is often used, with veneers, as doors and other exterior cabinetry parts.

The inherent stability, good machinability, and high strength of MDF create opportunities for it to be used as an alternative to solid wood. It will become more popular as prices for solid-wood plywoods increase. You may wish to check for the following features when ordering your MDF:

No formaldehyde emissions. Resin glue must not be a formaldehyde type.

Greater moisture resistance for limited swelling.

Light weight: Okay, two out of three isn't bad. MDF is denser than solid wood, and thus is much heavier than plywoods, and even OSB and particleboard.

Generally, MDF is an engineered, wood-based panel product manufactured from softwood fibers combined with an exterior formaldehyde-free synthetic resin. It is well known for its smooth surface, ease of workmanship, and ability to take a precise machined edge or panel route.

The smooth surface of good-quality MDF ensures a good surface for modern laminates and liquid coatings. Its consistent core produces excellent results with precision routing and shaping techniques. Sanding thickness tolerances are accurate to within 0.005 inch in many brands.

Some MDF is manufactured from more than 90 percent wood residuals for maximum use of timber resources.

MDF can retain the light-tan color of the wood fiber used in its manufacture; this is brand-specific, and you need a light color, for example, with three-ply laminating birch, maple, ash, and other light-colored hardwood veneers so there is minimum core show-through. Darker types are also common.

> **Store MDF indoors on a flat, level surface with enough support to prevent sagging, just as you store plywood.**

Store MDF indoors on a flat, level surface with enough support to prevent sagging.

Changes in relative humidity alter the dimensions of wood and wood-based sheet products. Condition MDF panels in the final environment for 2 to 3 days before fastening and finishing.

Look for a type of MDF that gives good performance when dadoing, shaping, and mitering on both faces as well as edges.

For best results use a carbide-tipped combination blade and a uniform feed rate through the saw. Back panels to prevent chipping along kerf on the sawtooth exit side or use scoring saws.

For drilling, a high-speed drill is best; this is one application where Forstner bits aren't useful. To avoid chip-out or breakage on the exit side, back the panel with scrap material.

For routing, a speed of 20,000 rpm is best, using double-fluted router bits, Z type. Hook and clearance angles of the bit are most important. Square-cut MDF edges can be slotted to receive metal or plastic T-strips; or they can be taped and painted, or banded with lumber strips. Unexposed edges can be machined for strong, accurate tongues, grooves, rabbets, and dados.

For fastening, use asymmetric threaded particleboard screws in predrilled pilot holes, at least 1 inch from the edge on the face, or $2^{1}/_{2}$ inches from the corners when used in the edge of the panel. Nails or staples work, provided that they are not less than $2^{1}/_{2}$ inches from the corners. They should be spaced at least 6 inches apart to reduce the risk of splitting.

Otherwise, all wood fastening methods work at least reasonably well, including nails, staples, rivets, screws, bolts, glue, or a combination. Type A or AB, sheet-metal, twin-fast type, and fully threaded screws designed for use in particleboard offer better withdrawal resistance than wood screws. Predrilled pilot holes are recommended for the size screw used. If nailing, use spiral or ring shank nails for extra holding power.

Commonly used adhesives for laminating wood veneer are urea and urea/melamine resins, both hot-pressed. Generally, all thermoplastic and thermosetting adhesives work well, as do contact cements. If you are using water-based adhesives, prepare with minimum allowable water content to reduce fiber-lifting problems.

MDF faces, composed of fine fibers, are very smooth and ideal for all kinds of finishing. Good-quality MDF panels can be filled, sealed, painted, or varnished with most commercial finishing materials including primers, fillers, lacquers, and synthetic base coats and topcoats, and high-temperature bake systems. The panels should be at stable room temperature (70°F or higher) when coated. To minimize fiber raising of the surface, solvent-based coatings are preferred over water-based emulsions.

We've covered many of the wood and wood-based materials useful for cabinetmaking at this point. It's now time to take a look at some other types, the materials used for facings and for countertop surfaces (we'll spend more time with them later, when we also discuss solid-surface cabinet materials such as Corian and Swanstone).

Laminates

The use of laminates as covers for kitchen and bath countertops goes back at least to linoleum many years ago. Since the 1950s, plastic laminates of many kinds have come into existence, and installation and durability factors have become easier and greater. Today's laminates are available in a dazzling profusion of colors and styles, and there are even metallic surfaces to be had. Probably postformed laminate countertops dominate the market at this point, and if that's your customer's choice, or the one you wish to present to your customer, then there are enough varieties to do any job. Simply put, laminates are formed around a substrate of flakeboard of some kind. (This varies with the manufacturer, and there are small makers all over the place; like CNC machinery, the units needed to make postformed countertops are not economically sensible for the start-up cabinet shop. This is particularly true in light of the cost of finished postformed countertops per linear foot, and the likelihood these days that another countertop selection will be a particular customer's choice.)

I've worked extensively with Micarta, Formica, and Wilsonart and gained some experience with Corian and Swanstone, so I will stick to those types of materials in this book. Corian and Swanstone come later, as they're solid surface and not laminates. Laminates are covered in the countertop chapter.

Tools for the Real Work

I am going to assume here that you know enough about tools to make your own selections as you work, gathering all those tools you expect to need, while forgoing those you won't need, or won't need with enough frequency to consider buying. If you don't, finish reading this book, and then go out and get a few more years' experience. Still, some economies of style and scale are easily achieved when you are first starting out in business, and because I have often tested tools for books and articles, some further comments may be helpful as we go along.

Very few small cabinetmakers bother with cabinet lifts and similar useful but costly items at the outset; you can do just as well renting special-duty tools—lifts and so on—until business demand forces you to buy something that expensive. General cabinetry tools are covered in the following list, along with some tools that often aren't classified as such—things like extension cords. For some jobs, you may find a need for a generator, at least for a few days: If the need is short-term, this is a rental item, especially early in your business operating career. I've not covered the tools that cost less than $10, such as utility knives, awls, center punches, laminate trimmers, and similar small items that are essential to doing a good job.

> Keep high-cost, specialized purchases few at the start, but don't skimp on quality or features on essential tools. Buy the best hand tools and table saws you can afford to get.

Not low-cost purchases, but really essential, are top-quality blade guards and out-feed tables. For the small shop where tools must be moved, roller bases are superb. *(HTC)*

Keep high-cost, specialized purchases to an absolute minimum, but don't skimp on either quality or features on any needed tools. Buy the best hand tools and table saws you can afford. Do the same with levels, pry bars, screwdrivers, planes, and drills. For twist drills, the new titanium-coated types are a marvel of durability for most cabinetry jobs, but cobalt is too costly. You may or may not share my biases. As a long time woodworker, I hate spade bits, so prefer to use auger bits and twist drills when I can. I quickly switch to spade bits, however, when there is a chance the bit is going to be used in sections with nails or other metal embedded. Much of that is going to depend on your installation needs, and if you aren't drilling into walls, except for pilot holes, then spade bits are completely unnecessary. Too, shop work is going to require brad point bits and Forstner bits more than other kinds.

Buy as Needed

Larger shop power tools, such as miter saws, can be bought as needed, although a top-slide compound miter saw is almost an essential today, as is a straight miter saw to cart along for installations. You can buy a decent straight-chop miter saw for less than $200, but the top-of-the-line compound miter saws cost twice that and up, mostly up. The choice is yours. I've used most of them and liked many. It does pay to remember that you're not fooling with a circular saw here: The unit is not easily adjusted back to square when knocked about too much, and factory repairs are very costly. If you paid $500 for the saw and then bought a top-of-the-line blade, you will have an investment of more than $575 in one tool. Some brands run appreciably ($100 plus) more. Treat your saw with care.

Equipping yourself is a simple matter. Staying equipped with accurate, working tools is another. Too many workers do not care for themselves or their tools particularly well, and the overall cost over time is high. Shortly, I'll discuss tools I believe are essential to the starting cabinet shop and give my opinions of the categories and their uses, taking a look at some of the brands I've used extensively enough to form opinions. Table 1 lists these tools and others that the cabinet-maker should own.

Table Saws

The basic power saw is used in almost all woodworking shops, and it is a mainstay of all cabinetmaking shops. The table saw comes in many variations today, far more than ever before, combining some technology along with the simpler, but no less important directives of many years of experience. The overall possibilities range from a 10-inch benchtop table saw that weighs under 50 pounds to huge 14-inch blade models that weigh well over 10 times that. For cabinetmaking, all the lightweights are completely out.

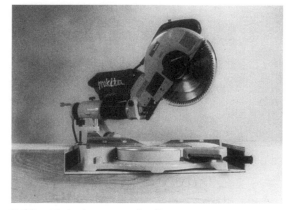

Regardless of what is said by many experts, wood is best worked to within $1/64$ inch, and not in micrometers or thousandths of an inch. Thus, you need a saw that will work to within $1/64$ inch for greatest accuracy, and one that will repeat such accuracy as often as you need it to. In even the smallest cabinet shop, repeatability and power are essential, as are large

Makita 12 dual compound miter saw.

TABLE 1 Cabinetmaker's List of Tools

Tool	Essential	Nice	Luxury
Circular saw	X		
Clamps, all kinds	X (as many of each kind as you can get)		
Reciprocating saw	X (for cabinet installers)		
Power miter saw	X		
Air nailer		X	
Air compressor		X	
Pickup tool box		X	
Truck	X		
Computer		X	
Printer		X	
Stepladders, 6-foot (2)	X		
Stepladders, 8-foot (2)	X		
Tool bag	X		
Bucket boss	X		
Sawhorses (2)	X		
Calculator	X		
Folding rule	X		
10- or 12-foot measuring tape	X		
25-foot measuring tape	X		
33-foot measuring tape		X	
Speed square	X		
Try square	X		

TABLE 1 Cabinetmaker's List of Tools (*Continued*)

Tool	Essential	Nice	Luxury
Combination square	X		
Drywall square (4-foot)		X	
Miter box	X (even with power miter saw)		
Chalk line, 50-foot	X		
13-ounce hammer	X (for molding work)		
16-ounce hammer	X		
20- or 22-ounce hammer		X	
Engineer's hammer, 42-ounce	X		
Cordless drill (12-volt or higher)	X		
Extra battery pack	X		
Corded drill, $3/8$- or $1/2$-inch	X		
Hammer/drill			X
Drill bits, 15-bit set	X		
Long-shank drill bits, $3/8$-, $1/2$-, $5/8$-, $3/4$-inch		X	
Hole saw, 4- or 5-inch	X		
Hole saw set		X	
Table saw	X		
Radial arm saw		X	
10- or 12-point handsaw	X		
8- or 10-point handsaw	X		
5-point ripsaw		X	

TABLE 1 Cabinetmaker's List of Tools (*Continued*)

Tool	Essential	Nice	Luxury
Compass or keyhole saw	X		
Hacksaw	X		
Scroll saw		X	
Pry bar, 36-inch	X		
Crow bar		X	
Nail puller		X	
Screwdrivers, 6-piece set	X		
Safety goggles, or face shield	X		
Dust and mist respirators, disposable	X		
Respirator mask	X		
Steel-toed shoes	X		
Shop vacuum	X		
Power plane			X
Plane, block	X		
Plane, bench	X		
Laminate trimmer	X		
Biscuit joiner	X		
Jigsaw (power)	X		
Belt sander	X		
Random orbit sander	X		
Extension cords (two 50-foot no. 12, with ground; others as needed)	X		
Tool belt	X		
Gloves	X		

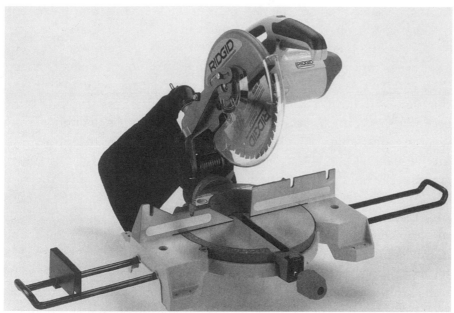

Ridgid 10-inch compound miter saw.

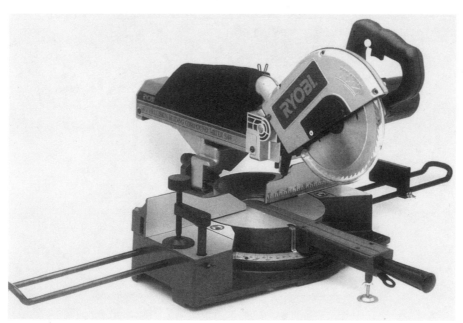

Ryobi 8^{1}/2-inch sliding compound miter saw. *(Ryobi)*

fences, because much of the cutting is going to be in either hardwood or sheet goods, often both on the same saw in the same day.

As a cabinetmaker, if your need is hundreds of cuts weekly, week after week after week, with $1/64$-inch accuracy and a wish for more, then you have a smaller, and more costly, range from which to choose. But you still have a range. Thus, repeatability and durability are essentials in your base shop saw; that gives two types to choose from, one less desirable, but also much less costly, than the other. I've known cabinet shops that opened up with a good-grade contractor's saw, but the light production saw was the first new addition. The contractor's saw simply will not maintain accuracy under daily pounding, regardless of how good it is as a contractor's saw; too, it is underpowered compared to the cabinet saw of 3 horsepower and more, is difficult to collect dust from, and often is light enough to skitter around when feed rates are high. So contractor's saws fall in the range at the bottom end; they are useful as immediate start-up tools, but lack the power and durability for long-term use in a cabinetmaking shop.

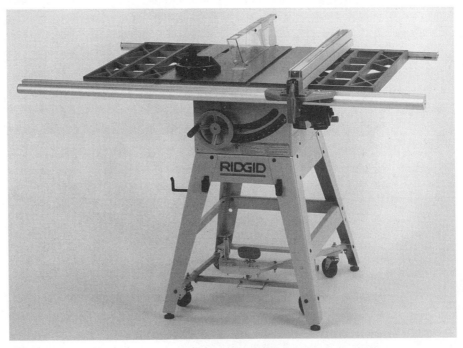

Ridgid 2424 contractor's saw.

Powermatic 64A contractor's saw, with Freud rip blade. The Powermatic has the best standard blade guard of standard saws.

I like to have one on hand for times when the standard shop saw is busy on other jobs. And it's handy to have a relatively low-cost tool set to one side of the shop with a dado head in place, so that specialty cuts may be made without breaking down the setup on the big saw. In fact, at today's prices, the cabinetmaker might well think about having two of these saws ready, one set for dadoing and one for use of a molding head (short-run stuff, where setting up the shaper isn't economical).

Saws in this category include the Bridgewood and Craftsman models that have open-work side tables. I've heard a number of people say the grids in these tables need filling, as you're likely to catch your fingers in them. I think the second table saw I ever used had such a grid, and many others since then have had grids; and, to date, I've not knocked a finger, scraped or otherwise messed with a knuckle, or had other problems. Like everything else in a woodworking shop, use of this tool requires some thought, and you will find that dragging your fingers along a table saw's table, with or without an open-work grid, is not really a good work habit. Unless you're buying other brands, it no longer matters, as the current Craftsman contractor's saw has a solid table. The insert is a pain, because you have to use a screwdriver to remove and install it. The Unisaw-style insert, used on production saws and many contractor's saws, is easier to level with the table surface and is less likely to tilt up on one end and down on another. Too, the insert is heavy enough that it will never warp or twist. There are some basic differences in table saws here, and for the two types that might be of interest to beginning cabinetmaking businesses, a short look at features may help. A contractor's saw sports a cast-iron table and either cast-iron or stamped-steel wings. It is powered by a $1^1/_2$-horsepower induction motor through a single V-belt, and the motor is mounted behind and below the saw.

A cabinet saw sports an enclosed base, a 3-horsepower or larger motor with the power delivered with three belts. A properly tuned contractor's saw can produce cuts that can't be distinguished from those of a cabinet saw, but cannot do it consistently over a long time and cannot do it in heavier materials (6/4 and up).

There are essentially three key elements in a good saw: the blade, the fence, and the saw itself. A well-tuned contractor's saw with a good fence and blade may, and usually will, outperform a cabinet saw with a poor stock fence and crummy blade.

At this point, we step up to that next level of table saw, the cabinet, or light production, model. These saws were designed with cabinetry and light production in mind, and for many decades they have filled the bill beautifully. For most small cabinet shops, these are all that will ever be needed. The Delta entry in the field is the Unisaw, and it more or less sets the standard, as it has for many decades. These saws are definitely not portable; once set up, they are best left in place, as their exceptional accuracy is better maintained when the saw isn't jerked around. The Unisaw weighs about 415 pounds in its most basic form, and each step up carries more weight. Motor covers are not standard equipment on the Unisaws, but other features are far more important: The miter slot is a T slot (with the horizontal slot on the bottom), so the miter gauge is held more accurately. There is a magnetic motor starter available—and I recommend it, as it keeps the motor from restarting without permission if power is cut off from any place other than the switch. The trunnions that hold the arbor assembly together are truly massive, and rack and worm gears elevate and tilt the blade mechanism.

Power? It's pretty much up to you, but standard power is a 3-horsepower motor that is probably called a 5-horsepower motor in standard advertising lingo. There is also a 5 horsepower, 220 volt only. For factories, there is a 5-horsepower three-phase unit, and three-phase power is a help in cabinetmaking shops, if it's available without major

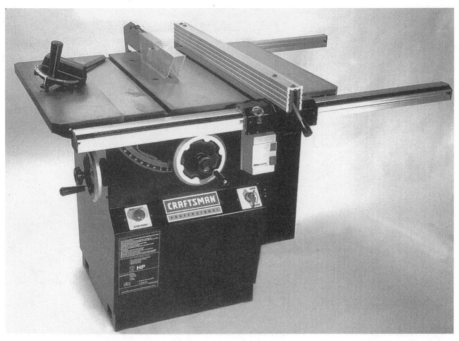

The Craftsman 22654 professional 10-inch cabinet saw (there's also a 12-inch version) is a bit pricey, but it has many features.

Craftsman 22652 12-inch table saw.

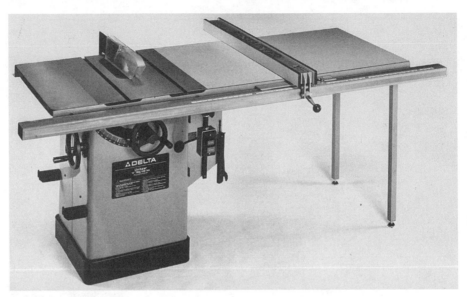

Delta Unisaw. *(Delta)*

cost. It is too complex for some applications—if you don't believe that, get three professional industrial electricians, or electrical engineers, talking about it, and the confusion and expense elevate things to an industrial level very quickly. I mention this in part because there is a continuing temptation to buy discarded tools from industry and school sources. Such tools are sold long before the useful life is finished, but almost all are for three-phase power. This can be a good start-up tool supply for small cabinetmaking shops, assuming the availability of three-phase power. Otherwise, motor replacement is too costly, although in some cases converters work decently.

Of course, Delta is not the only maker of light production table saws. The list is longer than it used to be, with the addition of overseas models from Jet, Grizzly, and others. And Powermatic has always provided a superb saw that is strong competition for the Unisaw. Too, a Canadian company, General Manufacturing, turns out a lovely northern version of the 10-inch tilting arbor table saw.

As a rule of thumb, Powermatic saws are likely to cost a touch more than the Delta products (about 10 to 15 percent), but offer even heavier trunnions and include their version of a T-square fence for that extra cost. Quality is on a par with the Unisaw (and both companies will be angry at that statement—there is a great deal of deserved pride

JTAS table saw with sliding table. *(Jet)*

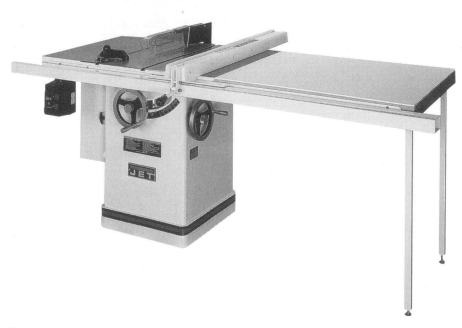

JTAS table saw. *(Jet)*

of product). General matches the pride of product, and the features, but presents a few problems with pricing. I have given up trying to figure Canadian pricing, and distribution tends to be a bit iffy outside the northeast United States. The saw is worth it if you can find one.

The Taiwanese saws, such as the Jet and the Grizzly, present most of, often all, the features of the American trio, but at a lower price and a slightly lower quality level. In the case of the Jet JTAS-10, in fact, there is a very, very slight difference in quality level. I've been using a left-tilt model for some months now, and I find it the equal of my old Unisaw, plus it was much easier to assemble.

It is not just labor that is cheaper in the countries that produce these lower-cost tools. In general (and like all generalized statements, this one is not always true), the gray cast iron used is of lesser quality and treatment. Machining may also be less accurate.

The point of import saws is a simple one: You get a great deal of tool for the money—often the cost is about 65 or 75 percent of that of an equivalent-size U.S. model. Currently, Grizzly's 3-horsepower 10-inch saw runs about $1100 versus about $1600 for the Unisaw, comparably equipped. For the import saws you must spend extra hours getting the setup just right and making sure the tool is accurate across a broad range. The three U.S. brands, and the Jet, can usually be taken from the shipping carton and put to work, although I'd recommend a complete check of all nuts, bolts, belts, and so on. Delta's units are just about spot-on as delivered, as are those of the Jet, Powermatic, and General. Bridgewood and Grizzly may need a little work to get to the same state. There is a wide availability of table saws in the light production style today as well as a good number of heavier saws; many companies now sell 12- and even 14-inch table saws for use in production shops. Generally, all a small cabinetmaking shop is going to need to start is the 10-inch light production saw and a backup top-quality contractor's saw.

These are the features of the top saws:

Delta Unisaw (right or left tilt)

- Table depth: 27 inches
- Table width: 20 inches (with wings: 36 inches)
- Weight: 440 pounds (varies with motor, accessories)

General 350

- Table depth: 28 inches
- Table width: 20 inches (with wings: 36 inches)
- Weight: 465 pounds (varies with motor, accessories)

Grizzly 1023ZX

- Table depth: 27 inches
- Table width: 20 inches (with wings: $36^{1}/4$ inches)
- Weight: 435 pounds (30-inch Shop Fox fence, motor cover, 3-horsepower motor)

Jet JTAS-10 (right- or left-tilt)

- Table depth: $27^{1}/8$ inches
- Table width $20^{1}/8$ inches (with wings: $40^{1}/4$ inches)
- Weight: 450 pounds (3-horsepower, Exacta 50-inch shop fence, 36-inch laminate side table)

Powermatic 66 (left-tilt)

- Table depth: 28 inches
- Table width: 21.5 inches (with wings: 38.25 inches)
- Weight: 450 pounds

These saws are not the only ones suitable for small cabinetwork, but are about the lowest-cost table saws that will take the day-in, day-out grind that sometimes requires blade changes every second day, with hardwood in 5/4, 6/4, 8/4, and more being fed through hundreds of feet at a time. When the saws get a break, they're usually slicing hardwood plywood. We've all seen the Powermatics and the Unisaws and the General 350s that have been in place, and seen daily use, for a dozen or more years with nothing more than belt and blade changes. That's the goal: minimum, or even no, maintenance beyond normal lubrication and blade changes, plus alignment checks every so often. Any of these saws, properly set up, will meet that goal, although setup may take longer in some cases.

Some other features of the saws include the following: All saws use two handwheels with locks to adjust the blade height and angle. The angle handwheel is at the left on right-tilt saws and on the right on the left-tilt saws.

The angle scale is at the front of all machines, and the depth handwheel moves in an arc as the blade is tilted. The scale is graduated in degrees on all. A simple pointer (adjustable) is used on all.

All the saws have adjustable stops for 90° and 45°. Powermatic claims its stops don't clog as readily as Unisaw's when used at high depths of cut. General's stops are harder to get to than Unisaw's, but the saw usually arrives with them set properly.

The General has heavy-duty painted metal handwheels and locks. The Powermatic and Unisaw have metal handwheels (not as heavy as those of the General) and plastic locks (the Powermatic has plastic cranks on the metal wheels). The Powermatic's handwheels are not highly finished.

The Jet and the Grizzly use metal handwheels. Both types are large and well finished. The Grizzly has very heavy metal handwheels that operate smoothly, but the lock is not positive. The General's locks are positive, while the Unisaw's are less so. The Jet locks nicely.

The table saw is the most important shop tool in most cabinetmaking shops. A very, very few shops make do with lower-quality table saws, or no tablesaw at all, but for the general custom cabinetmaker, a top-grade light production cabinet saw is essential. The above gives some general specifications and tips about the most frequently bought models. There are others; there are much more sophisticated and expensive European versions, and there are some hybrids. Given a choice, I'd go for one of the ones I've listed if only because the parts availability and service, including customer service, of all the companies is at the least good. Delta (maker of the Unisaw) has an almost legendary parts availability setup, while the others are a bit lower on the scale. Companies such as Grizzly work exceptionally hard at reaching goals, as does Jet, so their customer service is often spectacular: Grizzly even has a program that will get you in touch with a local buyer of the same tool (assuming one is available) so that you can get comments directly, without company intervention. To do that takes either good nerves or lots of confidence.

Band Saws

This is one of the saws that tends to fall under the "other tools" category in most shops, and that's probably, for most of us, where it is best viewed. But it offers capacities and capabilities that many of us need to remember: You cannot cut, no matter how slowly, 6-inch-thick wood on your table saw or radial arm saw in a single pass, but you can on most 14-inch band saws. It's extremely difficult to cut curves on table and radial-arm saws; but gentle curves, and with special techniques some fairly short-radius curves, are readily cut on the band saw.

The band saw lends itself, because of the thickness of cut, to production cutting of complex parts in a process known as *pad sanding.* So the band saw becomes very important in any kind of production scheme—even if what you're doing is cutting identical parts for toys for half a dozen different grandchildren. Most band saws have available accessories to make ripping and cutting circles easy: The band saw is just about the only tool of choice for cutting round countertops of any size. Blades are available that make near scroll cutting easier, as are blades that make ripping operations easier. Band saw blades in resawing operations (long rips on the wide part of the board) have a tendency to follow the grain in many woods, and special teeth help reduce this problem.

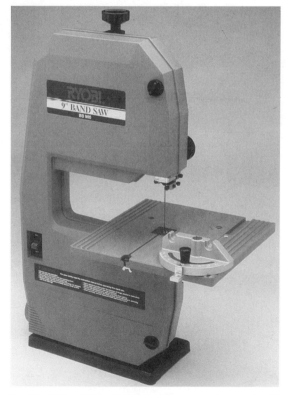

The traditional band saw runs its blade around two wheels, one above the other, giving a pretty specific neck depth for wheel size. More or less standard are the Delta and Powermatic 14-inch light industrial models and the Craftsman versions. Additionally, Grizzly has introduced a 16-inch model that appears to be a welcome size advance at low cost. The rigidity of the back frame, and the need for aligning only two axles, tends to make the two-wheel band saw the strongest and most accurate—as well as the most expensive. (Usually Inca has a more costly three-wheeler that is durable and accurate.)

Both the Delta and Powermatic 14-inch models come with closed-base versions

Ryobi's little 9-inch band saw is a surprise, and will do much of the work needed in a start-up cabinet-making shop. It is an amazingly accurate saw, too. *(Ryobi)*

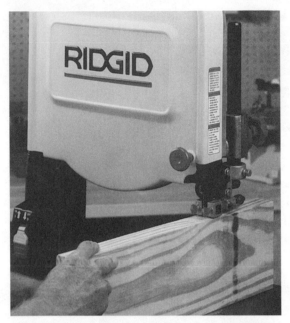

Ridgid 14-inch band saw shown here resawing pine. I have one of these, and often use it to resaw oak, where it does an excellent job.

with more powerful motors to go with their high price tags; but if you don't need the extra accuracy, and that power, you may be tossing money out the window. If you do need it, then buying the less costly versions is a waste of money, too. If most of your work is with hardwoods 1 inch thick and thinner, or with softwoods 2 inches thick and thinner, the lighter saws may do a fine job for you. But if you work with thicker hardwoods, or work with the band saw more, then one of the more powerful units merits serious consideration. And if you have the patience to tune such an import properly, give thought to one of the Taiwanese models such as Grizzly. They are lower in cost, require more care in setup, but are usually durable and useful afterward.

Band saws are versatile tools that, I believe, need to be in every cabinet-maker's shop in one size or another. Sooner or later we all need to cut a curve, and it is often at a time, and in a place, where a bayonet or jig-saw will not work or will not work well enough, so we have to have a band saw. Generally, while the 14-inch band saws are the most common, models that are 16 inches and larger are more versatile for the cabinetmaker, although the cost is much higher and the blades are more expensive, too.

Radial Arm Saws

Back in 1922, when Ray DeWalt invented the radial arm saw, there was no electronic wizardry, nor was there much beauty, but the saw did the work for which DeWalt intended it. In recent years, Black & Decker has transferred the DeWalt name to its line of high-quality portable power tools and dropped radial arm saws from its line, whether under the DeWalt name or the Black & Decker name. That leaves fewer makers, with Delta, Sears Craftsman, and Ryobi bearing the brunt. Generally, the Craftsman and Ryobi models are not up to daily work in a cabinet shop, so we won't take a detailed look at them.

The basic advantage of the radial arm saw is seen in cross-cutting. It will rip, but it is not as safe or as accurate in rip work as the table saw. Today's radial arm saws may also accept router bits on an accessory shaft; and, of course, they do all sorts of dadoing and similar work, with the cut actually visible to the operator. Table saws do the same work, but the cut isn't visible until it is finished, which tends to complicate setup a little bit. Radial arm saws also shine brightly when squaring up the ends of long stock, and when making miter and bevel and compound cross-cuts in long stock. Placed correctly, the radial arm saw has no limits on the length of the material it can cut. In practical circumstances, it isn't all that difficult to make sure you can trim either end of a 16-foot-long board, about the longest standard length you're going to find today.

At one time, the radial arm saw was touted by some experts as the perfect saw for the one-saw shop. It isn't, but then no saw is.

Jet 18-inch band saw. *(Jet)*

Radial arm saws can be finicky to tune and to keep in tune, so keep that in mind when selecting the basic saw for a shop. I would not use a radial arm saw as a basic saw. It is a superb supplementary tool that can do a great many things more easily than a table saw or a band saw can, but it is not a replacement for a table saw and most certainly will not replace a band saw. With some quick adjustments and a few accessories, the table saw—basically a rip sawing machine—can be made a superb cutoff machine as well, but it is never going to be truly fantastic, except with a sliding table, at making miter cuts.

I would not use a radial arm saw as a basic saw. It is a superb supplementary tool that can do a great many things more easily than a table saw or a band saw can, but it is not a replacement for a table saw.

I'm not even going to mention ripping operations and capacities with any radial arm saws. Suffice it to say that such work is better performed with a table saw—or, in a pinch and when you're not in a rush, with a band saw. I'll do ploughing and similar less-than-full-depth rip operations on the radial arm saw, but with extreme care, and on long pieces with a helper standing at the switch to cut it off if I yell.

I love the radial arm saw for two things: You can get really extreme depth in dadoes, if such is needed. A 10-inch radial arm saw will operate an 8-inch dado set, while a 10-inch table saw will not run more than a 6-inch dado set at full depth: A 12-inch radial arm saw will run a 10-inch dado set, assuming you have enough power. Then, in the same area, using a dado head or a molding head, you can get some

This Craftsman 10-inch radial arm saw features a "control cut" mechanism powered by a separate motor and steel cable to the cutting head, allowing the user to control the feed rate.

really fine multiple cuts and shapes because you are able to view the cuts. And it is fairly easy to set up to make multiple dadoes as you do when creating dentil molding.

Today's most prominent radial arm saws in professional shops are those made by Delta, in a wide array of sizes, and older DeWalt saws that may be rebuilt. (Original saws are still made, but may be hard to find.)

The cheapest professional radial arm saw around appears to be Delta's 12-inch model; this comes as a single-phase 2-horsepower saw, or as a three-phase 2-horsepower unit with a weight of about 300 pounds and maximum cross-cut of $14^3/8$ inches. Bevel stops are at 0°, 45°, and 90°, and at $22^1/2$° care is needed to keep the blade from striking the turret. This is a good, solid saw that is as accurate as most radial arm saws, and more accurate than many.

For real production, and countertop work, Delta produces the 18-, 16-, and 14-inch 33-400 series (401 through 423). The single-phase 5-horsepower 420 is a monster saw, with an 18-inch blade, 24-volt push-button magnetic starter, and a cross-cut capacity of 29 inches. It will miter to 22 inches at 1-inch depth of cut, and can cut to a maximum depth of $6^1/2$ inches—even at 45° it cuts to a maximum of 3 inches. You've got a possible dado width of 2 inches, too, and there are bevel stops at 0°, 30°, 45°, 60°, and 90°. The baby of the 400 series is the 33-400, the 14-inch single-phase 3-horsepower unit which has the same cross-cut, 45° cross-cut, and other figures as the 33-423, but which lacks the absolute power to do production work in material 3 inches thick and more. It's a pretty hefty baby, with an in-place weight of 725 pounds—which won't shift easily—and a height of 6 feet 4 inches. Table size in all the 400 series is 44 inches by $36^1/2$ inches, but that's barely relevant, because everyone lines these up along a wall with lengthened infeed and outfeed tables, according to need. In fact, Delta makes roller tables in 4-foot lengths for that purpose, although around here lots of cabinet shops build open-slat tables, without rollers (the open slats, usually placed about 4 inches apart, make it easy to drip and flip long workpieces).

These babies, though, are going to deplete your cash stash. The 12-inch 33-890 single-phase saw sells for around $1600, while the 14-inch 33-400 goes for upward of $4500 (that 29-inch cross-cut is expensive, as there is a lot of cast iron in that head).

Jointers

The average shop probably can get by with an 8-inch jointer when first starting out, though many seem to try to go with the much less costly 6$\frac{1}{8}$-inch version. (As an example, Grizzly's G1182Z 6-inch jointer currently is shipping at $395, with shipping free. This is a sale, so it won't be around when this book comes out, but it does illustrate the price ranges with one maker.) The 8-inch professional model (G1018) ships for $709, including shipping, but the 12-inch professional model is $1895, plus $150 for truck freight (shipping weight is 840 pounds). Delta's 12-inch DJ-30 can be had with a single-phase 3-horsepower motor, or a three-phase 3-horsepower, and weighs 706 pounds. Both offer three knife cutterheads, with the Delta 3$\frac{7}{8}$-inch diameter and the Grizzly 4-inch. Delta's DJ-20 8-inch jointer has a 1$\frac{1}{2}$-horsepower motor, weighs 335 pounds, and does 16,500 cuts per minute from its 3$\frac{3}{8}$-inch-diameter cutterhead, while the Grizzly has a 3-inch-diameter cutterhead and a weight of around 375 pounds, stripped. The Delta is currently selling for about $1400, shipping included, which shows the price differential between the two companies pretty well. Delta gives the older name, the slicker finish, along with probably greater durability and reliability, but for a start-up shop, there is nothing out there that beats Grizzly's combination of durability, decent quality, and customer service. Jet's 6-inch (JJ-CSXW) enclosed-base jointer at $500 is a good buy for a small jointer, while its larger 8-inch model, the JJ-8CS, costs around $1200. The 8-inch one is a solid machine, with a 2-horsepower motor, and handwheels of a size that makes infeed and outfeed adjustments easy. The CSXW has its handwheels on the front of the machine, which makes use of the small wheels much easier. The Powermatic 8-inch long-bed saw, model 1610050, runs about $1850, versus $500 for the 6-inch enclosed-stand model.

Almost any of the above jointers will get you in shape to do all you need to do at the outset. The larger models will last and be useful for a much longer time, but if the smaller tools are selected to start, a considerable money savings is effected; too, at a later date, those tools can be sold for about one-half their retail cost.

Planers

Planers are essential to the cabinetmaking shop. You'd think it would be easy to buy wood in all the thicknesses needed, but if you try, you pay so much more per board-foot that the cost of any planer is shortly looking much lower. Too, you'll find that every so often there comes a job that suddenly requires a few hundred feet of 5/8-inch lumber, when all you have on hand is 3/4- and 1/2-inch. Getting exact fits is always nice, and being able to bring a piece of wood to the size and smoothness you desire, when you desire, is a great step to more efficient working. Planers wide enough for cabinetmaking shop use start at 15 inches, but 18, 20, and 24 inches are generally more useful, if exceptionally more expensive. Jet's 15-inch JWP-15HO costs about $1100, while its 20-inch JWP-208-1 is $1000 more. Delta's 15-inch planer costs just over $1120, and prices rise rapidly for the other models, though features and capacities increase, too. The Delta DC-580 planer is a 20-inch planer available in single- or three-phase 5-horsepower models, with a three-knife cutterhead and a 5000 rpm cutterhead speed. The weight is 840 pounds, and controls are large and easily handled.

Grizzly's line of planers suitable for professional use starts with the G1021 15-inch model. This 2-horsepower unit has dual table locks to reduce snip from table movement, and the drive gears run in an oil bath. Shipping weight is 440 pounds, and delivered price is currently about $795. Grizzly's next step up is the G5850 planer, with a segmented-steel infeed roller and a maximum cutting depth of cut of 1/8 inch on stock as much as 73/4 inches thick and 20 inches wide. Cutter-

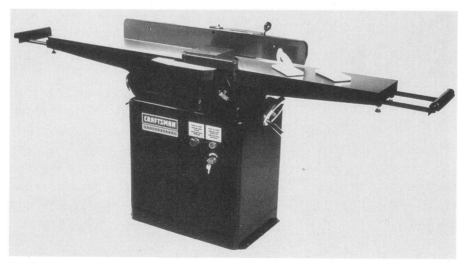

The Craftsman 20651 jointer is an 8-inch model that is excellent for the start-up shop.

Craftsman 21512 sanding station, with 6- × 48-inch belt and 12-inch disk. *(Sears)*

head rpm is 5200, and the power is from a 5-horsepower (25-ampere) 220-volt single-phase motor. Prices start to climb here, as the delivered price, including $150 freight, is $2599 for this 900-pound package. Grizzly's G7213 24-inch $7\frac{1}{2}$-horsepower planer is a three-phase-only tool, with a maximum material height of 8 inches, a weight of 1030 pounds, and a delivered cost of about $3600, including freight.

As you can see, going huge at the start-up adds considerably to business expenses, but also adds considerably to business capabilities: There is nothing other than a 24-inch planer that will quickly reduce stock size for such things as countertops of solid wood, desktops, and similar items where width is an essential to the projects. Some wide sanders will do almost as well, but they are exceptionally slow.

Shapers

Shapers come in an even greater variety than planers and jointers, so we're only going to glance at those that are suitable for the kind of short production runs most likely to be important in a cabinet shop. Although J. R. Burnette of Burnette's Cabinets in Huddleston, Virginia, keeps his major shaper ready for longer runs on custom work—I recently watched and photographed J. R. set up and start a 2000-foot run of 5-inch base molding for a home—most shops seldom need more than a very few hundred feet of molding at any one time. Generally, a dual spindle will work to expand capabilities ($\frac{3}{4}$-inch and 1-inch; or $\frac{3}{4}$-, 1, and $1\frac{1}{4}$-inch half-inch spindles are generally too light for real millwork). The larger

The Jet JJ12W 12-inch jointer is more costly, but it works better in a larger or more advanced shop. *(Jet)*

SCM 12-inch jointer, in use at J. R. Burnette's cabinet shop. This jointer was bought used some years ago, and has served exceptionally well since.

The Delta 8-inch jointer has a very long bed, is of moderate weight, and has plenty of power. *(Delta)*

spindles give smoother work in large runs, but also do smoother work when you're making raised panels and arched panels for cabinet doors. And that brings up a by-the-by: You'll note in photographs of J. R.'s run of molding that he uses a power feeder. Using a power feeder, with stock held from above with a featherboard as he does, is by far the most effective way to work any long run of stock. It gives both a smoother, more consistent finish and a safer job, since you're not getting important bits of you anywhere near the cutters and kickback is literally impossible. I agree with many others: Shapers scare me. They're essential for cabinetmaking, and I first used one in the dim mists of age 16, but they still scare me. Kickback can be horrendous and not always directional, at least as expected, and the machine is the only one I can think of offhand that tends to pull the operator into it. All these problems are eliminated with a power stock feeder.

Because the number of profiles needed is apt to be near infinite, the varieties of planers, planer-cutters, and ways to arrange those cutters on the spindle needs to be large. In the true professional range, Grizzly starts with the G1026, a good, basic 3-horsepower planer, with a magnetic switch, reverse, and two speeds (7000 and 10,000 rpm). There are three interchangeable spindle diameters—$1/2$-, $3/4$-, and 1-inch—giving a great variety of possible cutter uses. The spindle travel is a nice 3 inches, allowing many setup options. Spindle openings on the table go up to $5^1/2$ inches for those doing really large raised panels. At a 360-pound shipping weight, this model is easily shoved to one side when not needed, and its price of less than $900, shipped, makes it a good deal for the small shop. Grizzly's next single-phase step up is the 5-horsepower G5921Z. When Grizzly sticks a Z on its machines, you get a slightly higher price for a much more feature-laden unit. The 5921 has a one-piece cabinet-style stand; spindle sizes of $3/4$ inch, 1 inch, and $1^1/4$ inches; with spindle travel of $3^1/4$ inches. The different

Ridgid 13-inch planer.

spindles will take cutter heights (under the nut) up to 5$\frac{1}{8}$ inches on the 1$\frac{1}{4}$-inch spindle. The table counterbore is 7 inches in diameter and $\frac{5}{8}$ inch deep, taking a maximum cutter diameter of 5$\frac{7}{8}$ inches, which should provide just about all anyone can want in a raised panel. The 5-horsepower motor is totally enclosed and fan-cooled, and it draws 14 amperes. This machine isn't going to be easily shifted: Shipping weight is about 700 pounds. Prices climb, too: The G5921Z currently sells for less than $2100, delivered. For those who want to go even further into developing shapes and profiles and other types (most shops have no real need for

Ryobi 12$\frac{1}{2}$-inch planer. This portable planer is wonderful for job-site and installation changes.

The Craftsman 23374 15-inch planer is about the smallest in-shop size that is useful.

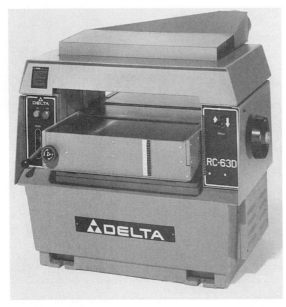

The Delta RC-63D planer is a 24-inch model. *(Delta)*

this kind of versatility), there is the other Z model, the G5913Z 5-horsepower tilting-spindle shaper. With a spindle tilt of 0° to 45°, the 5913Z changes the possibilities considerably. Spindle travel is the same as for the 5921Z, $3^1/4$ inches, as is the maximum cutter height under the nut, $5^1/8$ inches. Spindle speeds of 3600, 5100, 8000, and 10,000 rpm are also the same. The primary difference here is the tilting spindle that opens up creativity. Shipping weight is again around 700 pounds, but the cost is higher at nearly $2600 delivered.

From Grizzly, we step to Delta, which presents its 43-370 two-speed heavy-duty shaper that handles spindle sizes of $1/2$, $3/4$, and 1 inch. There is an extra-long $3/4$-inch spindle that accepts $4^3/8$-inch cutters under the nut, and there are two spindle speeds. This is a fine light-production shaper, and probably most all shops will need at the outset. The motor is a 3-horsepower single-phase (three-phase is available, as is 5-horsepower three-phase), and shipping weight is about 425 pounds. Controls are grouped and sized nicely, and the fence operates cleanly and easily. The table is 28 inches by 27 inches and can take two extension wings that bring it out to 36 inches by 27 inches. Currently, it sells for about $1400, without the wings. Delta's step up from there is a large one, into its five-speed shapers that require three-phase operation. The 43-791 and 43-792 differ only in that the 792 has a cast-iron sliding table. Spindle diameters are $3/4$ inch and 1$^{1}/_{4}$ inches, with a 5$^{3}/_{4}$-inch spindle travel. The 1$^{1}/_{4}$-inch spindle has a 4$^{7}/_{8}$-inch capacity under the nut. Table size for the 791 is 31$^{3}/_{4}$ inches by 39$^{5}/_{8}$ inches, huge in this tool class. Insert opening diameters go up to 9$^{1}/_{2}$ inches. Sliding table size (792 only) is 10 inches by 39$^{3}/_{4}$ inches, with a travel of 37$^{1}/_{2}$ inches. Machine weight is high, at 876 pounds for the 791 and 990 pounds for the 792.

Powermatic presents several nice spindle shapers, the 2-horsepower model 24 and the 5-horsepower model 27. The model 24 takes $^{1}/_{2}$- and $^{3}/_{4}$-inch spindles, has two speeds, and costs about $1000.

Massive chains and gears drive the RC-63D. *(Delta)*

Model 27 accepts $^{3}/_{4}$-, 1-, and 1$^{1}/_{4}$-inch spindles and comes in several versions. The basic model 27 is a four-spindle unit (it also accepts $^{1}/_{2}$-inch spindles), with a 28- by 29$^{1}/_{2}$-inch table. Spindle openings go up to 5$^{1}/_{2}$-inch diameter, and spindle travel is 3 inches. The largest spindles (1- and 1$^{1}/_{4}$-inch) have 4$^{7}/_{8}$-inch space under the nut. Spindle speeds are 7000 and 10,000 rpm. Machine weight is

470 pounds. The Super Shaper 27 is similar to the regular model, but features a 30- by 40-inch table, with removable 3-, 4$^{3}/_{16}$-, and 7-inch inserts—if you can't cut your panels on this one, you need to work with hand planes. Spindles include $^{1}/_{2}$-, $^{3}/_{4}$-, solid $^{3}/_{4}$-, 1-, and 1$^{1}/_{4}$-inch, for machine speeds of 7000 and 10,000 rpm. The 1$^{1}/_{4}$-inch solid spindle and the $^{3}/_{4}$-inch solid spindle are the main spindles for this machine. Machine weight is 657 pounds, so you're not going to push this one aside easily. The model 27 line extends further into the Powerstack model 27PS. Table size on the 27PS is 28 inches by 29$^{1}/_{2}$ inches with an 8-inch

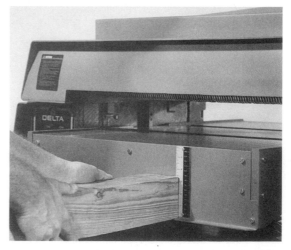

Quick size measures are easily done on the RC-63D. *(Delta)*

The tables are powered on the RC-63D. *(Delta)*

Truly massive tables make for easy, clean cutting: The RC-63D weighs 1675 pounds net. *(Delta)*

extension available. Spindle size is $1^{1}/_{4}$ inches with $4^{7}/_{8}$-inch capacity under the nut, and a crated weight of 485 pounds. Further up or down the line, there is the model 28. The 29- by 40-inch table is huge and has a maximum opening of 8 inches. The fence is a larger version of the one used on model 27, and spindle speeds are 3000, 4500, 6000, and 8000 rpm. This version takes only a $1^{1}/_{4}$-inch spindle, with a 6-inch capacity under the nut and a $4^{1}/_{2}$-inch spindle travel. The unit has an electronic brake and spindle table openings of $2^{1}/_{2}$, 4, 6, and 8 inches. At a weight of 699 pounds, this machine goes in one spot and stays there.

Jet has several spindle shapers, with the JWS34L being the basic professional model. It takes $^{3}/_{4}$- and 1-inch spindles and can be adapted for $^{1}/_{2}$-inch. Spindle speeds are 8000 and 10,000 rpm, with a table opening diameter of $6^{3}/_{16}$ inches and table size of $21^{5}/_{8}$ inches by 25 inches. The 1-inch spindle has an under-nut capacity of $3^{3}/_{16}$ inches, and spindle travel is $2^{7}/_{8}$ inches. Controls are nicely grouped, and the adjustment wheel is large enough for very easy handling. The motor is a 2-horsepower, single-phase, 230 volts only. Machine weight is 317 pounds. Price runs around $1070, delivered. The WSS-3-1 is a larger version, again for $^{1}/_{2}$-, $^{3}/_{4}$- and 1-inch spindle sizes, with a $29^{1}/_{4}$- by $25^{1}/_{2}$-inch table. The table opening is again $6^{3}/_{16}$ inches, with an insert opening of $3^{1}/_{4}$ inches. Spindle travel is $3^{3}/_{4}$ inches, and under-nut capacity for the 1-inch spindle remains at $2^{1}/_{2}$ inches. Spindle speeds are 7500 and 10,000 rpm, and the net weight is 400 pounds. The sliding table version of the previous unit is the SWSS-3-3, with many other features being the same: Table size is larger at $30^{1/4}$ inches by $26^{1/4}$ inches, with a sliding table $30^{1/2}$ inches by 9 inches riding on ball bearings. The table has a working slide of $28^{1/2}$ inches and has 3 horsepower available in single phase. The machine weight is 550 pounds. Next up is the TWSS-3-3. Available only as a three-phase tool, this unit has limited utility in the start-up shop, but has a huge $35^{1/2}$- by $27^{5/8}$-inch table, with an $8^{15/16}$-inch table opening and a spindle travel of $3^{1/8}$-inches. The tilting spindle speeds are 3000, 4500, 6000, 8000, and 10,000 rpm. The motor is a 5-horsepower three-phase unit, and the

Knives are strongly secured against flex, feed rollers are beefy, on the RC-63D. *(Delta)*

spindle tilts 5° forward and 45° back. Undernut capacity in the $1^{1/4}$-inch spindle is $4^{11/16}$ inches, while it is $3^{15/16}$ inches with the 1-inch spindle (a $^{3/4}$-inch spindle is also available).

You'll note that I've not included pricing on many of the larger machines. There are a number of reasons for this, including the fact that the machines themselves are sold primarily through major tool distributors and often are not readily available where prices are easily discovered. Too, with such dealers, you can often do a deal, working the original price down quite a bit—markup on larger machines is considerably higher than that on the more popular smaller machines that share marketing with hobbyists, so deals are apt to be more available. (A dealer who is marking up, for example, 15 percent on a $1000 item has no room to discount; if the markup is 30 percent on a $4500 item, there is some room to play, and you should aim to get a romp or two in, or go elsewhere.)

With the table saw, radial arm saw, jointer, planer, and shaper, you've got the basics of a totally equipped cabinetmaking shop. In fact, of course, you need sanders of several kinds, drill presses, and

The Jet JWP208W is a 20-inch, full-featured pro planer. *(Jet)*

possibly many specialized machines such as overarm routers, multihead drill presses (line boring machines), horizontal boring machines, hollow chisel mortisers, and so on, depending on the type of work at which you wish to excel.

Panel Saw

Another tool that deserves consideration is the sliding panel saw. This is one way to save back injuries when you are working with large plywood panels. The various brands each have great features, but I've worked with Milwaukee's panel saw and like that a lot, so I show a shot of it on p. 179. There are a number of other companies that also produce excellent versions. Prices for these tools seem to range from a low of about $750 on up past $2000; but if you do extensive work with sheets of material, and particularly with the newer engineered stuff such as medium-density

Shop-made dust collection connectors often work better than purchased types.

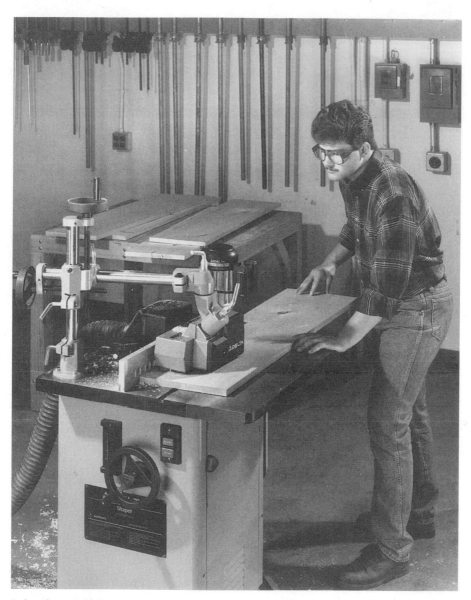

Delta shaper. *(Delta)*

fiberboard (MDF), which is heavier and harder to handle, then the panel saw is a great time and travail saver.

And you'll need hand drills, belt sanders, random orbit sanders and a multitude of other, smaller tools, bits, cutters, and clamps—you never have too many, and probably never have enough. There are so

(*Text continues on p. 177.*)

Setting up molding cutters.

Note home dust collector feed and home-made applewood fence.

Shaper is ready for final setup.

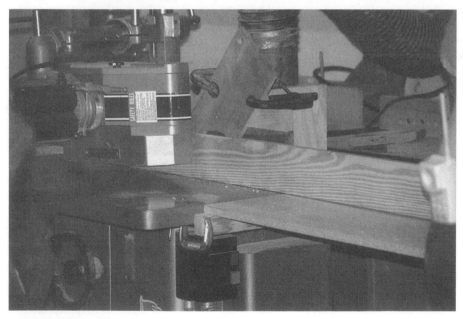

Fitting the featherboard.

Setting up for extension tables.

Extension table holder in place.

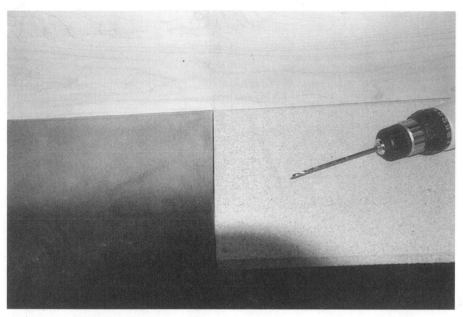

Extension tables are screwed to holders.

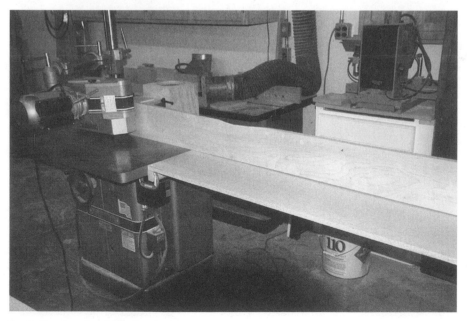

In-feed extension table in place.

Both in-feed and out-feed tables in place.

Shaper is set up. Note how power feeder covers the cutting edge of the bit. Safety is increased a lot.

Test piece runs.

J. R. Burnette runs the first full piece.

The first piece reaches the outfeed table.

A catcher can be a big help, saving lots of traveling from one end of the machine to the other when there are many pieces to run.

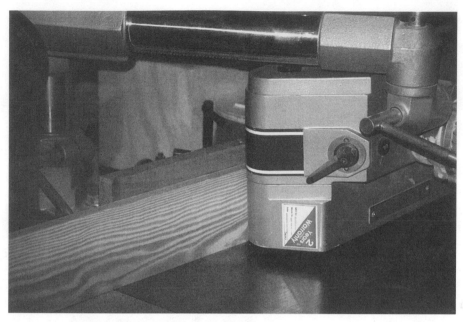

Power feeder is a massive aid to smooth finishes.

many different kinds of clamps, it's hard to even make comments about what a cabinet shop should have, but generally being heavy on bar clamps doesn't hurt.

Clamps

Woodworking clamps fall into categories: bar clamps, hand screws, C clamps, and band clamps. Included are pipe clamps, picture frame clamps, and miter clamps. The largest number of types fall in the bar clamp category. The American Clamping Corporation (ACC) catalog shows nine types of bar clamps, without sizes and without accessory assemblies (ACC has assemblies that hold edge pieces, etc.). Adjustable Clamp Company has even more, and other companies chime in with different versions.

The list of sizes for the bar clamps varies with the weight of the clamp. The heavier the clamp, the longer it can be, so normal-weight clamps reach 48 inches, light-duty stop at 32 inches, and Klemmy clamps, or super light-duty notched cam and tongue bar clamps, stop

(*Text continues on p. 186.*)

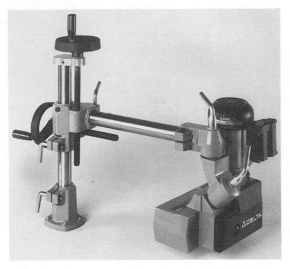

Delta lightweight feeder. *(Delta)*

The feeder wheels press down on the work to drive it steadily through the tool's cutting edges. *(Delta)*

Gear changes change feed speeds. *(Delta)*

Shaper-mounted feeder. *(Delta)*

Doing the run.

Delta 5-speed shaper. *(Delta)*

Milwaukee panel cutting saw.

Craftsman 21512 belt/disk sander.

Makita random orbital sander.

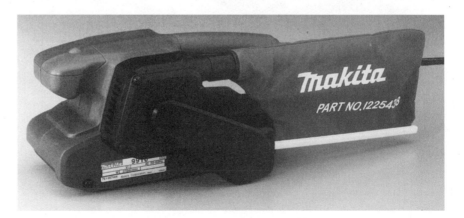

Makita 3 × 18 belt sander.

Bosch detail sander.

Craftsman stapler/brad nailer.

The Accuset brad nailer is also very useful in the shop.

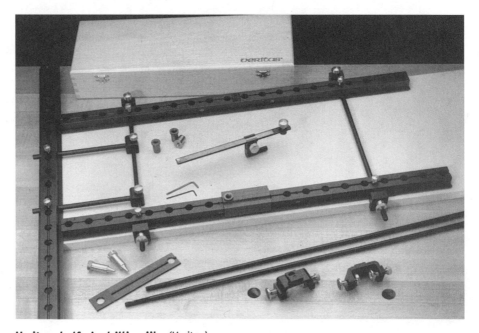

Veritas shelf pin drilling jib. *(Veritas)*

Airmate 3 Airstream dust helmet.

Ridgid radial arm saw.

Bosch laminate trim router.

Stacked performance is available from this Craftsman Excalibur Elite adjustable dado. It has a center blade and two outside blades with Dyanite carbide-tipped scoring teeth. This dado produces flat bottoms and eliminates chipping and tear-out in the workpiece. Dado widths are infinitely adjustable between $1/4$ and $13/16$ inch and are attained by simply dialing the desired dimension.

Tenoning jig.

Very old Parks planer, here doing yeoman's duty.

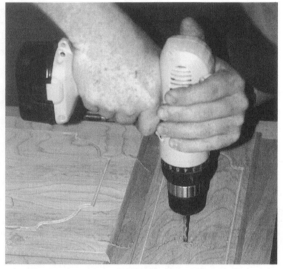

Cordless drill used to make holes for hardware.

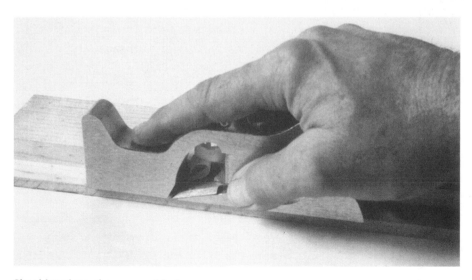

Shoulder plane cleans up rabbets.

Pipe clamps hold wide glue-ups.

at 24 inches. The heavy-duty Bessey K body clamps extend to 98 inches; they are costly, smooth-working clamps, superb for heavy work.

Band clamps are also variable. The Stanley Tools version is about 1 inch wide and 10 feet long, and Vermont-American's is 2 inches wide and 14 feet long. Band clamps are for clamping odd shapes, whether octagonal frames, chair leg assemblies, or picture frames.

Hand screws are for uneven, and even, clamping when marring is not allowed. Tips for preparing hand screws help preserve the screws, and probably the finish, on some projects.

Hand-screw bodies are wood, easily ruined by glue and other problems should you get messy. To minimize glue buildup, you might coat the entire outer jaw with paste wax (don't polish it). Over the years, though, the wax creates enough slickness to make the use of the hand-screw tips quite difficult—small objects tend to slide out. Get 2- or 3-inch-wide masking tape and cover the front of the hand screws. Strip off when loaded, and replace.

> **Probably the best advice for clamps has two parts: Buy the best you can afford; buy more. No cabinet shop ever really has enough clamps.**

Hand screws work well to clamp non-parallel surfaces, and they do not creep.

The above is only a glancing look at the world of clamps. There are so many out there that keeping track is a job, and the needs of the basic cabinetmaking shop are

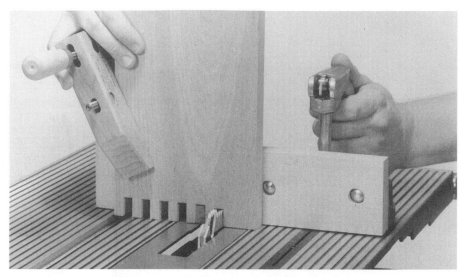

Handscrews do many jobs well. Here, they're holding material in a finger joint jig. *(Shopsmith)*

Handscrews can also serve as stops and fences. Here, the handscrew acts as a fence as a piece is drilled for a mortise.

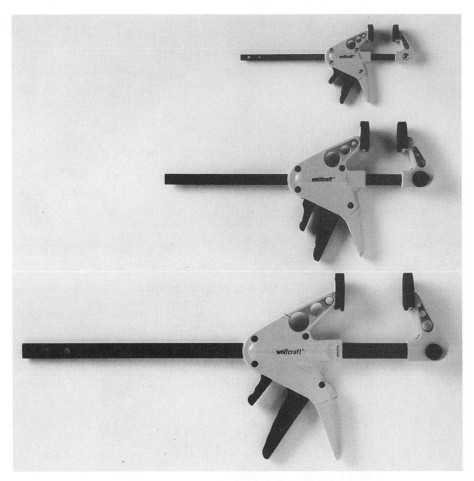

Wolfcraft one-hand fast clamps are exceptionally useful and quite sturdy.

so wide as to force the purchase of many types. For the beginning cabinetmaking shop, I'd suggest as large an array of Bessey K body clamps
as you can get, along with as many Jorgenson pipe clamps as possible,
and a short ton of different-length pipe for the clamps. Toss in at least
a dozen edge clamps, and you've got the beginnings of something.
That's a start that you'll expand over the years.

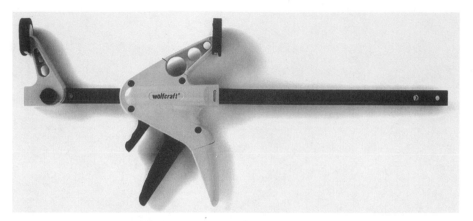

Wolfcraft clamp set to spread.

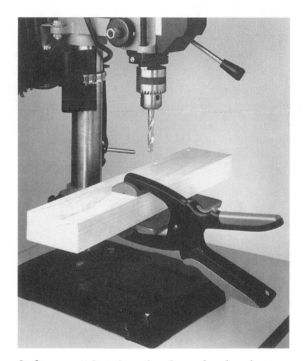

Craftsman ratchet clamp is a form of spring clamp.

Steve uses C clamps to hold guide in place.

Basic Cabinet Joinery

For the most part, semiproduction joinery for cabinets is precise, but not fancy. Dovetailed drawers are common, but they are machine-made and half-blind mostly, and in most shops they are done with jigs that are readily available. There is a lot of lap and half-lap joinery, and plenty of use of butt joints and biscuit joinery. Miters, too, are common. Beyond that, joinery is not very advanced, and it doesn't need to be. That doesn't say it can't be. If you have the capabilities, the desire, and the customers willing to pay for it, all sorts of fancy joints, from through and pegged and foxed mortise and tenon to through dovetails of varying sizes, are possible and desirable. But we're not considering them here, because for the average cabinetmaker, especially the average start-up cabinetmaker, they have no real place.

Butt Joints

Butt joints are the simplest joints any woodworker uses, whether or not cabinets are being made. There are variations on the theme, but the basic butt joint is one straight side placed against another straight side and glued or otherwise fastened there. Butt joints bring two pieces of wood together to form a junction, which may be an L, a T, or a flat board. Common uses include all sorts of framing (with large timbers,

> Common uses include framing, producing wide and flat boards, and joining carcass pieces in cabinets.

the butt joint becomes a framing joint), producing wide, flat boards and joining carcass pieces in cabinets, chests, and other projects.

Butt joints are often splined to increase strength and ease alignment problems. Splining requires a groove in both pieces, the insertion of the spline, and gluing. (Most butt joints are glued for cabinet construction.) Developments in splining include biscuit joinery, which uses short splines shaped like flat footballs to make strong butt joints in any of the styles. Biscuit joiners make alignment of the splines easy, while installation of the biscuits is quick. The biscuits swell when glued, which aids strength and alignment.

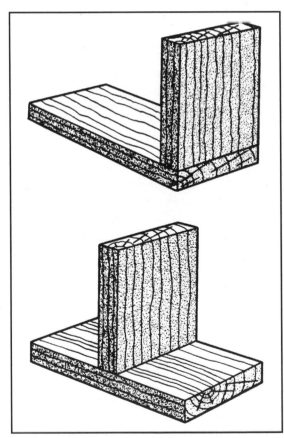

Butt joints.

Miter Joints

Miter joints are angled butt joints. While most people think of door frame and picture frame molding when miter joints are mentioned, these joints are useful in a number of ways for the cabinetmaker. They provide the simplest method of joining plywood so that plies don't show. Miter joints may be splined or biscuit-joined.

Square miter joints form a 90° angle, with each member cut at 45°. There are many variants, some with 60° angles joined to 30° angles and others with different angles of cuts.

Stopped miters are sometimes used when one piece is thicker than the other, with both pieces visible. The larger piece has the miter stopped and a straight cut made from that point, while the smaller piece is mitered and fits into the miter and lip thus created. Miter joints are often keyed. A slot or other shape is cut through

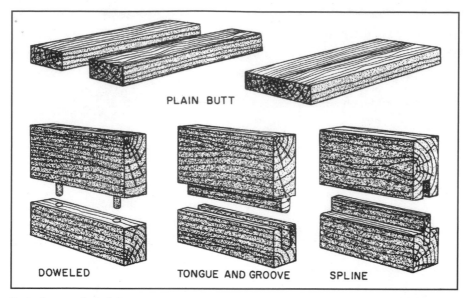

PLAIN BUTT

DOWELED   TONGUE AND GROOVE   SPLINE

Variations on butt joints.

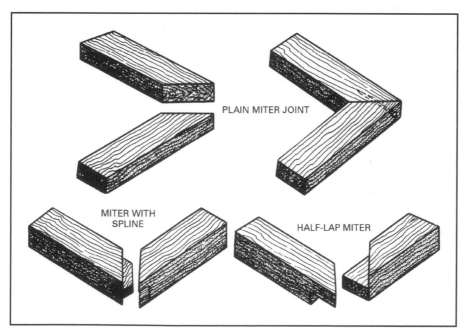

PLAIN MITER JOINT

MITER WITH SPLINE   HALF-LAP MITER

Miter joints.

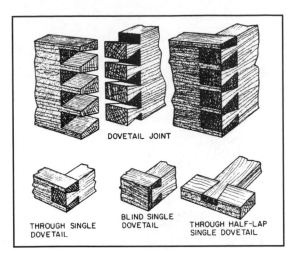

DOVETAIL JOINT

THROUGH SINGLE
DOVETAIL

BLIND SINGLE
DOVETAIL

THROUGH HALF-LAP
SINGLE DOVETAIL

Dovetail joints.

both pieces of the joined miter, and a piece to fit the cutout shape is inserted to aid strength. Most such keys are simple splines, added from the outside, and are called *slip keys.* Dovetail keys are great-looking, especially in contrasting woods.

Tongue-and-Groove Joints

Most wood flooring is joined with tongue-and-groove joints, but for those special cabinet jobs, a version of tongue and groove makes a great backer board when solid wood is the desired material: There is much less of a problem with flatness, and no problem with glue-up, as you relieve one side in a mild V to form a specialty joint that is not glued. The tongue-and-groove is a simple joint, made easily on table saws with molding heads, with routers, and on shapers. The groove is cut in one piece, while the tongue is cut on the other side. Pieces are then assembled one to the other, tongue to groove, and nailed, through the tongue and into the joist (for flooring), or into the end piece for cabinetry. Tongue-and-groove joints are also excellent alignment aids for producing large, glued flat surfaces from narrow boards. Too, they add strength by adding glue surface.

Dovetails

Dovetail joints form shapes like a bird's tail, narrow at one end and fanning out in width as the wood reaches its endpoint. The dovetail is the sign of strength in drawers, and today it is mostly machine-made. (Dovetails cut readily by hand, once you learn the techniques, but are time-consuming to make. And the thought of doing 30 or 40 or 50 drawers, or, for that matter, a dozen, with dovetails all around is enough to make anyone's head ache.) For semiproduction cabinetmaking, there is one router dovetail jig that surpasses all others in simplicity and basic ease of use:

> **Dovetail joints are narrow at one end, fanning out in width as the wood reaches its endpoint. The dovetail is the sign of strength in drawers, and today it is mostly machine-made.**

Out-of-the-box setup takes 15 minutes, and setting up for use after that takes about 1 or 2 minutes, most of that spent clamping the device to the board and inserting the bit in the router. I'm writing here of the *Keller dovetail jig.*

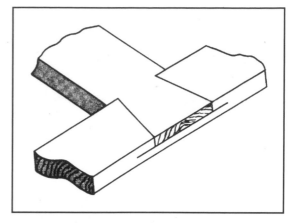

Dovetail half-lap joint.

The through dovetail joint shows the end grain of both pins and tails. The tail is the portion that gives the joint its name, looking a lot like a spread bird's tail. The pins are the smaller pieces present on each side of the tail. Half-blind, or lap, dovetails show the end grain of the pin pieces only. Mitered dovetails (called *secret dovetails* in a more dramatic age) show no end grain. Within these three basic categories are a number of subcategories, some useful, many not.

The through dovetail is a showy joint and adds to cabinet appeal. It is useful for showing off combinations of light and dark woods, thus is suitable for the more showy work you might do. Any dovetail resists both pull and draw forces, so is great in drawer construction, with through dovetails most commonly used on drawer backs. The through dovetail is also useful in general carcass framing for chests, where you wish joint design to be obvious. Hand-cut or large-template dovetails are used for larger chests. Both the Leigh jig and Porter-Cable's Omni-Jig work well on such joints, but both are a great deal more work and bother than the Keller jigs.

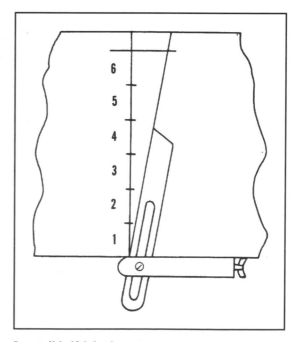

Dovetail half-joint layout.

For drawer fronts, the half-blind dovetail resists both pull and draw forces, but does not show the joint style to the world. It is a fine joint for more formal pieces, where it was traditional until recent years.

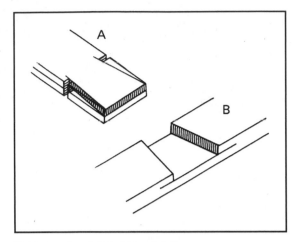

Making a dovetail half-lap joint.

Mitered dovetails are seldom used today and were seldom used in the past. Utility is the same as for other dovetails, but layout and cutting are absurdly complex. To my knowledge, there is no power tool jig that reproduces this joint with any degree of ease.

Finger Joints

The finger joint is stronger than the dovetail joint when used in boxes, drawers, chests, and other carcass construction because of its large glue area. The box joint has replaced the dovetail in much fine furniture in the past century. Joint variations abound.

Finger joints are more easily cut on mitered surfaces than are dovetails, and the joint is easier to match up and produce with rapidity, without the need for a great deal of practice, beyond learning the basic use of a table saw.

The usual method of producing finger joints is to make a jig that gives the needed width cuts with a dado blade, with a spacing bar setting the distance between finger slots. The jig is simple to make and use.

The Porter-Cable Omni-Jig offers a template that lets you make finger joints using a router. This is simple to use, though costly and limited to 1/2-inch finger joints. Special bits let the Leigh and Keller jigs be used in a similar manner. These jigs, while relatively costly, are the small shop's answer to computer numeric–controlled (CNC) router machinery and, as such, speed setup and cutting of larger numbers of joints. As already noted, the easiest to use is the Keller jig, though it is also the most limited in that each jig produces only a single joint size and style.

> The box joint has replaced the dovetail in many uses. Finger joints are easily cut on mitered surfaces, and the joint is easy to match up and produce in production settings.

Corner Locking Joints

This is a machine-made joint that is simple to produce with a router and special bit. It isn't as pretty as the dovetail, but is

Cutting dovetails with the Keller jig.

Cutting dovetails with the Keller jig.

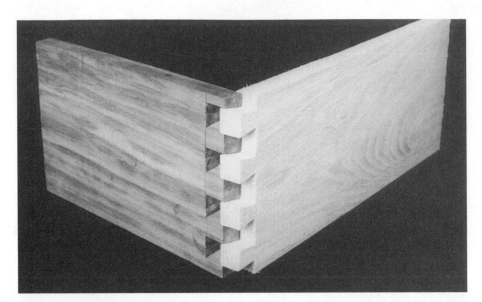

Walnut and ash dovetail.

Redheart and oak dovetail.

Leigh D3 jig clamp.

Cutting a dovetail with the Leigh D3.

Porter-Cable's Omni-Jig.

Cutting dovetails with the Omni-Jig.

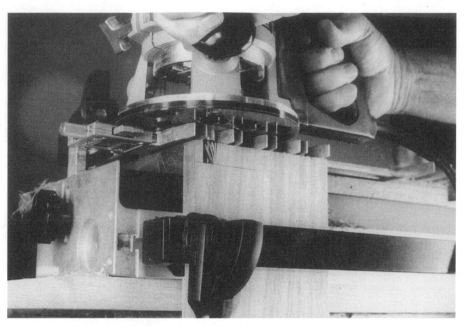

Using the Leigh D3 to cut finger joints.

much faster to produce. Two passes, one for each mating side, and you have the joint. Most bits are designed to do this from a single setting.

Framing Joints

For a lot of cabinetry, the mortise-and-tenon is the primary framing joint. Tenons are the parts that fit into the mortises, or holes, cut in other frame members. For joining cabinet and other frames, and for joining such things as furniture legs to stretchers and aprons, there is nothing stronger than a good, tightly cut mortise-and-tenon joint.

There are many variations on the mortise-and-tenon themes, largely because the joint is so useful and strong in holding larger pieces together. Most tenons have

Corner locking joint bit.

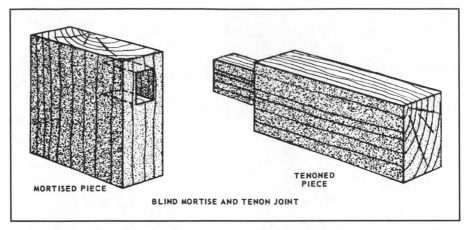

Blind mortise and tenon.

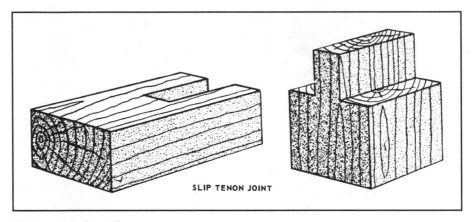

Slip tenon and mortise.

shoulders (the wood from which the tenons extend) on four sides of the piece, though not all do. Shoulders may be on one side, two sides, or three sides as well, depending on construction needs. Common use indicates a tenon about one-third the thickness of the members being joined, if both members are the same thickness. This leaves enough material on the mortise side to help keep twisting forces from snapping the tenon through the mortise sidewall. No single tenon should exceed, in width, 6 times its thickness. That means, for example, that a $1/4$-inch-wide tenon shouldn't be more than $1^1/2$-inches long for maximum strength.

There are many different forms of mortises and tenons. A stopped, or blind, mortise does not go through the mortised wood. The stopped mortise is matched with a stub tenon. The through mortise takes a through tenon, exposing the end grain of the tenon. This may also be foxed, or wedged, into its end. A stub tenon may also be wedged. A tenon that is wedged spreads against the mortise sides. The best protection against pullout is a slightly spread mortise to accept the wedged end of the tenon, whether it is a stub or through tenon. The gunstock tenon adds a slight curve to a shoulder of the tenon so that it matches the design of

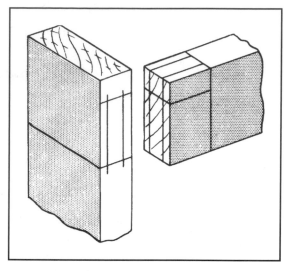

Layout of stub mortise and tenon.

a curved rail and provides a continued design arch. Often it is necessary to design a supporting notch on the mortise member so that

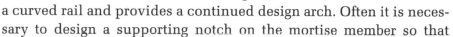

there is no unsupported piece, in the arch, with short grain. Such short-grain pieces are weak and break off easily.

Doubled mortise and tenons in a single joint do not produce greater strength, but are sometimes used when material requirements demand. Experienced joiners look on doubled mortise and tenons as necessary evils, because two thin tenons are not equal in strength to one thick one.

Haunched mortise and tenons are specialized joints sometimes used on the outside members of a frame. They are a tenon with a section cut away to allow resistance to wedging forces, while the mortise is left filled in the area where the tenon is cut away. Usually, a short stub, or haunch, is left on the tenon, which is received in a special stub, or haunched,

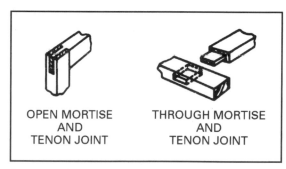

OPEN MORTISE AND TENON JOINT   THROUGH MORTISE AND TENON JOINT

Open mortise and tenon and through mortise and tenon.

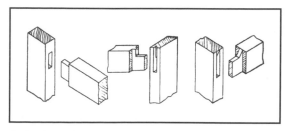

Stub, haunched, and table-haunched mortise and tenon joints.

mortise. With the haunch down, the resistance to twisting forces is very high for tables and similar structures.

Other Framing Joints

Cabinet frames tend toward thinness, so may be joined with joints other than mortise-and-tenon joints. In fact, some of the joints used are mortise-and-tenon joints with slightly different names. A *bridle joint* is a mortise-and-tenon joint that has three sides of the mortise open, thus forming a bridle. That usually means the tenon only needs two sides trimmed to fit, and the joint is a simple slip fit. The term *bridle joint* is, of course, related to a horse's bridle and its open end; thus the outer, or mortise, section of the joint bridles the tenon, holding it in place.

A common description of a bridle joint is a reversed mortise-and-tenon joint. It isn't. Nor is it a mortise-and-tenon joint that has had the wood normally left solid in a mortise cut away. It is a variant of the mortise and tenon, usually with one piece of the mortise cut away to allow the bridle to form. The bridle joint is not as strong as a mortise and tenon, but is stronger than lap and half-lap joints and is useful in light framing. The cradle you see being constructed on p. 205 uses bridle joints at the ends of the side frames.

Other joints are versions of mortise and tenon with changes. A splined and doweled butt miter is a good example. Here the miter is made, and the joint is checked for fit. Then the center thirds of the miters are removed, as if each were an open mortise joint. The length of material, or depth, removed at the back of the mortise is determined in a simple manner. Measure down the cut face of the miter, and divide by 5. Measure below the miter heel that one-fifth, and mark. Carry straight across the board with a square. That provides the depth of cut for removing the mortise. Once the spline is cut and inserted, you can drill the dowel holes. This may be done in two steps, before the spline is inserted or after the spline is inserted. Fitting such splined miter joints may prove difficult. Check the joint as the spline is being fitted, and plane the spline in small steps. Splined joints can be designed and repeated around any basic corner joint, and they will give quite a fancy look if different-colored woods are used as keys. The keys may be cut in different shapes as well, giving an even fancier look. Obviously, this joint is only going to be used in decorative areas of very high-end cabinetry.

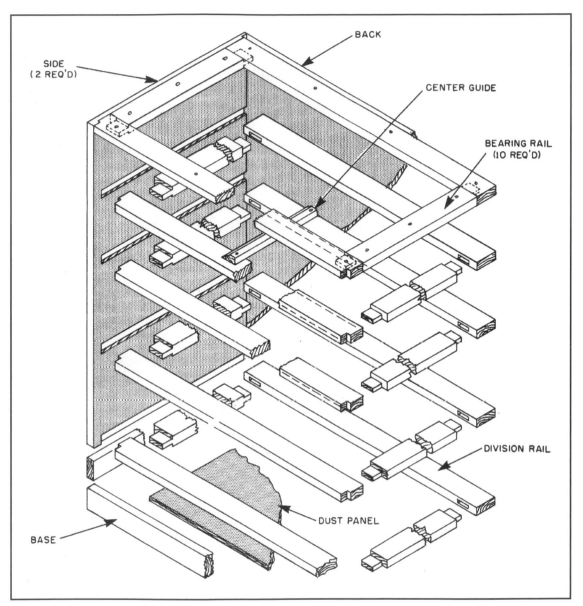

Cabinet joinery: With all the mortise and tenon work here, you can consider this high-end work.

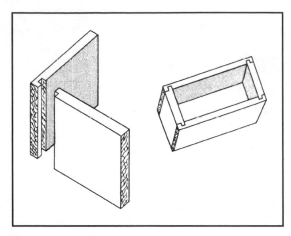

Dado and rabbet joint.

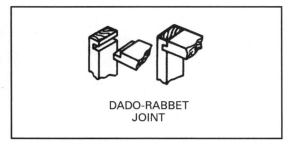

Dado-rabbet joint.

Dowel Joinery

Because I feel that one of the most difficult and overrated types of joinery is doweling, coverage here will be very light. It isn't difficult, just tedious, with massive alignment problems that eat time that might be better spent in designing dovetails or finger joints. Biscuits can often, though far from always, take the place of dowels. Dowels provide good support and help for a number of kinds of joints, but add complexity to what are often simple joints and may present alignment problems that are more difficult to handle than the more difficult joints would be. Dowels are substitutes for more complex, harder-to-cut joints such as mortises and tenons, and dowels are not as strong.

Too, dowels tend to present gluing problems. The need for gluing grooves and that hair of extra depth needed in the drilled holes are things some woodworkers forget, much to their chagrin when the joint fails early.

Rabbet joints.

Cutting a rabbet on a table saw.

Dowels are virtually all made of hardwood rod, often birch, turned to exact diameters. Many are now grooved to allow glue flow, and some even come with beveled ends. If you want off sizes or lengths, you'll have to form them yourself.

Biscuit (Plate) Joinery

The biscuit joiner has done a lot to replace dowels these days. These flat, football-shaped pieces of compressed wood give a means of aligning and supporting joints that is far simpler than fooling around with dowels and doweling jigs. The small saw blade in the plate, or biscuit, joiner slices slots on both sides of a joint; the biscuit is fitted in after glue is applied; and the joint comes together, with alignment needed just in a single plane.

For most uses where the available stock is large enough—the slot requires about $2^3/_4$ inches for even the smallest biscuit—biscuit joinery is replacing both butt and dowel joinery.

> The biscuit joiner has done a lot to replace dowels these days. These flat, football-shaped pieces of compressed wood give a means of aligning and supporting joints that is far simpler than fooling around with dowels and doweling jigs.

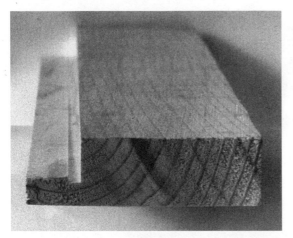

One reason that two passes on a standard saw blade is not the ideal way to make a rabbet is the difficulty of aligning two blind cuts.

Biscuits.

Fastening Joints

There are dozens of variants of most cabinetmaking joints, and nearly as many ways of fastening the joints to prevent separation. Glue is the standard for most kinds of cabinetwork and furniture work, but is not a solution for items that need to be taken apart. Nor does glue work well where joints must fit more loosely than is customary for furniture. Such loose-fitting joints are used in outdoor furniture, where wood movements are more extreme in shorter periods of time. Speed of joinery also determines the type of fasteners. Nails are used for house framing instead of fancy joints, screws, or glues, because nails are faster and cheaper. Screws allow pieces to be taken down and moved without destruction, while nails do not. Specific fasteners, called *knock-down fasteners,* make takedown and movement of furniture and other items easier, while providing a more solid joint than simple screws. A lot of cabinetmakers these days rely on a combination of brads, or staples, driven by pneumatic tools, with glue for permanent fastening. This offends some purists, but is a production speeder that also cuts down on the cost of clamps, because the glue-nail process eliminates the need for many (but far from all) clamps.

Joinery combines joint design with fastener or adhesive type to best suit a particular purpose. Care in design, cut, and assembly ensures long life for any project. Production time is an important measure of joint design effectiveness for the cabinetmaker.

Making Joints with a Router

One of the best tools we have for making joints is the router, once a rare tool, but now commonly found in the cabinet shop. I'd venture to say that there isn't a cabinet shop today that uses power tools that doesn't have a router or two or three or more.

Precision-built, the router is similar to a drill in that it has a high-speed motor spinning a chuck (collet) that will hold any one of many cutting bits. All this is held in a frame that is part of the base. This base rests on the wood surface, the bit projecting beyond it. The depth of the cut made by the bit is controlled by raising and lowering the motor in the frame. This is done by loosening a locking ring, then turning an adjusting ring that links the motor and the frame. It is possible to learn the basics of using this tool in minutes. Wood being worked on should be held firmly by clamps or on special surfaces, so the router can be operated with both hands.

A plunge router allows the user to plunge the bit into the workpiece without moving the router base. This means work can be started without tipping the bit into the work, or running it in from the side. The plunge router is handy, but far from essential. Many of the best routers do not offer a plunge feature. Under heavy use, plunge routers wear out a tad sooner. This is normally not a problem, unless the router is a cheap one.

Routers on the market today offer more features and power than ever before. Power ranges from 1/2 to 5 horsepower. For many workshop uses, a good-quality router offering 1 1/2 or more horsepower will serve for years. Lighter-duty routers are too limited for some work, while the heavier-duty models, those of 3 horsepower and more, are heavier and harder to handle, though they can turn out heavy work and really last under heavy use. Safety is important: There is no guard on the bit, which turns at 15,000 to 30,000 rpm. Change bits or work on the bit only when the router is unplugged. Let the bit reach full speed before starting work, and let it stop on or in the work before you lift the router.

Routers can help produce many types of joints, especially when used in combination with jigs.

The Keller Jig

The Keller dovetail jig takes the simplest approach to dovetail joinery that I've ever seen, offering three sizes of templates, along with carbide

The Keller dovetail jig takes the simplest approach to dovetail joinery, offering three sizes of templates, along with carbide bits. This rig is the only one to provide unlimited dovetail width, and the speed of setup has to be experienced to be believed.

bits. While the cost is relatively high, this rig is the only one to provide unlimited dovetail width (not too important unless you get to doing side jobs of reproduction blanket chests and similar items). The unit is sturdy, of $1/2$-inch machined aluminum plate. Templates provide for fixed pin spaced dovetails and come in three versions. For most cabinet shop uses, the model 2401, at 24 inches wide, is best. It works with stock $3/8$ to 1 inch thick.

To use this jig, clamp the stock in a vise, after mounting the template to flat wood blocking. Make adjustments and trial cuts until the fit is right, then cut the tail board. Use the tail board to mark the pin board, then clamp it with its appropriate template and cut it.

The jig is of high-quality construction, great precision, and durability. And the templates are aluminum. With aluminum templates—or plastic templates, or steel templates—it is essential that the router bit come to a stop to prevent nicking the templates as the tool is raised from the work. If you strike a steel template, the bit may be damaged or metal pieces may fly.

The Keller dovetail templates offer a more restricted line of joints than do the Leigh jig and the Omni-Jig, but are easier to set up. The Keller units make basic through dovetails, with fixed pin spacings, in three sizes. The templates come in pairs. One of the pair makes tails; the other produces pins. Keller includes dovetail bits of appropriate size and style with each unit. The models, with overall width indicated by the first two digits, are the 3601 (36 inches), 2401, and 1601. They differ in stock thickness acceptance and power requirements as well as width, with the smallest useful with lightweight routers of at least $3/4$ horsepower, and the others requiring at least $11/2$-horsepower routers. The big unit will accept stock from $5/8$ to $11/4$ inches thick, while the middle unit, probably the best buy for most woodworkers, takes stock from $3/8$ inch thick. The small model is limited to stock $3/16$ to $5/8$ inch thick.

The Keller system uses fixed-pin units, thus is simple to set up. The use of two templates simplifies this even more, though it increases cutting time minimally. Pin spacings remain at 3 inches for the big unit, $13/4$ inches on the middle unit, and $11/8$ inches on the small set.

Mount the templates to a backing board of your choice. Mounting holes (already drilled) are slightly oblong, and adjust until a test joint fits perfectly. I would suggest using a fir or pine backing board with the 2401 and 3601. Hardwood looks good, but because of size, weight is a problem when changing from one template to another.

The workpiece is clamped, upright, in a bench vise. You will need a solid bench and bench vise for this work. I've mounted a Jorgenson 10-inch model to my assembly bench, and it works beautifully. Set the tail template on the top of the board, and center it to get your tails where you want them. Clamp the backing board to the workpiece, and rout carefully, moving from left to right. Once the tails are cut, use them to mark the pin board. Set the pin template on the marks. Clamp and rout.

With any soft metal template, rout carefully. These units will never wear out, unless you strike the template with the router bit, which chews right through the aluminum. Keller supplies carbide-tipped bits which will never even notice the softer metal. In most cases, make sure that you are engaging the template with the template guide at the start, and that the bit is no longer rotating at the end, and you'll have no problems. Slight touches can readily be fixed with Bondo.

Finger Joints on the Table Saw

Making a finger joint jig is simple, requiring a few pieces of wood, accuracy, and a dado blade set. The dado set is needed for the finger jointing and other table saw work anyway. Shop-made jigs take different-size cutouts for each size of finger joint. You need a separate jig for each joint size, but you are not limited to three or four or five joint sizes. If you need 1-inch finger joints, a jig can be made to produce them, though most dado heads will require two passes per joint cutout with such wide fingers. (The maximum cut with many dado heads is $13/16$ inch, good for single-pass finger joints up to $3/4$ inch wide.) More modest finger joints give a better appearance, with $1/2$-inch being the largest size used for wood up to $1^1/2$ inches thick. The thinner the stock, the lighter the finger joint that gives top appearance.

The simplest jig is usually the best. Start with a miter gauge extension that passes the blade, or dado head, on the side where you prefer to work. Care in measuring is imperative with any jig. Sloppy jig measurement and assembly means joints won't fit. Use oak, maple, ash, or a similar hardwood. Softwoods may be used, but they do not retain accuracy for long. The extension must be at least 3 inches high and 16 inches wide, with both height and width increasing as the size of boards to be joined is increased. The extension is securely screwed to the miter gauge, and set so the miter gauge moves freely in its slot. Don't get the extension so tight to the table that the gauge can't slide with it mounted.

Make a pass over a $3/8$-inch dado blade, set $3/4$-inch deep. Check dimensions by measuring, then make a practice cut in scrap stock first. Measure over $3/8$ inch from the first slot and cut a second $3/8$-inch by $3/4$-inch slot.

Into the first slot, insert a $3/4$- by $3/8$- by $2^1/2$-inch stop block. Secure the entire jig to the miter gauge, making sure that the second, or open, jig slot is directly over the set dado blade.

Make a guide strip the width of the dado blade setting. Use that to offset the board being cut. The guide strip is placed alongside the board edge. Both are held upright, butted against the stop block. Spring clamps work well here. Making the first pass over the dado blade gives a cut on the board. Remove the guide strip, move the board over so the cut fits over the stop block, and place the mating piece in front of the already cut piece, its flat edge butted against the stop block. Clamp tightly, and pass through the dado blade. Move the entire assembly over (notch moves onto stop block), and repeat the cut. Continue until the entire board is ready. Pieces will then mate, and you will have cut two sides at one time, a solid time saver over other finger joint jigs.

Make a separate jig for each joint size you want to use. The four most popular are $1/4$, $3/8$, $1/2$, and $5/8$ inch. Where different sizes are needed, replace the $3/8$-inch in the jig design with the above fractions. For the larger two sizes, it is practical to increase the length (the $3/4$-inch measurement) to 1 inch.

For ease of fit, add $1/32$ inch to cut depth. This allows fingers to extend far enough to allow some light sanding to finish the joint. If depth is too shallow, the only way to finish the joint is to sand down the entire side.

It is possible to build a jig that will give size alternatives, not requiring a jig for each size. It is difficult to set up the jig, though it is not hard to make, and only rarely is it of real use. Great care in measuring and checking is needed for this jig to be of any use at all. Size limits for table saw jigs that make finger joints are not as restrictive as those for router jigs with dovetails. The widest dovetail jig is 3 feet across (the Keller jig can be set and used for wider boards, of course) and expensive, while the closer-to-standard 16-inch-wide Omni-Jig is also costly. Finger joints are limited in width by the working space available and by the support available for the wood. Limitations in height are another story. Dovetailing is often done on a flat board, laid flat; so while width is limited by jig width, length is limited by shop length and work supports, as is width with finger joints. Finger joints, though, are commonly cut on table saws, which take most of 3 feet out of the floor-to-ceiling distance, leaving a 5- to 7- or 8-foot length possibility for most of us. Boards get very hard to handle once they are much more than 48 inches long. Jig modifications must be made to help deal with widths greater than 4 feet. But most cabinetry does not require drawer joinery deeper than about 24 inches, so this is mostly a hypothetical problem.

More Table Saw Joinery

Table saws do fine work with dado joints and grooves. Grooves and dadoes differ. Grooves go with the grain of the wood, while dadoes go against, or across, the grain of the wood. Dado sets are simple accessories, most being stacked-blade assemblies that give a wider kerf, or cut. The outer blades are similar to standard saw blades, while the inner, or chipper, blades commonly have only two teeth, at opposing sides of the blade. Chipper blades clean out the area between the outer blades, producing a set-width groove, or dado. Most standard dado sets will accept enough chipper inside blades to give a dado width of $13/16$ inch in a single pass. Many accept more chipper blades, if your saw arbor will handle the extra width. Most arbors won't, even in cabinet saws. Single- or double-blade "wobbler" units are available. The one or two blades in the tool are set to wobble at a maximum specified distance, producing a cut of the same size as the wobble. Such dado sets leave more material in the bottoms of the grooves and are generally not as effective and efficient as stacked-blade dado sets.

Groove and dado joints are useful for setting in shelves in cabinetry, and in bookcases where shelves and dividers do not need to be adjustable for height. They are stronger than simple butt joints and neater-looking.

Make stopped dado and groove joints, using stop blocks mounted on the saw to keep the cut to a length. Using stop blocks on such operations on a table saw requires forethought, because the saw may have to be turned off as the stop block is reached—if the material does not allow a good, safe grip for removal from the cutting area. It is sometimes easier to use stop blocks on dadoes and grooves with a less powerful tool, one with less penchant for kicking back, such as a router.

If material is long enough—that is, if it extends far enough away from the blade—you can use stop blocks on the table saw, with a dado setup. Using such a block leaves an arced bottom in the cut. This may need to be cut away with a chisel, or the board going into the dado can be fitted to the dado. A router leaves curved corners, instead of an arc rising from the bottom. Those can be chiseled out, usually with a corner chisel.

Rabbet Cuts

Dado blades are great for rabbet cuts—simple shelves cut into the wood, into which another piece of wood or other material fits, making a rabbet joint. A rabbet cut is set up to cut into the auxiliary rip fence. Use a straight piece of wood (my preference is maple), and attach with screws through the holes in the fence, or use clamps. If there are no holes in your fence and it is a hard surface to clamp, drill two. Get the edge of the auxiliary fence down on the table, but not so tight the fence won't move with it in place. Now, cut a relief arc in the auxiliary fence facing. This arc will need to be different depths for different cuts, so start with a slight cut at two-thirds the depth of the facing width. If you used a $3/4$-inch-thick board, cut in $1/2$ inch, and so on. Raise the blade slowly to increase the depth of the relief arc, to a maximum height of 1 inch.

Using the dado setup to cut a rabbet, set the depth of the blades, after installing blades to give a width of about $1/16$ to $1/8$ inch more than the needed width. Set the auxiliary fence so that it gives the needed width, and lock all settings. Feed the material through, keeping a snug fit against the fence facing. Use push tools for any cut where the blade is going blind, which is the case with all rabbets and dadoes.

Tongue and Groove

Dado blades also serve to cut tenons, and tongues and grooves. The simplest of the cuts is the tongue and groove. Set the fence to center to proper width for cutting the groove—the distance that will be cut away on each side to form the tongue. The depth is fractionally larger than for the tongue, but no more than $1/16$ inch, no matter how large the tongue dimensions are. Make the dado width somewhat smaller than the actual requirement for the groove to be cut. Make one pass with the board, and reverse and make a second pass along the other side. The groove will be exactly centered if cut like this.

A single setup is made for the tongue. Set the dado blades to depth and width, and pass the board over so that one side is cut away. Reverse the board and cut the other side away, to finish the tongue. Do this for all boards to be tongued, but only after checking exact fit with the first two grooved boards. If the fit is good, make all the cuts needed. If the fit is off, make any needed adjustments. Cut the tongues last because they are more easily adjusted than the grooves.

Cutting Tenons

Tenons are readily cut with dado blades. One cut is made like cutting tongues since a tongue is, in essence, a long tenon. Run the dado blade up to the height to be cut, set the width on the inside, and pass the work through. This needs some form of carriage if the work is tall. You may wish to make your own carriage, or buy one.

Standard 8-inch dado blades can extend only about 2 inches above the table insert, so there are limits on the length of tenons cut in the upright position. Longer tenons are cut in a series of passes, with the dado set to the depth needed to clear material from one side of the board. Clamp the board in the miter gauge, and use a stop block on the rip fence. Make repeat cuts until the last cut comes off the stop block. You may also start with the cut from the stop block, but don't use that as one of the intermediate cuts. Flip the board over and repeat the process to cut the other side of the tenon. True tenons have the edges cut away. This needs another saw setup.

Stud tenons do not have the edges cut away, and are often used as parts of what the British call bridle joints—that is, the mortise has only two solid sides, similar to a horse's bridle, so insertion is quite easy. Strength is lower, but the entire joint is easily cut on the table saw.

Splined Joints

You can cut splined joints with the dado blade and cut the slots for the thin splines with a standard saw blade. The spline may run the length of the board or may be stopped (a blind spline joint).

Standard grooves for splines are made exactly as are grooves for tongue-and-groove work, but are usually not cut from thick stock (most splines are thinner than 1/4 inch). Splines are made of plywood, or of stock with grain running against the stress direction, regardless of how board grain runs. The spline grain runs 90° to the spline groove. Blind splines are cut with two stop blocks, one at the front of the fence and one at the back. Working at the front of the table saw, clamp stop blocks firmly to the rip fence, and brace work against the front stop. Lower onto the turning cutter, whether that cutter is a dado head or a single saw blade. Move it toward the rear stop block after it is flat on the saw table.

Form the spline to fit the groove, not the other way round. The ends of the groove will be arced, which means the splines are cut to fit that arc. It is far easier to fit the splines than it is to fit the grooves.

Lap Joints

Lap joints are rapidly made by notching, using a dado head to cut away stock. This is important for lap joints that are not at board ends, where a standard blade takes too long to make all the passes needed to clear the notch for the lap. The dado blade is used to set the end markings, cutting first at one end and then at the other. After that, multiple passes clear the center of the notch. The process may be repeated if both boards are to be center-lapped.

End laps work in much the same manner, but only the inside end is cleared first, to the marks, with multiple passes until the end of the board is reached.

Laps may be mitered, and made to match different board thicknesses, as well as cut to half-depth of two boards of equal thickness.

Lap joints are made by notching, using a dado head to cut away stock for speed.

More Miter Joints

The miter joint is a variant of the butt joint in that pieces cut at 45° angles are butted together and fastened in place. It is useful when working with plywood to leave only the surface of the plywood showing, covering underlying plies. The miter joint is a cabinetry joint for both moldings and frame carcass construction. The table saw is not perfect for creating smooth cross-cuts. The radial arm saw is better, but is limited in cross-cut capacity. To get wide cross-cuts, it's necessary to go to industrial radial arm saws such as Delta's 14-inch (you will not like the overall tool or blade prices on this one). Even then, the cross-cut capacity is 29 inches, just about what is needed to slice across a kitchen countertop.

Lock Joints

The lock joint, like the finger joint, is a U.S. invention designed to speed up joint making, though the lock joint is a lot harder to make with a table saw than the finger joint is. It requires accuracy of a fine degree, and great care, and in $3/4$-inch stock it works as follows: Make the first cut $1/8$ inch deep and $1/8$ inch wide, $5/8$ inch down from the board end (most standard kerf table saw blades offer a $1/8$-inch kerf, but check). Cut a $1/8$-inch by $1/8$-inch rabbet on the same side, top edge. Come back from the other side of the end and cut a $1/8$-inch by $1/2$-inch-deep rabbet, leaving a $1/8$-inch extension. Next, on the second side, cut a $3/8$-inch slot, $3/4$ inch deep. The bottom edge is now cut off exactly $5/8$ inch in from the end. Make a second pass to make a $1/8$-inch by $1/8$-inch-deep dado above the cutoff part; or set your waste removal blade at $5/8$ inch to make that dado cut at the same time as you remove the $5/8$ inch of waste. The resulting joint is tight and strong, and more easily made on a router table. Compared to hand-cut dovetail joints, it goes together very quickly with a table saw. To save time, stack pieces for multiple parts in order, and make cuts in sequence on all pieces before moving on to the next setup. You must be dead sure of your setups.

It is far simpler to make lock joints with a specific bit and a router table or shaper. Some shaper bit manufacturers—Jesada is one—make lock miter joint cutters. Thus, to save time, I strongly suggest you avoid using the table saw for this joint, though to save time in drawer construction, I strongly suggest you use it. Jesada's bits, the 380-261 and 380-262 ($3/4$- and $1 1/4$-inch) cutters are joined with a drawer lock joint cutter (380-117 and 380-118).

Splinter Protection: For the Work, Not the Worker

Regardless of the tool used, consider feed speed and direction when you think of preventing splintering. On the subject of feeds, consider a feature of wood related to feed speed and cut type and depth. That is the cross-grain decorative or other cut. At the end of a cross-grain cut there will be feathering and splintering of wood, no matter how much care is taken.

Among the methods used to get rid of this splintering is the use of a board slightly wider than needed for the finished project. It is then ripped to final width, and the splintered part is discarded. Too, the feed may be slowed down even further, reducing tear-out and splintering. When all four sides of a project need cutting, something that often happens with decorative molding and routing, work the cross-grain sides first. The cuts with the grain are used to clean up splintering.

Biscuits in the Shop: Using Plate Joiners

Plate joiners are also known as biscuit joiners, an insult to good biscuits. The "biscuits" or plates are flat and football-shaped, 0.148 inch thick, regardless of width (smaller sizes are available, but the standard is 0.148 inch). The saw blade in the joiner tools cuts a kerf 0.156 inch thick, to provide a loose fit. The plates absorb water from the glue and rapidly swell to more than 0.160 inch.

The plate joiner in portable form hasn't been around as long as many of our other tools, but has been around longer in Europe than in this hemisphere. The plates themselves trace back to 1956, when Steiner Lamello Ltd. (from the German for "thin plate") started making them. Within 13 years, the company was manufacturing a portable groove-milling machine for the plates. Lamello started the craze, which first reached the United States about a decade ago, but which stayed relatively dim until recently. When plate joiners first started receiving attention in national magazines, the use of the machines grew greatly.

The growth of Lamello lured other firms into the market, so now Virutex (Spain) and Freud (Italy) have portable models, while DeWalt and Porter-Cable provide U.S. models of general interest. Ryobi makes a nice little miniature biscuit machine. All these models work with three sizes of little plates, to come close to replacing dowels as an aid to wood joinery.

Accuracy

Accuracy of jointing is far better with plates than with dowels. The slot cut to accept the plate allows adjustment along the length of the biscuit, while a dowel pegs you to a point and keeps you there. If you've drilled the dowel holes a fraction of an inch off, your frame has to be a fraction of an inch off. With plates, it is unlikely you'll be much more than a small fraction off because of the way the joiners are designed; but if you are, you can move things around until the mate is perfect (or the glue sets up).

Buy biscuits only as needed. Buy them in small packages, and buy only a single package of each type. Do not open any of the packages until you need the biscuits, then reseal them as soon as possible, and store the packages in a dry area.

I can find no real difference between the no-name brands of biscuits and the more costly brand-name types (Lamello, Freud, Porter-Cable, etc.). It makes sense to go with the cheapest available when the only differences are cost and name. The biscuits are of solid beech which is stamped to size after being sawn into laths.

Preparing Biscuit Joints

Biscuits are used to join surfaces, replacing splines or dowels in the process, and should lead to a neater, quicker job. Gluing is simple. Glue is best dribbled down along the sides of the slots, after a screwdriver or knife has been used to remove chips from the slots.

As with all joint preparation, the wood must be cut accurately to size, with ends square or mitered as required for the junction to be made. The better the overall preparation, the better the resulting joint, as always.

The more practice cuts you make, within limits, the better you'll find the results of your finish cuts. There is something going round that seems to say written directions and a few markings on a machine will take the place of practice these days. Not so. *Nothing* takes the place of practice.

Cuts are aligned by using marks on the joiners.

Edge-to-Edge Joining

Mark where you need the plates to add strength to the joint. This is at 8-inch or wider intervals for edge-to-edge joining. For such joining, work with two boards at a time, no matter how many you're joining. Mark the boards in 2 inches from their ends, and about 10 inches apart between those marks. Place the cutting guide for your model so that the slot is cut halfway down the boards' thickness. Cut to the marks on both boards, insert biscuits, and check the joint. If necessary, clean out the grooves.

The final steps are to disassemble, place the glue, insert the biscuits a last time, and clamp the boards together, in alignment. When the glue has set, repeat the process on the next pair of boards, if needed.

Final preparation of the resulting wide board is easy, light sanding. This takes less time than working with most other edge-to-edge glued materials.

Corner Joining

It is in the corners that biscuit joinery really shines. The ease, speed, and results will have you flinging all sorts of doweling jigs in the trash. Again, align the pieces, but this time the top-piece board end is facing you, sitting flush on top of the end of the bottom board and at right angles to that board. Make your marks, starting 2 inches in from the ends, marking 4 to 6 inches apart. Cut the slots, starting with the face board, then going to the end of the other board. Test-fit with dry plates, after cleaning out the grooves if needed. Disassemble, add glue, and reassemble, with clamps.

Center Butt Joints

Joining internal parts with butt joints is easily done, but requires a few more steps. When marking to the center of a $3/4$-inch board, mark on the $3/8$-inch line, then always work to the same side of that line. When boards are marked for a particular place in an assembly, key mark them so they return exactly to that place for practice assembly, and for later final, glued assembly. If they're not marked and the piece is at all complex, pieces will end up in the wrong places.

Cut vertical slots first, then cut horizontal slots. Marking distances (thus plate insertion distances) are similar to those for corner joints:

Come in 2 inches from each end, then set other biscuits 4 to 6 inches apart.

If you are having trouble getting all the glued parts together in box assemblies within the 10 or 15 minutes of open working time of aliphatic resin wood (yellow) glues, do not just hold things together for 10 or 15 minutes until the glue sets. Instead, get Franklin's Extend and use the extra 5 to 10 minutes it provides. Extend works anywhere you're having trouble getting the time to assemble a glued unit, and it is the equivalent of Titebond II.

Miter Joints and Plates

In some materials mitered joints look far better because they provide a finished, all-wood corner appearance, without allowing plies to show with plywood, or to keep grain direction and general appearance more similar in solid woods and pressed boards. As a modified butt joint, a miter joint offers little strength beyond what any butt joint does, so it needs some form of support to provide the greatest durability.

Splines and dowels have long been popular for this type of support, with splines taking the prize. A miter joint is set on a table saw so that the blade rips a $1/8$-inch-wide kerf the length of the miter. The corresponding board is treated in the same manner. A spline is then fitted into the groove, glue is applied, and the unit is clamped together. Splined joints offer horizontal mobility during assembly so that accurate fit is simple, in that plane. Dowels offer a difficult fit in all planes, especially once the angle changes from 90° to 90° included—that is, 45° + 45°, 30° + 60°, and so on.

Plates offer adjustment along the same plane as do splines, while offering easy and accurate handling of the slotting so that there is less likelihood of a misfit. Produce a square miter, just as you would normally, in the size required for the job. Bring the two mitered faces together, and mark 2 inches in from the ends, and after that at 4-inch intervals. For very large carcasses, place the marks at 6- to 8-inch intervals.

Set the faceplate to 45° and make the cuts for the biscuits. Make your dry test assembly. This is the time to correct any out-of-square problems created by the biscuits, but if the cuts are accurately made, and the miter was done properly, there should be no problems. Disassemble, glue, and reassemble. Check for square as you clamp.

Plate joiners offer a number of options to increase strength with thicker stock. It is quite possible to offset cut marks even in fairly thin ($3/4$-inch) stock so that dual biscuits may be used, one above the other. It is seldom essential, and not a good idea. Internal pressures can cause an imprint of the plates on the surfaces.

In thicker stock, extra strength could well be needed. If we assume stock thickness of $1^1/2$ inches, the offset will normally be $3/4$ inch. By making it $3/8$ inch, as if for $3/4$-inch stock, we can make one cut, on a marked line, then turn the board over and make a second cut on the same vertical marked line. That produces two slots for plates and allows use of two biscuits per area, with a normal amount of stock outside the biscuits as well as between the biscuits.

Gluing and Clamping

For most of cabinetmaking history, hide glue was the epitome of wood-joining strength, even with its multiple problems. It also has multiple advantages, though not in the modern cabinetmaking shop, as we shall see. Things have changed considerably.

The primary use of wood glues is to hold joints together. There are other applications, but the major use is to make strong wood joints in furniture and other assemblies. Proper gluing will relieve some of the internal stresses of wood, particularly in laminating flat boards and reducing warping, cupping, and other forms of distortion. Too, adhesives allow us to use many parts of a board that are otherwise bits and pieces of scrap.

> Wood glues hold joints together. The joint must be cut correctly and fit tightly together. Proper gluing can also relieve some of the internal stresses of wood, particularly in laminating flat boards and reducing warping, cupping, and other forms of distortion.

Wood Selection

Successful joints always start with correct wood selection, which must be rapidly followed by tightly cut joints, well mated and coated lightly with an appropriate adhesive, firmly clamped, and permitted to dry. You must make sure that the woods being joined will successfully mate. Too great a difference in moisture content in woods creates problems, as may similar differences in wood structure. As an example, teak, with its high silicone and oil contents, does not bond well with other woods and is even difficult to bond to itself.

The best wide-area glue-ups (especially as sizable laminates or flat boards) result when the same species of wood is used. As examples, all the boards are pine or fir or oak or cherry. If part is cherry and part is pine, difficulties arise, though such difficulties are reduced by checking and following the classification below of various hardwood and softwood species according to gluing properties, adapted from a chart by the U.S. Forest Products Laboratory. The use of different species of woods in glued joints is still a problem area.

Group 1: Glues very easily with glues of wide range in wide properties and under a widened range of gluing conditions.

- Hardwoods: American chestnut, aspen, black willow, cottonwood, hackberry, yellow poplar
- Softwoods: Bald cypress, California red pine, grand fir, noble pine, Pacific silver pine, redwood, Sitka spruce, western larch, western red cedar, white fir

Group 2: Glues well with glues of fairly wide range in properties under a moderately wide range of gluing conditions.

- Hardwoods: American elm, basswood, butternut, magnolia, mahogany, red alder, rock elm, sweetgum
- Softwoods: Douglas fir, eastern red cedar, eastern white pine, ponderosa pine, southern pine, western hemlock

Groups 3 and 4: Glues satisfactorily. Requires very close control with good-quality glue of glue and gluing under well-controlled conditions, or special treatment, to obtain the best results.

- Group 3 hardwoods: Black cherry, black tupelo, black walnut, dogwood, pecan, red oak, soft maple, sycamore, water tupelo, white ash, white oak
- Group 3 softwoods: Alaska cedar
- Group 4 hardwoods: American beech, hard maple, hickory, osage orange, persimmon, sweet and yellow birch

Whenever possible, use the same species; but if different species are desirable, use solid sections of those species where glue is not necessary, if possible. Door construction provides a good example, with the inner panel floating loose in the stiles and rails. The glued inner panel may be of any species, while the stiles and rails may be of a different species. Because the two sections are not solidly joined, there are no joint problems.

For the same reason, use plain-sawn boards with plain-sawn boards, and quarter-sawn boards with quarter-sawn. Otherwise, the differences in grain directions will create distortion problems.

Allow boards to temper, or acclimate. That means leaving them at least 24 hours in the environment in which they will be glued. Two to three days is better.

Selecting a Cabinetmaking Glue

Regardless of claims, there aren't as many true cabinetmaking glues as one might expect. Even fewer are of interest to the general wood-worker producing strong joints. Contact cements are used for general cabinetmaking and are briefly covered here. The joint produced using them is not a true cabinetmaking joint; it is a laminate of dissimilar materials, and properties in the cement allow it to last even though expansion and contraction rates differ markedly.

Cabinetmaking adhesives are broken down into animal, or hide, glues and synthetics. Animal glues are far older, and are far less used today because synthetics offer certain properties that hide glues do not. Similarly, there are some properties offered by animal glues that synthetics don't offer, or don't offer as completely.

No one really knows when the first hides, hoofs, and bones of animals were boiled to produce glue. These glues are by-products of the meat and tanning industries, and are readily found in dry, granular forms or as ready-to-use liquids. The most common liquid hide glue today is Franklin's. The glue is nontoxic. Its slow drying time gives a long assembly time for complex projects, allowing adjustments and changes that can't be made with faster setting glues.

Hide Glues

Hide glues are thicker than white and yellow glues, resist solvents (other than water) well, and give a pale tan glue line. They sand well without gumming. The lack of gumming is important, because glues

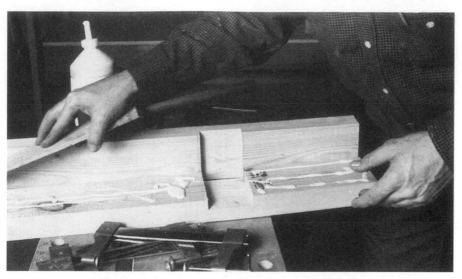

Glue must be spread quickly and evenly.

and adhesives that gum heavily will clog and ruin sandpaper quickly. Water resistance is poor for hide glues, which means it isn't the greatest stuff for kitchen and bathroom cabinetwork.

Dry hide glue is available. It must be mixed with water, heated, and then held at 140 to 150 degrees Fahrenheit. The process starts with soaking the granules in cold water. The granules soften after several hours, and the excess water is poured off. The glue is then heated, and the temperature is maintained while the glue is stirred until it is smooth and free of lumps.

Hot hide glue is applied with a stiff brush. The joint is clamped while the glue is still hot. Wood being glued needs to be warmed to at least 70 degrees in cold shops. Neither the wood nor the glue should be overheated, as that ruins strength.

Using this glue is a complex procedure that requires good advance planning and a costly heating bucket. As a result, the use of dry glue is not common among today's cabinetmakers, especially given its lack of water resistance.

Synthetic Adhesives

Most of the adhesives used in cabinetmaking today are synthetics, formulated specifically for different applications in the cabinetmaking fields—and some originating in other fields. Most are types of resin glues that gather strength by chemical reaction, or curing. Curing is dependent on the temperature of the glue. The strength of the cure and the speed of setting are increased by raising the glue temperature. Heating the glue to more than 120 degrees isn't a good idea, as the maximum cure temperature is about 110 degrees for yellow (aliphatic) resin glue, while others do best at 75 to 80 degrees.

Polyvinyl Acetate Resins

White glues (polyvinyl acetate resins) come ready to use, in squeeze bottles on up to 1-gallon and larger jugs. There are many brands, and many are somewhat acceptable for general cabinetmaking purposes. White glues don't use a chemical reaction type of cure. The water in the glue moves into the wood and the air, thus the resin gels. On unstressed joints, you can usually release clamping pressure within 45 minutes. Leaving the clamping pressure on for several hours is better. Stressed joints need at least a 6-hour set before clamps are released. White glues are not always dead-white. Some are dyed close to yellow and others are tan, to appear more like aliphatic resin glues. Aliphatic resin glues are more heat- and water-resistant.

White glues have poor sanding qualities because the glue softens with the heat generated by sanding, gumming up the sandpaper. The same characteristic causes a loss of glue strength at 100 degrees and up. Water resistance is low enough that a high-humidity environment can create separation problems.

Set is fast, limiting assembly time to 10 or 15 minutes. Pressure application must be fast, so preassembly of projects is essential. Fit all clamps within about a half-turn before disassembling, applying adhesive, and reassembling for clamping.

The glue line is close to transparent once the white glue has dried. In too cold a shop (under 70 degrees usually, though I've successfully used this type of glue at 65 degrees), the glue will appear a chalky white, and the joint line may be weakened.

White glue gives with the day-to-day movement of the wood, a process known as *cold flow,* so it should not be used on a highly stressed joint. Cold flow allows joints to move naturally without creating cracked glue lines and weakening the joint.

Liquid Yellow Glues (Aliphatic Resin)

Aliphatic resin glues were designed as improvements over the polyvinyl resin glues, and they really provide some worthwhile changes.

Heat resistance is higher, which makes sanding easier while also improving strength at 100 degrees and up. These glues set well at temperatures up to 110 degrees, which means they can be used on hot summer days in hot areas without air conditioning. Assembly is usually easier at lower temperatures because raising the glue temperature speeds the set rate and reduces open time. The glue line is a translucent pale tan or amber color.

Yellow glues are less likely to run and drip than white glues because their basic consistency is heavier. This makes for neater gluing jobs. Their greater moisture resistance means you can assemble projects for use in damp basements within reason.

Set is faster than for white glues. This can be a problem if you are gluing complex projects. Switch to Extend when project assembly will take more than 10 minutes.

The bottle of Titebond glue in the background is the most important feature of the photo, because Titebond yellow glue—aliphatic resin—is a standard setter in woodworking.

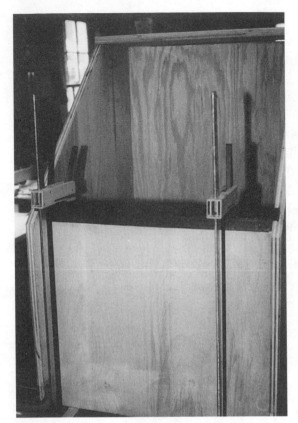

Clamp carefully.

Total cure is at least 24 hours. Use water to clean up before the glue sets.

Waterproof and Water-Resistant Glues

A number of glues are available for use where moisture is a problem, thus are extra handy in bathroom and kitchen cabinetry. Two of the older and most useful are plastic resin and resorcinol resin glues.

While plastic resin adhesives are highly water-resistant, only the resorcinols earn a waterproof rating. Where there is some tradeoff possible, it's more economical to use the plastic resins, since resorcinols cost 3 or 4 times as much as the plastic resins. [And don't forget that the fortified aliphatics such as Titebond II (and Extend), which can be used down to 55 degrees, have the sanding qualities of aliphatic resin glues and are highly water-resistant—they aren't for total submersion, but otherwise are great in wet environs. This has become an exceptionally important category of cabinetmaking glues in the past few years, though the aliphatics in general still hold the real place of honor.]

Resorcinol resin glues are dark-red liquids (the resins) to which a catalytic powder is added before use. Resorcinols have a reasonable working life, depending on formulation, after mixing from about ¼ hour to 2 hours. It is best to work with the longer life, so check the labels first. Resorcinols are expensive and fairly difficult to use, but where a really long working time along with very good water resistance is demanded, only epoxies—even more costly and difficult to use—are in the same class.

Before starting, get well set up. Make sure wood moisture content is below 12 percent, joints are tight and precise-fitting, and heavy-duty clamps are on hand.

Brush resorcinol on or spread it with a spatula. Tongue depressors, available at any drugstore, make great glue spreaders. If dense hardwoods are glued, watch out for glue starvation—the lack of glue in tight-fitting joint areas. Lightly coat both surfaces with glue, and leave the joint open for the maximum time before clamping. Increasing shop temperatures helps, especially in colder shops. The project must be clamped as quickly as possible. Clamping pressure is high, about 200 pounds per square inch (psi). The pressure must also be uniform. That means more clamps.

The glue line is ugly, a dark red or reddish brown.

Urea (Plastic) Resin Adhesives

At one point, and for a long time, good water resistance and low cost could only be found in the urea resin adhesives. Today, there are many competitors: Titebond II and its competition—polyurethane glues (these are expensive and hard to store, and of little use to the cabinet-maker, which is why this is the only reference to them). Plastic resin adhesives are dry powders that are mixed with water just before use. The resin is urea formaldehyde which is a highly water-resistant adhesive, best on wood with a moisture content of no more than 12 percent. Best use and cure temperature is 70 degrees. Formaldehyde lacks popularity today and is said to cause some people problems with its gassing off. Possibly, it does.

Plastic resin glues are superb for producing joints in projects that must withstand long-term dampness. Some of these glues do very well in true exterior applications. They make good general-purpose glues because they work easily in all situations, with the exception of high-density woods such as maple and oak. Precise fit of joints is essential, as plastic resin glues are not good gap fillers. The best gap filler, other than epoxies, is hide glue. Check with the manufacturer of the glue to see what its off-gassing performance is like under particular circumstances.

The setting of the glue is affected by temperature, so complex assembly jobs may be left to cooler weather or done in an air-conditioned area. Otherwise, working life ranges from 1 to 5 hours. Clamp pressure must be in place for at least 9 hours, preferably 12. Clamps may be taken off when the glue squeezed out is hard.

Clamp pressure must be moderate. The joint line appearance is good, a light tan color. Gumming is not a problem as the resin resists heat well.

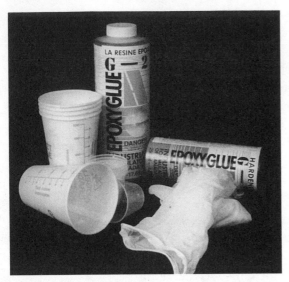

Epoxy, a two-part glue, is a problem solver that has a very adjustable open time and cure time.

Epoxy Adhesives

Epoxies weren't used in the cabinetmaking shop for a great many years and are seldom needed now. Like resorcinols, epoxies come in two parts, with a liquid hardener (or catalyst) added to a liquid resin. There is no powder. Curing is by chemical reaction. Heat is given off as the reaction takes place.

Mix only as much as you will use. The material is costly, and waste is expensive.

Epoxies are adhesives of choice for bonding teak and similar high-oil woods. Epoxies can be formulated to suit just about any bonding need, in moderate temperature applications, if you precisely follow package directions.

Epoxy doesn't shrink, so is a good gap filler. Some epoxies are available as putties. Thus the largest gap is readily filled, though precise joint fits are still better for long cabinet life.

Set time is an important factor in wood adhesives. One of the reasons epoxies missed early popularity was that virtually all of them are fast-set types, setting in less than 5 minutes. Such speed is superb for many jobs, including some small cabinetmaking projects and repairs, but is a horror story for larger projects. There is no way to get a large project together and squared up in the available time. Thus, epoxies were useful for repairs and useless for most other purposes. With slower set formulations, epoxies became more useful. Epoxies aren't useful, or financially possible, for general cabinetmaking jobs such as bonding strips of wood to form a butcher block, or bonding wood for a tabletop, or coating a lot of joints for strength. The cost is extreme, and the need simply isn't there.

Epoxies are quite toxic, too, which limits their uses in some shops. Epoxies are messy, though that problem is easily solved. Wear thin plastic gloves, available in packs of 100, to avoid the hand mess. Clean up quickly with acetone for other messiness, keeping the gloves on. Make sure all mixing containers and sticks are disposable. When you can't keep from making a mess, make the mess itself as easily disposable as possible. Check with the Occupational Safety and Health Administration (OSHA), or check the sheets the manufacturers put out, for possible dangers other than the mess.

If you're working with dense woods or with exotics such as rosewood, teak, and ebony, nothing surpasses epoxy as an adhesive. Clamping pressure is light; working time is adjustable, depending on the system, to as much as 90 minutes; gap filling is superb; strength is incredible; and the resulting glue line is either clear or an amber color, depending on the brand used. Some brands won't stain wood, others will. Check. Some may be used with a variety of catalysts, with the catalyst providing either a slow or fast cure. Fillers can be found, as can pigments.

Epoxies are not for general-use joints, but for special cabinetmaking uses, they've really come into their own.

Contact Cements

Anyone who ever builds kitchen or bath cabinets will work with contact cement. Contact cement usually doesn't go into the cabinets (Eurostyles, with laminates over wood substrates, do use contact cement over almost the entire cabinet), but into the countertop where plastic laminates such as Formica are glued to wood substrates. The wood substrates may be plywood or one of the formed boards. The latter are usually best because, over the years, plywood grains show through the plastic laminates.

Contact cements come in two basic types. One uses a water base, or water solvent, while the other uses a nonflammable solvent base. UGL's Safe Grip uses 1,1,1-trichloroethane as the solvent. Some types still use flammable solvents. It is probably best to not use these, because they also do a lot to contaminate the air.

Vapors are harmful with the newer solvents, so make sure you work with proper ventilation. Some water-based solvents are pretty rough in the fume field. A few may not be safe around an open flame. Check before using any of them. Avoid flammable contact cements, because they are truly wildly flammable.

Contact cements give a quick bond that allows cleaning up and trimming of the final project right away. Using them is simple. Coat both surfaces with the cement, using a brush or a roller or a spray gun. Let the surfaces become dry to the touch. Place the laminate on the substrate.

UGL contact cement.

When the laminate is positioned correctly over the wood, it is pressed onto the contact cement, but to get the position, the laminate must remain mobile. A slip sheet of Kraft paper or waxed paper may be used, covering the entire surface. Leave enough to grip outside the two pieces being joined, bring the top piece down, align the two pieces, and slowly start slipping the paper out. Once the paper is out 3 inches, roll or tap over the cleared area to ensure a bond. Pull the paper the remainder of the way out, being careful of alignment. Roll or tap to ensure the bond, and you're done.

You might also use $3/4$-inch square wood stickers at intervals. Placing stickers across the full width of the base material works well at 1-foot intervals. The coated laminate is laid on the stickers and carefully aligned with the base. Remove, first, the center sticker so the laminate touches the base material coating. Tap or roll, and then remove the remaining stickers, tapping and rolling as you go.

Choosing Glues

The selection of the appropriate glue is important, but so are the application of the glue and the clamping of the parts. Equally important is that you are working with a tight-fitting joint so there are no problems with gap filling or joints weakened by wide expanses of nothing but glue.

Make your glue selection based on the qualities you need most. If assembly is complicated and time-consuming, and moisture is not a problem, choose Extend or urea resin glue. If moisture is a moderate problem, choose a urea (plastic) resin glue instead of yellow glue.

For general cabinetmaking uses, liquid yellow glues give the best combination of moisture resistance, sandability, and thicker-spreading qualities that provide better gap filling.

For great water resistance, select Titebond II, Extend, epoxy or urea resin—Titebond II first, unless the epoxies fill some other specific need such as great unsupported strength or filling a gap. Epoxies cost too much for general use.

For total waterproofing, select resorcinol resins. These are expensive and difficult to apply properly since they require very precise-fitting joints, and they leave an ugly reddish glue line, but they are totally impervious to water.

The type of glue chosen influences the method of application, though most can be applied with a brush, stick, or roller. Joint surfaces must be checked first. If the joint surface is designed to be a tight fit, it should be. Clean off all dust, oil, loosened and torn grain, and chips. Any machining that must be done should be done as close as possible to the time of gluing and assembly. A test assembly is a good idea, because once the glue is added, correcting mistakes is, at best, very messy. If mistakes get in and glue sets, mistakes remain.

Before you apply the glue, decide whether the unit can be assembled within the time required for a specific type of adhesive. If a glue has a 10-minute open time, you must have the assembly completed before that time passes. The thicker the glue you spread, generally the longer the open assembly time. If wood is extremely porous or extremely dry, open assembly time decreases. If you can't make the test assembly within the allotted time, change either the conditions of gluing or the type of adhesive used so that enough time to complete and clamp the assembly is available.

Mix, where required, all adhesives according to the maker's directions, and as accurately as possible. Spread evenly over the surfaces to be joined.

Clamping and Clamping Pressure

You clamp a glued joint for three reasons: The wood surfaces must be brought into direct and close contact with the glue; the glue must become smooth, flowing to a thin, continuous film; and the joint must be held steady until the glue dries.

Clamping pressure varies with glue type, but generally coincides with glue thickness and wood type. The heavier the glue, the greater the clamping pressure. You want a thin, smooth glue line, not a joint that is squeezed dry, which occurs when too much pressure is used.

Most cabinetmaking glues fall in the intermediate-thickness range, requiring a clamp pressure of 100 to 150 psi. Some dense hardwoods may require pressures up to 300 psi. Softwoods do not require much clamping pressure and will not stand it. At the very least, some deformation will occur as clamping pressures approach 300 psi on softwoods.

A clamp may apply force ranging up to 1 ton. This pressure is divided by the overall size of the clamped area to determine the pounds per square inch being applied.

Resorcinol and urea resins require a great deal of pressure, while epoxy needs little pressure. Avoid excessive pressure in favor of even pressure over the entire area. It is better to get even glue squeeze-out over the entire joint than to rack the project up as tightly as possible.

Some cabinetmakers say one thing, others say another when the number of clamps to be used comes into question. I suggest clamps every 8 to 10 inches when possible, with some light clamps used as often as every 4 to 6 inches. In no case do I feel secure with clamps spaced out more than 16 inches apart. Even contact cements require about 50 psi for a proper bond. That's the reason you need a heavy roller or a 2 by 4 and a hammer. If you don't have a laminate roller, use a rolling pin with a sheet of moderately hard, heavy cloth wrapped around it. It can be used without the cloth, if a few gouges and some dirt are not a problem. If you use a roller as thick as a rolling pin, bear down hard to get the correct amount of pressure per square inch.

> Clamping pressure varies with glue type, but generally matches glue thickness and wood type. The heavier the glue, the greater the clamping pressure. The thicker the wood, the greater the pressure. You want a thin, smooth glue line, not a joint that is squeezed dry, which occurs when too much pressure is used.

Clamps and Tools

Many styles of cabinetmaking clamps are available, and few have changed much in recent decades. Of the new items, nothing is really earthshaking. Cabinetmaking clamps fall into one of these categories: bar clamps, hand screws, C clamps, and band clamps. Among these you find pipe clamps, miter clamps, and a slew of others. The largest numbers of variations fall in the bar clamp range, which makes sense since those are used for case framing, large assemblies of multiple-piece tops, and sides.

Sizes for the bar clamps vary with the weight of the clamp. The heavier the clamp, the longer it can be, so normal-weight clamps reach about 48 inches, while light-duty units stop at 32 inches. Superlight-duty notched cam and tongue bar clamps stop at 24 inches. The heavy-duty Bessey K body clamps that ACC offers extend to 98 inches. These are costly, silky working clamps superb for heavy work that also do very well in medium work. Edge clamp accessories are offered in single and dual models, and they allow a great many variants that standard edge clamps, such as the Jorgensons I have, do not. The edge clamp accessory fits onto a standard bar clamp bar, allowing a great

Plan carefully when cross clamping so you know which clamps fit where.

Clamp the shortest distance possible that will let your assembly stay square and tight.

deal of positioning leeway. My edge clamps are fine heavy-duty units, but do not offer as much leeway. They were not designed to do so, and yet they do their particular job very well. Band clamps vary. One of mine is about 1 inch wide and 10 feet long. Another one is 2 inches wide and 14 feet long. The first one is from Stanley Tools and the second from Vermont-American. Both are excellent tools. Band clamps are useful for clamping odd shapes, such as octagonal frames, chair leg assemblies, and picture frames. Some are made with angle sections that slip onto the bands to fit different-angled corners.

Short, lightweight bar clamps are a help in some areas that are usually thought of as C clamp territory, such as gluing two sheets of plywood. It takes just as many clamps, but there is more depth of throat, and somewhat thicker overall clamping possibilities.

Hand screws in this country usually bear the Jorgenson brand name. Tips for preparing hand screws help preserve the screws, and the finish on some projects. Hand-screw bodies are wood, thus susceptible to glue buildup and other problems. To keep this glue buildup from becoming a problem, coat the outer jaw with paste wax and don't bother to polish it off. This means virtually nothing will stick, including glue. Hand screws work well to clamp nonparallel surfaces, and they do not creep as some clamps will. They are available in more sizes than are listed in catalogs. Prodding local suppliers may get you a chance at the larger sizes you need and cannot find otherwise.

This is far from an exhaustive look at clamps, missing a lot of types including corner clamps, useful for mitered work, spring clamps, and the wide variety of levered hold-down clamps.

The Basic Box:
Cabinet Carcass Construction

It would be fantastic if I could tell you there is a perfect material, and a perfect method, for putting together carcasses, or cabinet boxes. There isn't. Each cabinetmaker forms his or her own opinion of what works best and goes on from there. J. R. Burnette likes medium-density fiberboard (MDF) in various thicknesses and laminates for most of his construction. Others, particularly those who work specifically and specially with 32-mm construction, prefer melamine-covered particleboard. Some use plywood and will use nothing else. It is a matter of choice and one's own attitudes toward what wood product constitutes quality of a degree to enhance the reputation.

So what's the best way to build the carcass for kitchen and bath cabinets? I can't even tell you that. As you'll note, we have chapters on face frame and frame and panel construction, as well as 32-mm construction. On a personal basis, I find the 32-mm stuff goes together more rapidly, but I also find it doesn't look as pleasing, to me, as do traditional face frame cabinets. In a lot of cases, though, that makes me wrong, because the smoother-looking Eurostyle 32-mm look goes great in lots of specific instances, in particular designs, and even in some classic homes (it can be used to produce a more retro 1930s look than actual 1930s cabinet designs can).

Melamine-coated particleboard (MCPB) is the material of choice for many cabinetmakers building Eurostyle cabinets. I'm always leery of MCPB because of its tendency to disintegrate when wet—you have to live a time in a mobile home to appreciate leaving a hall window open while away, to come back to no floor, except where it is glued to the joists, so that repairs are incredibly difficult. The stuff, at least older materials, just melts. My preference is for MDF, with whatever laminate is preferred, making sure the MDF uses water-resistant (at least) resin adhesive. MDF with standard adhesives disintegrates almost as quickly as does particleboard. Plywood may also be used, as may solid woods, though solid woods are seldom used today because the various manufactured products provide practical answers to old problems of wood movement. The reason frame and panel doors exist is primarily to allow the panel, solid wood (though not always), to move. Solid wood, on average, can move as much as $1/8$ inch in 1 foot as humidity changes. It doesn't seem like much, but on a 23-inch-deep cabinet, that's nearly $1/4$ inch of movement across the width, enough to bust something. Thus, for traditional cabinetry, a panel in a frame may be used, even in carcass construction, though today the simple box is most often used, with a manufactured wood base.

If butt joints are not used in a traditional carcass, then almost any joint type may be used, including various types of tenons, miters, dowels, rabbets, dadoes, and others. All this sounds horribly complicated; it can be for really super deluxe high-end, take-no-prisoners work, but today most people cannot pay for such work, and wouldn't anyway, in areas meant to be used for a decade or two and then changed over.

> There is nothing particularly complex about the average cabinet. The base cabinet boxes have two sides, a bottom, and a back. The upper cabinets have two sides, a bottom, top, and back board. There may be a nailer strip. There is a toeboard assembly.

There is nothing really complex about the average cabinet. The base cabinet boxes have two sides (if this were a house, they'd be gable ends), a bottom, and a back. The upper cabinets have two sides, a bottom, top, and back board. Countertops generally cover the top of base cabinets, so no top is used. Usually, in 32-mm carcass construction, bottom and top boards are

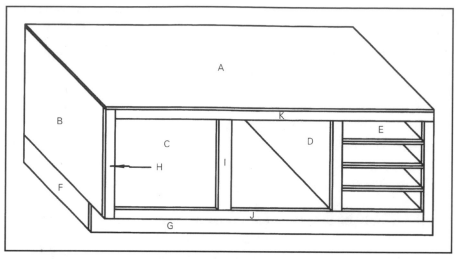

Basic face frame cabinet drawing. This pattern is set for 24-inch-wide drawers and two sets of 15-inch-wide doors. A, countertop or underlayment; B, finished end panel; C, bottom; D, inside divider; E, dust panel; F, end toeboard; G, toeboard; H, end stile (full length); I, interior stile (inside rail fit); J, bottom rail; K, top rail.

attached to the sides. The widths of the bottom and top boards determine the carcass interior width because the sides are attached to these boards with butt joints. Use the back board to cover all edges of the bottom, top, and sides.

The back board is sometimes a choice when you are making cabinet boxes. There are options including backing strips, 1/4-inch full backer boards, attachment strips, or full-width full-strength material. You can save money with either of the first two, but I don't like them. Generally a minimum 5/8-inch backer board works to strengthen the box—stiffens it—and makes for an easier installation job, though the cabinet is heavier and costs a few dollars more. Mounting to the wall is very easy, because you don't have to hunt for a stud that matches an exact spot on a strip of board. The cabinet also resists the twisting force that sometimes ensues if you mount it to a less-than-plumb wall, or a wall that has other imperfections. (I once lived in a Hudson Valley home built in the late 1830s, and the downstairs walls literally bulged in the middle by several inches. I've no idea, in retrospect, whether that was deliberate or not—I tend to think it was—but it made all sorts of furniture placement difficult, and cabinet hanging nearly impossible.)

MDF cabinet boxes using butt joints and fastened with at least $1^1/_2$-inch-deep thread screws are very strong. Your parts thicknesses may vary, and if they do, make sure the screws are at least 3 times, in length, the thickness of the material being fastened. That's a rule of thumb, of course, and $5/_8$-inch materials don't insist on $1^7/_8$-inch screws: use 2-inch. Other joints are even stronger, but take far more time and equipment to complete. And you may be building in totally unneeded strength and durability. A good cabinet can last as long as a house, but it is seldom required to do so. That isn't an excuse for building in lesser quality, but added to the necessary price shopping that even upper-end customers must do, it is a rationale for not going overboard. It would be nice to do a Chippendale and fill a shop with Cuban mahogany, dozens of apprentices and assistants, and revert to the work of yesteryear. But Chippendale mostly built furniture and had an exclusive clientele. Those of us not working for Joe and Jane Sixpack are apt to be working for Joe and Jane Corkscrew, who would prefer to spend at least some of that disposable income elsewhere.

Corner blocks of hardwood (poplar, soft maple, etc.), and cross-bracing, may be used to provide both resistance to racking forces and fastening points for countertops when you build base cabinets.

Once your basic box is built, you can go one of two ways: Tape can be applied to the exposed front edges, which is the basic Eurostyle 32-mm cabinet. The North American standard is a $3/_4$-inch solid-wood face frame covering exposed edges which gives you a traditional-style cabinet. Both systems are very popular; overall, Eurostyle is faster to build, thus lower in overall cost. It also doesn't use solid wood, or at least doesn't have to, so the cost is again kept down. Most traditional cabinetry uses solid wood at least in the face frames, and sometimes in the paneling as well. Seldom is solid wood used on sides and backs of kitchen and bath cabinetry, regardless of style.

Door Sizes and Hinges

After you've built the cabinet box, you need to figure the door size. Although that's a relatively simple matter, a couple of issues must be considered. The first is the finished type of cabinet: Eurostyle cabinets use much different hinges than do traditional face frame styles. The hinges in the 32-mm system are hidden, never visible with the cabinet doors closed. Oddly enough, face frames may use almost any type of hinge, including the Eurostyle.

Figure the door size for the Eurostyle frameless cabinet with hidden hinges as follows: The door size equals the outside dimensions of the cabinet box, in width and height, less $1/16$ inch on all edges (for single doors). That means your basic 16-inch-wide by 32-inch-tall cabinet needs a single door $15^7/8$ inches wide by $31^7/8$ inches tall. Two door cabinets need an additional $1/32$-inch space for the center gap between the two doors, so a 24-inch by 32-inch cabinet would have two doors, each $11^{27}/32$ inches wide by the same $31^7/8$ inches tall.

Face frame cabinets using 32-mm hinges use the inside width, plus 1 inch to the combined width for double-door cabinets and plus $1/2$ inch for single-door cabinets. Thus, your 32-inch cabinet that's 30 inches inside takes two $15^1/2$-inch-wide doors. Door height depends on a number of factors, including door mounting (flush is what works with Eurostyle—overlay doors go on with regular hinges). It also depends on the width of the rails and stiles, so needs to be figured independently as you work.

Figuring door widths for traditional hinges depends largely on the hinge style. Measure the inside cabinet width, to make sure the overlay door covers the opening as well as the required mounting surface for the hinge. Test-fit the hinges with a single test door for a series of cases to see what you come up with. You can use cardboard templates if necessary, or a simple cutout of door size from scrap MDF or other material, to test-fit in one size. Then note the additional space needed on the door compared to the inside width of the case, and work from there. It will fit, but a lot depends on the door style and cabinet style.

I like overlay hinges for several reasons, including the fact that many look really great. A properly done overlay door also looks good and provides leeway in fit. If the door is a bit wide after the original overlay is cut, it's a simple matter to cut one side or another, or the top or the bottom, of the overlay to better the fit. It is a true sign of clumsy cabinetmaking to cut the overlay so the door is too small, though.

> **Overlay hinges look good, and the overlay door provides leeway in fit. If the door is a bit wide after the original overlay is cut, cut one side or the other, or the top or the bottom, of the overlay so that the fit is better. It is a sign of inept cabinetmaking to cut the overlay so the door is too small, though.**

Once you've got the basic boxes on hand, there are a number of ways to work out what goes where, as far as doors are concerned. Even if you go the easy way, and slip into buying your doors ready-made from a supplier, making sure of fit is imperative. (And if you start out without a shaper, or at least a very, very good router table, you might find this not such a bad idea. It saves time which can translate to money if there is no shaper in the shop, and no time or money to get one and learn to use it.) The doors are there for you, and you've built the carcasses. It is not a really fine idea, but one that happens. Finish building the cases, and when they're done, have the special-order outfit on alert. Get your door sizing done with samples and checking, and then order the doors. That's the order in which you'd build them, so that's the way to order them.

No real complexities are necessary, just a general order of march, so that items get built in the proper sequence, are checked for fit, are assembled and ready to go, and can then be transported to the site where they are easily installed because everything fits.

Layout

That dreaded word again. This time, though, lay out on the sheet material; and if you're in a position where you have to cut face frame material to length and width, this can save you time and money. Essentially, you want to lay out the parts of the cabinets on the faces of your sheet material in a manner that does two things. First, and of utmost importance when grain is visible and natural finishes are used, the wood grain must match or otherwise fit into the look of the piece. Second, you want to get as many pieces as possible from a single sheet of plywood, MDF, or other material, so as to prevent waste and control costs.

One day, I decided to build some office book shelves. I laid out the $11^1/2$-inch-wide sides, top, and bottom and the $10^1/2$-inch-wide shelves to save the greatest amount of material, using birch plywood.

> Lay out the parts of the cabinets on the faces of sheet material in a manner that does two things: First, when grain is visible and natural finishes are used, the wood grain must match or otherwise fit into the look of the piece. Second, you want to get as many pieces as possible from a single sheet of plywood, MDF, or other material, so as to prevent waste and control costs. There are now programs that help with laying out for economy, but none that can match figures yet.

That meant that some of the shelves had surface grain running straight, and others had surface grain running across the board. Do not do this if you're producing cabinets, shelving, or anything else for a customer. Given the fact that now it's almost impossible to tell what direction the grain runs, at least without removing 80 or 90 books, the original look of the shelving was not all that great, because of the screwed-up grain directions. This kind of thing works just fine for your office equipment. It works not at all for the customer's product.

For fancier overlays or veneers, you may even wish to match or book-match veneers to give a cleaner, neater, more attractive overall look. Whatever you do, unless it is a part of the actual design, do not run grain in vertical and horizontal directions on the same face, or adjacent faces, of the same cabinet. This is less of a problem in 32-mm design, of course, since most of that cabinetry has a plastic laminate overlay and is based on MDF or another engineered substrate. For face frame work, the stiles of the cabinet are vertical and tend to dominate the front, even though rails and drawers tend to the horizontal. For that reason, most grain direction in kitchen and bath cabinetry is vertical—not always, of course, and variants are used for design purposes all the time. Just be sure it doesn't look like a mistake!

Layout and Parts Lists

Without a good layout of each cabinet, you're drifting in space when it comes time to estimate materials. Once you've got a handle on the number, style, and sizes of cabinets, it's time to draw up a parts list. The list can be a problem until you know how you're going to lay out the parts for the various cabinets.

Standard plywood or MDF or other engineered panels are 48 inches by 96 inches (give or take factory allowances, which sometimes see 97-inch-long by 49-inch-wide MDF sheets). Other sizes up to 60 inches wide and 192 inches long are available on special order, but prices rise quickly, both for production and for shipping on those panels.

Work to get the best use out of each panel; without figuring for grain, you can easily lay out any panel for optimum economy. Each 48-inch panel divides into two $23^{15}/_{16}$-inch-wide panels (assuming you use a standard $1/_8$-inch kerf saw blade), three $15^{15}/_{16}$-inch ones (or very close: number three may be a shade wider or narrower depending on how you set up your cuts); and four pieces $11^{15}/_{16}$ inches wide.

Lengths are readily divisible by multiples of 2 inches, so you can slice off 16-, 24-, 30-, and 36-inch pieces. Generally, plywood strength is greatest along its face grain, but also the difference in strength for most cabinetmaking purposes is not really significant (for heavy-use shelves, yes, but otherwise, not likely). MDF is supposed to be just as strong in one direction as in the other, which is nice, because there is no way to tell which way the grain runs—there isn't any. As I stated earlier, it's better to run grain vertically in doors, sides, and end panels, whenever possible. Grain for horizontal surfaces such as shelves should run along the shelves for greatest strength. It also looks better that way, as the eye tends to follow the grain smoothly, where cutting cross-grain interrupts the vision and gives a choppy look.

Toeboards in natural finish may have the grain running in either direction, but many toeboards, even on natural-finish wood cabinetry, are painted or covered with laminate. The laminate is more durable than most materials, while paint is easier to renew than clear finish.

Drawer sides may be made out of solid material—poplar is popular for this—and grain then runs the long way. Plywood drawer sides may have the grain running in either direction, but if you go one direction in one drawer, you should go in the same direction in all drawers.

Grain in drawer bottoms, which will almost always be plywood, should run parallel to the long direction, or in almost all cases front to back.

Drawer front grain usually is run vertically, to match panels and stiles. There are some opinions that may or may not fit with mine, but Dave Fleming, a retired boatbuilder, has the following to say:

I have no "axe" to grind nor am I trying to impose my limited aesthetic on anyone: These are my observations and opinions, based upon my own experience. There is nothing wrong in the sense of workmanship or effort in any of the projects. But in my eyes, I see a lack of awareness of the material used. By that I mean, for example, the way the grain of the wood in a piece is laid out. Some use very prominent grained woods and fail to lay out the framework so that grain flows or follows the direction of the piece. Or the wood is so heavily stained that the character of the wood is lost. Pieces are made for a purpose and are so heavily decorated that the piece is obscured in all the twirls and twists of the decoration. My view is that a piece should speak of the wood first and it, the wood, should flow with the intent of the piece. For example, a kitchen cupboard is first a utilitarian piece, not a Bombay chest. So let the wood define the use, and then if needed a bit of sweep in the baseboard or a delicate finial can be tried. For the beginner perhaps a wood that is not as dominant might be experimented with. See how that small chest looks in common pine and then determine whether walnut or cherry is appropriate for that design and use. Play with finishes that enhance the natural grain of the wood before moving on to the stains and heavy Poly-U (polyurethane) type looks. A simple oil, sanded and slurry-filled, might be all that that piece of cherry needs to stand well on its own. At first, concentrate on the design for its use and the construction that will satisfy that requirement. Then (concentrate on) the wood as it should fit those requirements and the finish that meets the needs of the piece.

Don't rush to splash on some stain or whatever, leave the piece without anything for some little time and come back to it at different times of the day and view it in the different lights and see how each time the piece takes on a different aura or impression. Most of all, don't rush, be patient, be observant and take photos and make notes of how you feel viewing the piece at these different times. We should all learn and grow with each piece we complete. And we should all keep prominent in our minds that we can and will do better with each succeeding project.

Dave is right, in my opinion, though his finishing comments are primarily aimed at amateurs who will not be trying for 15- or 20-year durability in a kitchen or bath environment. And the above quote is used with Dave Fleming's permission. The different viewing times might be significant with your clients, too. Ask them to view the entire job, as you've finished it, at two or three times of day and see what they think of it with just natural wood appearance.

Making Parts Lists

The drawing at the top of p. 252 shows a face frame cabinet that serves to illustrate, not define, what is needed to make an accurate and useful parts list. Label parts on the cabinet starting with A, and determine how many parts in each cabinet fit that label. Then go on and do the rest. There are major similarities, and should be no major differences, but you may end up with wildly different-shaped parts as you go through your listings. Corner cabinets, for example, have different needs than do sink base cabinets (even if the corner cabinet is a sink base cabinet).

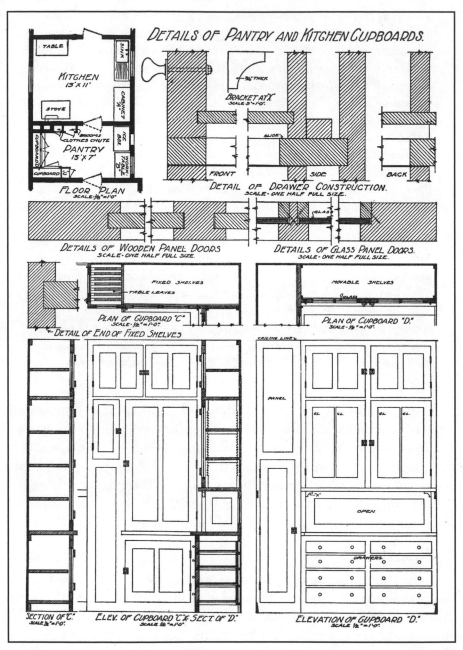

Details of pantry and kitchen cupboards, pre–Great War style (circa 1911). Note the detail in the drawings.

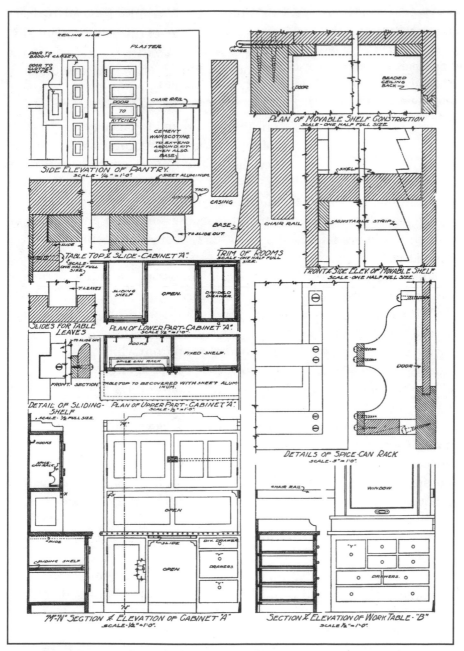

Details of pantry and kitchen cupboards, pre–Great War style (circa 1911). Note the detail in the drawings.

> Stile and rail widths depend largely on the design of the cabinets, but anything under 2¹/₂ inches demands the use of something other than biscuit joinery, which tends to complicate construction. Pocket screws can be used in narrower face frames, too.

Stile and rail widths depend in large part on the design of the cabinets, but anything less than 2¹/₂ inches demands the use of something other than biscuit joinery, which tends to complicate construction. Dowels are more difficult and if misaligned, remain misaligned. Pocket joinery works, but requires extra machinery that tends to cost considerably more than a biscuit joiner. Add the charts to the drawing, and it becomes even simpler to get an accurate, clear-cut list from the layout.

Tables 2 to 6 are for illustrative purposes only; do not take them, and the drawing (at the top of p. 252) that accompanies them, as indicative of a finished product or a plan. They do show what is needed, and can be readily adapted to almost any situation and cabinet, with a minimum of thought and effort.

With a completed series of story sticks, and your cutting lists, you're ready to cut the materials to size: I have a strong preference for laying the parts out on the faces of the wood from which they'll be cut. That gives you a last chance to correct any missed alignments of grain, change wrong grain directions, or shift cuts to miss knots and other wood problems.

As a quick note on measuring, the old adage of measure twice and cut once is absolutely wise, except for those of us who are better off measuring three times and cutting once. If you use a measuring tape, do not measure from the hook. Move down to the 1-inch, 2-inch, or other mark and use that as a starting point (don't forget to add that figure to the other end).

If you use a soft pencil, then you have a decent cut line that will disappear under cutting, but a lot of cabinetmakers prefer to do the rough layout (adding about ¹/₂ to 1 inch extra in both width and length to parts) with a chalk line and then to finish cut. A top-grade combination blade with at least 60 teeth works best for final cuts. Use a full thick blade, not a narrow kerf type. Keep the good face of the plywood up on the table saw, which keeps the chipping to the back of the plywood, where it is less likely to be visible.

Carcass parts are cut first. But if you have enough cabinets to use more than one sheet of plywood, it pays to do the rough layout of all

TABLE 2 96-inch Base Cabinet: Plywood Parts

Part number	Part name	Thickness (inches)	Width (inches)	Length (inches)	Material	Notes	Number needed
1	Left end	$5/8$	$23^3/4$	32	Hardwood plywood		1
2	Dividers	$5/8$	$23^1/2$	$31^1/4$	Hardwood plywood		2
3	Half dividers	$5/8$	$23^1/2$	15	Hardwood plywood		1
4	Subsurface	$3/4$	24	96	Particle-board	May be MDF	1
5	Substrate for laminate countertop	$3/4$	24	96	MDF	May use $1^1/4$ inch	1
6	Dust panel	$1/4$	24	24	Softwood plywood		2
7	Drawer bottoms	$1/4$	24	24	Hardwood plywood		3
8	Cabinet bottom	$3/4$	24	96	Softwood plywood		1
9	Cabinet back	$1/2$	32	96	Hardwood plywood		1
10	Drawer fronts	$3/4$	6	24	Hardwood plywood	Run grain vertically	2
11	Drawer fronts	$3/4$	10	24	Hardwood plywood	Run grain vertically	1
12	Door panels	$3/8$	29	18	Hardwood plywood	Run grain vertically	1

TABLE 3 96-inch Base Cabinet: Toeboard Parts

Part number	Part name	Thickness (inches)	Width (inches)	Length (inches)	Material	Notes	Number needed
T1	End toeboard, finished	$3/4$	$4^1/2$	23	Hardwood plywood	May use hardwood	1
T2	Toeboard stretchers	$3/4$	$4^1/2$	$21^1/2$	Softwood plywood		5
T3	Hidden end toeboard	$3/4$	$4^1/2$	23	Softwood plywood		1
T4	Toeboard (front, finished)	$3/4$	$4^1/2$	96	Hardwood plywood	May use hardwood	1
T5	Toeboard (rear, hidden)	$3/4$	$4^1/2$	96	Softwood plywood		1

TABLE 4 96-inch Base Cabinet: Drawer Parts

Part number	Part name	Thickness (inches)	Width (inches)	Length (inches)	Material	Notes	Number needed
D1	Drawer side	$5/8$	6	$22^1/2$	Poplar		4
D2	Drawer side	$5/8$	10	$22^1/2$	Poplar	May be hardwood plywood	2
D3	Drawer end	$5/8$	6	23	Poplar		2
D4	Drawer end	$5/8$	10	23	Poplar	May be hardwood plywood	1
D5	Drawer dividers	$3/8$	5	22	Poplar	May be hardwood plywood	6 (or as desired)

TABLE 5 96-inch Base Cabinet: Solid Hardwood Parts

Part number	Part name	Thickness (inches)	Width (inches)	Length (inches)	Material	Notes	Number needed
H1	Top rail	3/4	2 3/4	96	Oak		1
H2	Bottom rail	3/4	2 3/4	96	Oak		1
H3	Stiles	3/4	2 3/4	27	Oak		5
H4	Drawer rails	3/4	2 3/4	24	Oak		3
H5	Door rails	3/4	2 3/4	14	Oak		8
H6	Door stiles	3/4	2 3/4	31			8

TABLE 6 96-inch Base Cabinet: Cabinet Hardware

Part number	Part name	Width (inches)	Length (inches)	Material (inches)	Notes	Number needed
A1	Overlay hinge	1 1/2	1 3/4	Brass		8
A2	Door handles	NA	4 1/2	Brass		4
A3	Door latches	NA	NA	Magnetic		4
A4	Drawer slides	NA	22	Steel	Full extension	3 pairs

plywood parts at one time. Doing so lets you work to get the best choice of grain patterns for every part, including the last-to-be-cut pieces such as drawer fronts.

Cut the rough cuts. There are plenty of needs for cutting plywood, but the biggest one is for a knowledgeable helper. Your helper has to know how to help without hindering; you do not want, or need, input in feed direction or speed. The helper helps. You control. Basically, the helper is an animate material supporter.

> If you have enough cabinets to use more than one sheet of plywood, do the rough layout of all plywood parts at one time. You can then work to get the best choice of grain patterns for every part, including the last-to-be-cut pieces such as drawer fronts.

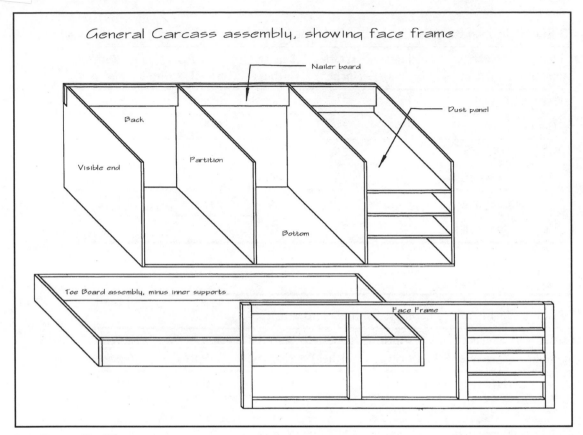

Face frame, off cabinet.

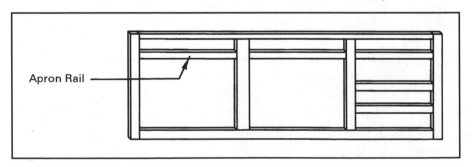

Face frame with apron rails.

You also need a good outfeed table. Any smooth table works well, but it must be at the same height as, or ⅛ inch or so lower than, the table saw table. Roller supports, I've found, tend to like to add their own directions to the material, and that is not helpful.

Once you've been doing this for a time, you'll eliminate the rough cuts and go straight to finish cuts. But at the start, rough cutting gives you the basic shape and size, and lets you make finish cuts in smaller pieces of wood that are easier to handle.

Depending on the fence you use, you may have a possible cut width as small as 24 inches or as much as 50 inches. For sheet use, the 50-inch fence width is usually a great deal handier, because you can then rest most of the uncut width on the table, and then the outfeed table, instead of having to support, for example, 32 inches of ⅜-inch oak plywood while you cut off 16 inches to get your final 32-inch width. It's lots easier, and usually more accurate, to run the 32-inch width flat on the table, and up against the fence.

When you have the narrower fence setup, always allow for waste. Taking our example above, you must have a 32-inch final width. The cutoff is 16 inches, but includes the saw kerf. Thus you measure from the fence to the outside edge of the blade, instead of the inside edge, as is usual. Measure to the tip of a sawtooth set to the outside.

Final cuts are always made with a 60-tooth or smoother cutting 10-inch blade. With great success, you may eventually

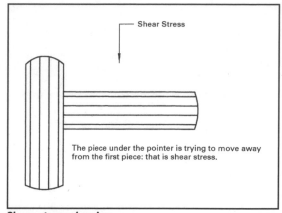

Shear stress drawing.

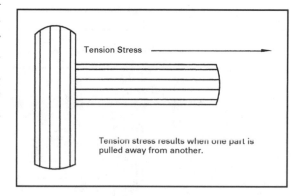

Tension stress drawing.

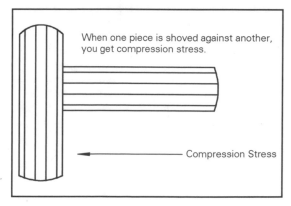

Compression stress drawing.

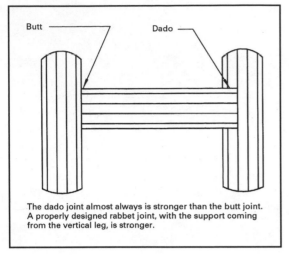

The dado joint almost always is stronger than the butt joint. A properly designed rabbet joint, with the support coming from the vertical leg, is stronger.

Dado versus butt joint strength drawing.

bring in the scoring saw setup, but for a start, using the best possible blade is your quickest and lowest-cost solution for smooth cuts.

Joints That Hold

Assembling the case requires a few distinct joints. We've covered general joinery already, but for the average case, you'll select very few of those joints. For the basic carcass, or case, you should need only rabbet, butt, and dado joints. Face frames and toeboards require more joinery skills, but generally rely on biscuits, or

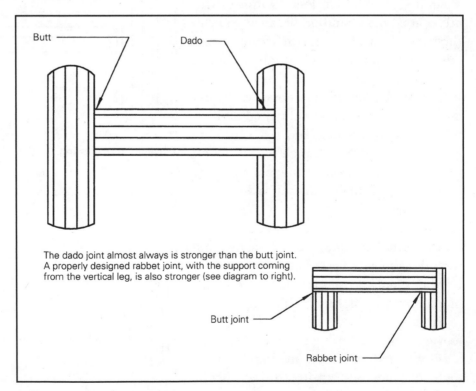

The dado joint almost always is stronger than the butt joint. A properly designed rabbet joint, with the support coming from the vertical leg, is also stronger (see diagram to right).

Dado-butt and rabbet-butt joints.

pocket screws, while cope and stile joints, a form of mortise and tenon, work for door frames.

At this point, you've got everything cut to size, at least for the case, so you're ready to cut the joints for the case. Typically, you'll place the back as a butt joint, inside the finish or other sides. If you wish greater strength, place the back in a rabbet, as deep as the back material and one-half the width of the material used for the sides. The problem with this method is that you are dependent on totally square floors and walls. For the normal house, leaving a scribing allowance on the side widths is recommended. At the top of the back, use a strip of ³/4-inch by 3-inch plywood as a nailing rail. This rail can butt against the insides of the sides, but should fit into notches cut in any dividers inside the carcass. The sides may be done in a stopped rabbet, too, if you wish, which results in a stronger case, more resistant to wracking forces, if glued up that way.

Bottoms of the sides are rabbeted to take the bottom of the case, and they may be dadoed for greater strength and for a wider bottom rail on the face frame (rabbeting limits the face frame bottom rail width to about ³/4 inch, while a dado can be set at any desired height).

All fixed shelves fit into dadoes in the dividers and the sides. If a divider is to take fixed shelves at the same height from both sides—something best avoided because it weakens the divider—the divider must be ³/4 inch thick to accept

Unisaw with Unifence cutting plywood. *(Delta)*

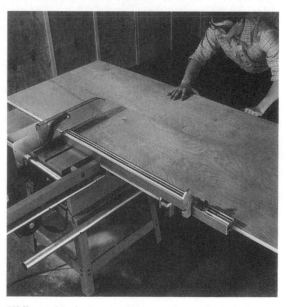

Sliding table being used to cut plywood. *(Delta)*

> Select joints for strength as well as speed of cutting and assembly. Learn the directions of forces that pull and push against the carcass, and design your joints to be strongest in the direction of greatest stress.

$1/4$-inch-deep dadoes. Otherwise, dadoes may be $1/4$ inch deep in $5/8$-inch-thick dividers, and $3/8$ inch deep in $3/4$-inch dividers. (Deeper dadoes are acceptable, but it's usually best to keep the dado depth at one-half or less the depth of the divider.)

Joint Strength

Select joints for strength and speed of cutting and assembly. Learn the directions of forces that pull and push against the carcass, and design your joints to be strongest in the direction of greatest stress. The earliest sign of failure on most cases is in the joints. The drawings on pp. 253 and 254 show the direction of some forces against joints and show how the joints are stronger in different ways. Except in drawers, there is very little tension force, the pulling away of one part from another, but cabinets often get a combination of forces, with shear stress at tops and shelves being the most common. Bending, twisting, or wracking stresses combine compression (push) and tension stress, often in unpredicted planes. It is best to design the carcass so it doesn't yield readily to wracking forces. This is most easily done by making sure all joints are solid, tight, well glued, and lightly nailed. The addition of nail rails to the backs of base cabinets really helps because the glue surface is increased a lot and there is an extra member in place to serve as a brace. A well-glued heavyweight bottom also works well to help resist wracking forces.

Cutting of dadoes and rabbets can be done with a router, or with a radial arm saw, or with a table saw. Within its limits, I believe the radial arm saw is the easiest to use. Next is the router. The table saw presents more difficulties, but is still a good tool for the two joints, if you use auxiliary fences to keep from eating into your regular fence face.

The radial arm saw is tops because you're watching the joint as it is being cut. The material is laid joint side up on the saw table, and the saw is pulled through it. All kinds of joints and decorative cuts (dentils, especially) are more easily done this way. The basic problem arises in the width of cut: Most base cabinet shelves are about 15 inches wide, and that's the absolute maximum limit for many radial arm saws, including the 12-inch models. They may run out to 15¾ inches on some brands, but if your rabbets or dadoes exceed that length, you're in trouble.

There are two ways around the problem, both costly. Delta's turret series, starting with the 14-inch radial arm saw, have maximum cut widths of 29 inches. Similar cut widths are available with older DeWalt radial arm saws. In either case, a used saw is probably going to cost you about $2500, and a new, or reconditioned one, considerably more than that, which explains why other methods are often recommended.

Sliding table being used to cut solid, long stock. *(Delta)*

Routed Dadoes and Rabbets

The operator gets almost the same view of the cut being made with a router as with a radial arm saw. Guides are necessary but are easy to make. Bits are relatively cheap. The router itself, because only a top-quality 1½-horsepower model is needed, generally is going to cost well under $200, with an array of different-sized bits costing possibly $100. Stopped dadoes are easy, too, because the operator is looking at the cut and can see where to stop, even if the jig isn't set up to stop him.

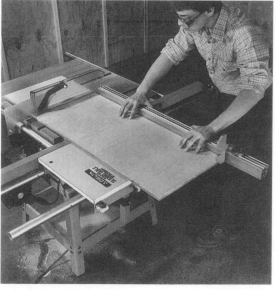

Easy finish cuts are made with a sliding table and a stop, on a Delta contractor's saw. *(Delta)*

Router Bits

Sizing router bits to agree with the plywood you are using can be a problem. Straight bits are generally available in $1/2$-, $5/8$-, and $3/4$-inch cutting widths. In other words, 9 chances out of 10, your $3/4$-inch plywood is actually $23/32$-inch thick. Your router bit, though, is a full $3/4$ inch. The result is a sloppy, weak joint. Companies are now producing bits specifically designed to address this problem: Eagle lists a series of two-flute (don't bother with one-flute) router bits that are designed to work with plywood, thus all bits are undersized by about $1/32$ inch. The same bits are also provided full-size, for other uses. Thus, bits for $5/8$-inch plywood

> Straight router bits are generally available in $1/2$-, $5/8$-, and $3/4$-inch cutting widths. Usually, your $3/4$-inch plywood is actually $23/32$-inch thick. Your router bit, though, is a full $3/4$ inch. The result is a sloppy, weak joint. Companies are now producing bits specifically designed to address this problem. Use them for best results.

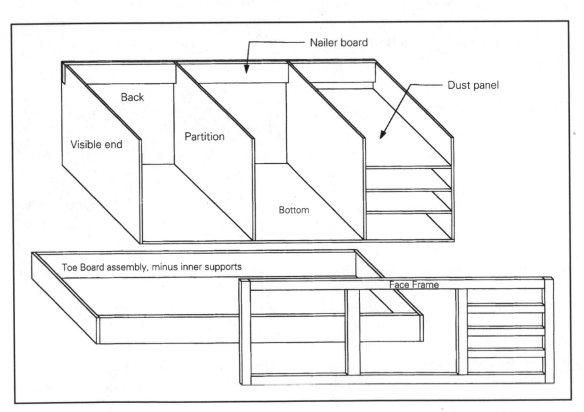

General carcass assembly drawing.

are $^{17}/_{32}$ inch, and those for $^1/_4$- inch plywood are $^7/_{32}$ inch, to complement the $^{23}/_{32}$-inch bits for the $^3/_4$-inch plywood. Jesada offers a similar bit line, as do others. The only other solution to the sloppy fit is to use a smaller bit and make two passes, which is a time waster.

Jigs for router use abound, especially for simple, straight-line dadoes or rabbets, and most are simple. For once-in-a-while use, simply clamp a board to the side, back, divider, etc., in a position that lines up the bit, and run the router along the board. Standard router bases on $1^1/_2$-horsepower routers are 6 inches in diameter, so it is easy enough to work from the center of that unit to get the rough location. Then do a more precise location with the board lightly clamped and the router sitting in place, but off. Tape the board to get exact alignment.

Some cabinetmakers prefer to replace the round bases with rectangular bases made of any of a number of plastics. This actually makes centering the bit and using jibs simpler and a bit quicker. Polycarbonate and phenol plastics work well for such bases and should be installed in thicknesses to match the base being removed.

You'll note throughout the router discussion I've emphasized the $1^1/_2$-horsepower router as the way to go for dadoes and rabbets. There are a number of reasons for this, including cost. A really topnotch Porter-Cable model 690 with knob handles costs just about $150. And it offers a choice of $^1/_2$- and $^1/_4$-inch collets, thus accepting bits that do not deflect too easily. Always, when the bits are avail-

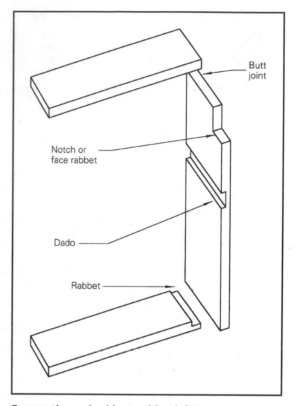

Frequently used cabinetmaking joints.

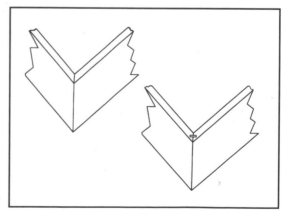

Miter joints.

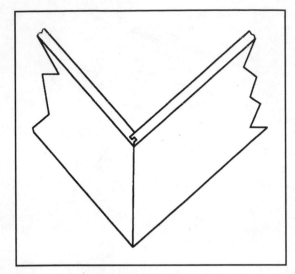

Lock miter joint.

able, use $1/2$-inch diameter shank bits. They are much more stable than $1/4$-inch. In my opinion, $1/4$-inch-diameter bits are best used for trim work, such as light round-over of edges. Serious groove and dado cutting tends to lose accuracy quickly with such a shank diameter. The slight extra cost of the $1/2$-inch shank is well worth it; identical $3/4$-inch bits cost $26 in $1/4$-inch shank as well as $32. Actually, those aren't totally identical, because the $1/2$-inch shank bit has an overall cutting length of 1 inch versus the $3/4$-inch cutting length of the $1/4$-inch shank bit.

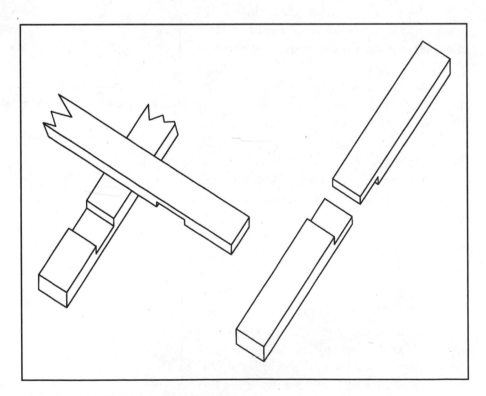

Cross (left) and half lap (right) joints.

In addition, such routers are light. The huge routers rated, inaccurately, at 3 horsepower and up also weigh in at upward of 14 pounds. These are great in a table, or for really heavy-duty freehand hogging out of material, long and wide and deep dadoes, and similar jobs. But for the day-to-day use that is modest to medium duty, the lighter routers are easier on the people using them.

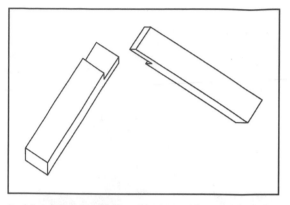

End lap joint. Basically, this is nothing more than a half lap joint pivoted 90°.

When you are making stopped shelf dadoes with a router, mark the end of the dado, and add 3 inches for the base (unless you've changed the base). You can place a stop block on the workpiece, or you can rely on not missing the mark. I prefer stop blocks.

Cut miter joint in cherry.

Easy-to-view cuts make the radial arm saw excellent for dadoing and similar grooving work that must be done blind. *(Delta)*

The easiest way to set up a routed dado accurately, instead of using center-to-center distances, is to make a guide in the necessary width for the board being routed, with a permanently fixed guiderail (the router runs against this, so a cleanly, and accurately, jointed and ripped piece of ⅝-inch-thick oak or maple is best: length as needed, width at least 5 inches). Align the straightedge with ¼-inch tempered hardboard cut to the width, and cut at least 1 inch wider than one-half the base of the router. Glue the tempered hardboard to the straightedge. Clamp the partially made guide tightly to a surface that either overhangs or is on a destructible (waste) surface without any projects. Use the router, pressed tightly against the straightedge, to carefully cut off the tempered hardboard. You now have a bit edge guide, par excellence. Align the edge of the guide with the top edge of the needed dado or groove, clamp in place, and run the router on through, at the correct depth, and you've got your dado. The same guide can be used to produce rabbets, though it is just as easy, and probably faster, to do standard rabbets with rabbeting bits which are available in a wide range of sizes and types.

Shelving

Any shelving that is to be installed outside dadoes now needs to have the holes drilled for the shelf supports, or slots dadoed for the standards. In general, the holes are by far the most professional-looking method of adjusting shelves that we find in general use today, but are far more readily drilled by gang drills in factories. They are time consumers in a small shop, because most drills, and drill presses, can only drill a single hole at a time. And there is always the possibility of drill

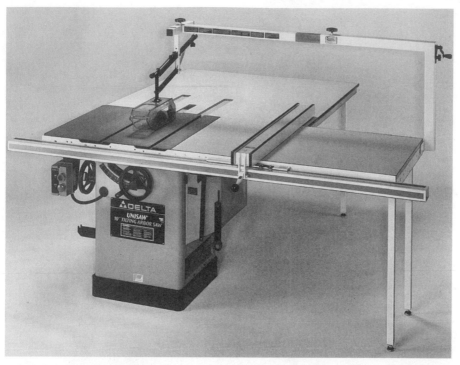

The free-floating Biesemeyer guard makes many cuts not only easier, but safer. *(Delta)*

This feed unit for a particular type of label printing press is one item J. R. uses to help smooth out slack periods. It shows the kind of versatility of production that helps the small shop stay afloat. And it also shows a quick and dirty use of clamps on templates for a router.

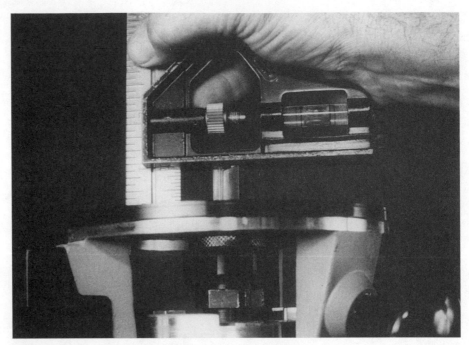

Setting up Porter-Cable D-handle router.

through. Probably one of the simplest ways of helping move things along is to use a plunge router for this job. Make a template to place the holes where you wish them inside the plywood carcass side or divider. That may be tempered hardboard, phenol, or a translucent acrylic. Simply make the template the size (in width) of the router base, plus 2 inches, and then drilling the number of holes you expect to place for adjustment works well. Make the drilled holes the same size as the template guide (usually a Porter-Cable type, two-piece) that screws into the router base. Then use a $1/4$-inch (or other size to fit the type of shelf support being used) two-flute straight router bit in the router, and set the depth for exactly what you need. Clamp the guide in place. Place the router on the guide, with the template guide in the hole you wish to use as a starter. Plunge the router. Repeat as often as needed, in as many holes as you wish. A single guide may be made to fit back and front hole spacing. (In many base cabinets, the front of the shelf will be about 6 inches inside the carcass edge, so the template needs to be much wider to that side: The rear of the shelf will be at the rear of the carcass, and the shelf support pins may be set as little as $2^1/2$ inches in from the back of the cabinet, while coming in as much as 8 inches from the front.) This system removes two worries: getting the holes square to the surface of the wood and drilling through the wood. Without going to the expense of multihead drill presses, this seems to me the best way for the start-up shop to do the drilling for pin-style shelf supports.

Failing any desire to drill pin-style shelf support holes hole-by-hole, look to your tool catalogs. Delta produces a 13-spindle boring machine, model 32-325D (electric) with a $3/4$-horsepower motor that will do the job in almost any cabinetmaking shop material. Thirteen holes will cover just about any shelf adjustment you will need to install in a kitchen or bathroom cabinet. This is not an inexpensive machine, with a base cost (no stand) of about $1000. To judge whether you'll find it worthwhile,

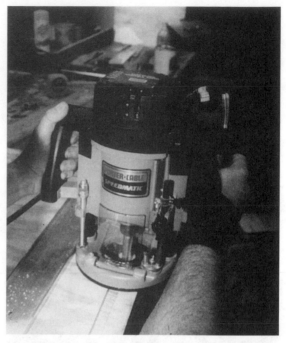

Use a large router when the work gets really steady or tough.

This router jig eases aligning dadoes and grooves, and works with any router with a 6-inch base.

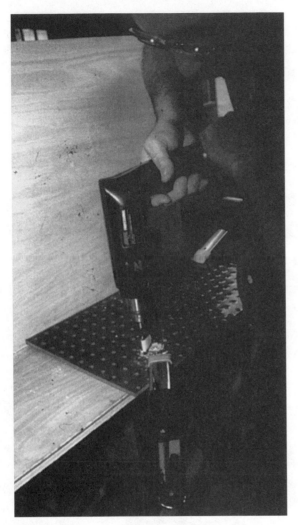

Use a pegboard as a low-cost guide for shelf support pegs.

figure out how many hours your workers are spending drilling individual shelf pin support holes. Then figure out how much faster they can go with the capacity to do 6 to 13 holes (the usual needs) at a time. It's only a few seconds per hole, but if your guys or gals are putting in hundreds of holes, the investment in the spindle boring machine may well be worthwhile.

Steel Shelf Standards

In many cases, the use of steel shelf standards is the fastest, simplest, and sanest method of providing adjustable shelves. These standards accept shelf supports that are exceptionally sturdy and are also very easy to install. You really should not, though, use them as back-to-back supports in thin stock (under $3/4$ inch). Dadoes to accept the shelf standards may start below the dado for the base of the cabinet if adjustability is to be useful down that low; there is no need to deal with the bottom end of the dado in such cases. The dado can be stopped at the top of the carcass side or divider, within 2 inches of the top. If that's done, leave it rounded, just as it comes from the router, or sloped just as it comes from the dado head on the table saw, because to all intents and purposes the nonperfect fit is invisible. Cutting standard dadoes to a depth of $3/16$ inch and a width of $5/8$ inch works well; remember, too, that on stock thinner than $3/4$ inch, these standards may be surfaced-mounted.

When shelves are cut to final length, cut off an extra $5/16$ inch at the ends to ease changes in shelf length for humidity changes. In other words, make the shelves $5/16$ inch shorter than the measurements say they need to be, because they're apt to expand and contract at least that much during the different seasons.

Aligning Standards

You must align standards in exactly the same manner. I always use matching pieces from a set, starting at the bottom when possible; but when cuts must be made, using matching pieces means lining up numbers correctly. (With 1 at the bottom on both pieces on a side, you're starting out fine, but remember, the divider that has two more standards must start at the same point, or else your shelves will be tilted.) Standards are cut to length, when needed, with a hacksaw. I've always liked a metalworker's vise for holding the standards, but a wood jig with rails just higher

An excellent tool for general hole drilling and even better for cabinet installation, where power may be a problem.

than the standard sides works as well, if you use a strip of wood to clamp the standard into the jig. The entire jig is then held tightly, or clamped, to the end of the workbench and the piece hacksawed off at the mark.

Try to avoid cutting through the holes in the standards; it leaves sharp internal pieces that can scratch finishes, catch cleanup rags, and maybe even cut people.

Assembling the Carcass

Once the parts are cut and the rabbets and dadoes placed properly, it's time to edge-band or otherwise cover plywood edges that will be exposed. That's most easily done with wood tape, with heat-sensitive adhesive, which may be put on with a machine. There are a great variety on the market—I've seen several small local shops using the small Freud edge bander, too, and that's a very handy, relatively low-cost tool that does a good job. On small jobs, I've been known to use a common steam

Once edges are covered, you begin
putting the cases together. Do not begin
by reaching for the glue bottle and brush,
hammer, nails, screwdriver, or pneumatic
stapler or nailer. Start by making sure
your assembly bench is clear and clean.

iron, without the steam feature switched on. Larger edge-banding units are faster and cost much more.

Once edges are covered, you begin to put the cases together. Do not begin by reaching for the glue bottle and brush, hammer, nails, screwdriver, or pneumatic stapler or nailer.

Start by making sure your assembly bench is clear and clean. J. R. Burnette uses an assembly bench that looks somewhat too low at first glance: It is about 20 inches off the floor, a sheet or 4-foot by 8-foot plywood on sturdy legs, or on a box. This bench serves as a spot to allow spraying of contact cement on laminates as well as an assembly bench for carcass parts. In my travels. I've come across another idea that looks good to me: For final-assembly benches, place a layer of carpet on the top. This, if heavy enough, may be just laid in place. You can then lay good-quality plywood face down on the bench without worrying an immense amount about unremovable scratches. You can, of course, use the carpet on any height bench. Use a tight-nap carpet and possibly even an indoor-outdoor washable type to prolong its life. (And unless major changes have come about in the past few years, the washable carpet is cheaper than fancy Berbers and such.) If the carpet is not fastened to the top, it's easy to whip off and set aside while finishing goes on, so that it lasts longer. Make it slightly oversized, slit the corners so you can make a fold, and use duct or other tape to hold the carpet in an upside-down box shape. This can then be easily lifted off, but stays in place nicely while assembly is going on.

Get Organized

Good organization is an imperative here, because once glue is applied and takes a set, to resquare a case or move a shelf or make any other changes becomes difficult to impossible. It's at this point that you check each individual part for fit in its dado, rabbet, or butt. Check for fit in the joint, and check for ease of getting the joint square. You need a tight fit, but one that allows room for glue, and square should be easily attained in seconds. Make any corrections at this point.

Also make sure the shop temperature is up in the range stated by the manufacturer of the adhesive you're using.

Lay out all necessary clamps for the various parts of the assembly. These naturally will vary in size and type depending on the carcass assembly type and size, but you'll need pipe or other bar-style clamps as your main source of clamping. Adjust clamps within a few inches of their final size needs. This saves a great deal of time in glue-up, which means you keep your working time well within the setup time of the glue (10 to 15 minutes for most woodworking glues—use Titebond Extend and you go out to 15 to 20 minutes).

> **Good organization is essential in getting ready for assembly, because once glue is applied and takes a set, resquaring a case or moving a shelf or making any other changes becomes difficult or impossible.**

Glue-Up

Lay out the bottom, ends, and any dividers as well as a rear nailing rail. Tilt pieces into place to check fits. Make sure you've aligned the insides of the ends to the inside of the carcass. Check to see that dividers are facing in the proper direction. Fit all dust panels into the dividers that form the drawer (or drawers) sections. Many cabinet-makers use nail-on strips to hold these dividers in place and use a nail-blue butt joint to form the joint. I prefer to set the dividers in a 1/4- or 3/8-inch-deep dado. That eliminates the need for nailholes in the base and also makes a stronger joint.

After checking all fits, assemble any internal dividers on dust panels; that is, slot the dividers into their dadoes, and nail and glue them in place (do a test assembly *first,* of course). Square up the panels as they go in place. This partial-assembly method allows more working time with the final, full assembly, because you allow the glue to set on the partial assemblies before installing them in the overall carcass assembly. Clamp carefully once the unit is square.

If there is a fixed shelf into one end, you can assemble that now, sliding it into its dado and fastening with glue and nails. To keep this square, you may want to take a brace onto the back edge; use a diagonal piece of 1 by 2 scrap pine as long as is needed. Again, many cabinet-makers use butt joints for such shelves. Don't. Use a dado. The overall increase in strength is quite large for the small increase in cutting time.

Canac kitchen and bath cabinets are the result of good design, proper assembly, and glue-up. *(Canac)*

Wellborn Waverly kitchen cabinets result from good assembly work. *(Wellborn Cabinet, Inc.)*

Please note that the glue-and-nails technique is not an essential. You will get a joint that is just as strong without the nails, but nails serve as assistants to the clamps. Adding nails to the glue, too, tends to help the assembly hold its shape as you move it around to add another piece, before the glue sets up. Leave the nails out if they take too much time and you don't feel you need them.

With one end subassembly done and the center assembly complete, place the base in a position to receive the parts. At this point, the wisdom of the size of J. R.'s assembly bench becomes obvious. Too, the height begins to seem more reasonable, because you're working below eye level and lifting only short distances.

Again, check fits. Make sure the subassemblies fit into their respective rabbets and dadoes. Check for back fit, if you've not already done so. Make sure the nail rail fits into its notches. Disassemble.

Add glue for the center assembly and slide it into place. Check square and clamp, top to bottom. Use two clamps. Slide the end

Amera cabinets benefit from good design and assembly characteristics, and generally good construction. This room off the kitchen has been outfitted with Dorchester cherry cabinets in a rich natural finish. *(Merillat)*

assembly, with shelf (assuming there is one), into place, with glue in the dado on the divider face of the center assembly, and on the rabbet on the bottom. Use nails, then check square, and clamp again, top to bottom. Use two clamps. Recheck square of both assemblies at all points where square can be checked. Make any adjustments before glue sets up. Turn the assembly over, so the base bottom is facing up, and use nails, if desired, to help hold dividers and end in place.

As a quick note, clamping pressure is medium on all these clamping jobs.

> **Keep on checking square during assembly. Make any needed adjustments as you go, while the glue is not set.**

Slip the end panel in place in its rabbet. Use glue and nails, and check square with the base. Slip the nail rail in place, using glue and nails to fasten. Clamp along the bottom (after checking square). Clamp also along the nail rail.

J. R. Burnette's assembly bench is behind him and ready to go. He also has a couple of even lower units, and a rolling unit or two for finishing and gluing.

Keep on Checking the Squareness

Take diagonal, corner-to-corner measurements at the front of the cabinet, making sure those measurements are equal. This ensures overall cabinet side-to-side squareness.

Lift the assembly (with a helper) off the assembly bench and set it to one side where the glue can set up; check square one last time after moving the unit. The unit may also be left in place for the addition of the face frame; movement is probably needed if you have only a single assembly bench and the job has a number of cabinets to be assembled.

Walk away from the assembly for at least 45 minutes to 1 hour to allow the glue to set up, at which time the entire unit may have the clamps removed.

At this point, you've got the carcass assembly techniques pretty well in hand, so it's time to go to the next part of the job, the construction of the face frame to cover the front, raw edges of the carcass, and to get ready to attach face frames to the carcasses, so that the finished cabinet begins to take on even more shape. At this point, you're almost ready to see what final-size drawers, drawer fronts, and door frames will need to be, but not quite. You've got the rough dimensions, but those may be as much as $1/2$ inch off in some directions. It is after the face frame is added that you get final dimensions.

Wall Cabinets

There are two basic differences between wall and base cabinets: size and the addition of a second nailing strip (or a back heavy enough to serve as a support). Other than that, construction methods are essentially the same for both except that you enclose both top and bottom— you enclose top and bottom with base cabinets as well, but the top is the countertop and is the subject of a separate chapter, where types and construction are covered more thoroughly than are the simple plywood (or MDF) top and bottom for the wall cabinets. Wall cabinet tops and bottoms may be of softwood faced with veneer, or MDF faced with veneer, or hardwood veneer plywood of the several types covered in the chapter on wood. Generally, only the inside is seen, and in most cases, it is simplest to provide an easy-to-clean laminate surface in a light color that contrasts with or complements the wood grain (with Eurostyle cabinets, the laminate may match or contrast). That prevents grain and color match problems with no hassle at all, and MDF is readily available already covered with such surfacing—it can also be surfaced with any of a multitude of materials, in the shop or by special order from your supplier.

Face Frame Style and Substance, Plus Doors to Suit

Traditional North American cabinetry uses face frame cabinetry, usually with doors that are frame and panel construction. When the cabinet's case, or carcass, is constructed, the face for that form comes next and provides some interesting ways to change designs and to present the customer's face to the world. The received word is that face frames reduce access to cabinet interiors as much as 20 percent compared to Eurostyle frameless construction. That's a maximum, I understand. No one seems to have done a test on just how much stronger and more durable face frame cabinets are, but in my opinion, they're at least 20 percent more durable, and 300 percent or more better looking—but the latter isn't something I'd care to tell a customer who insisted on Eurostyle cabinets. In essence, to me, Eurostyle frameless cabinets are simply easier and faster, thus cheaper, to build. They suit some tastes admirably.

Generally, face frames are simple, of solid wood up to 4 inches wide, or as little as ³⁄₄ inch wide, usually of ³⁄₄-inch thickness. In most designs, the face frame wood is of the same type as in the carcass plywood, though this is not an essential; it is very unusual to see different

> **Eurostyle frameless cabinets are easier and faster, thus cheaper, to build. They suit some tastes admirably.**

species of wood obviously mixed in cabinets. The wood may be hardwood or softwood, depending on the customer's desires, though today the hardwoods dominate the field, with maple making a resurgence as this is written. Cherry and oak are exceptionally popular, and hickory is also showing signs of being a wood that is desired in more and more kitchens. Walnut is seldom found these days; for today's styles, it's too dark.

There are two basic methods of forming face frames: one uses biscuits, or dowels, and glue, and the other uses pocket screws, with or without glue. At one time, dowels were the fasteners of choice, but over the past decade or so, biscuits and screws have replaced them, for several reasons. Biscuits allow greater speed in alignment and allow you to make adjustments for minor miscues. Strength is similar to or greater than that of dowels because of the increased gluing surface. Screws are record setters for speed, but are also exceptionally strong,

Canac Eurostyle frameless cabinets. *(Canac)*

Canac frameless cabinets in a more traditional American look. *(Canac)*

especially in resistance to wracking (twisting) forces. Mortise-and-tenon joinery could also be used (and cope and stick joints in door frames are a form of mortise and tenon), but it is exceptionally time-consuming and requires even greater precision than doweling. It is a joint saved for those custom applications where the owner must have absolutes, though for most cabinetry the extra strength is a total waste.

Canac frameless cabinets in a more traditional American look. *(Canac)*

Lap joints and half laps are also useful in forming face frames, and they are held together with glue.

Face frames are held to the case, or carcass, with biscuits and glue, nails and glue, or dowels and glue. For modern work, I'd ignore the last. Cheesy construction uses visible screws. I'd ignore that, too.

Note that we're not speaking here of fancy mortise-and-tenon and bridle and other joints. These are furniture joints and don't suit kitchen cabinet construction too well, as extra time to construct and assemble such joints means that cabinet prices would be totally out of sight. There is little need for the extra strength: properly made biscuit joints, and properly done pocket-hole screw joinery, hold together across a butt joint well enough to make the wood break before the joint itself separates. When a kitchen requires anywhere from 20 to 80 of a particular style joint, it is wise to use the fastest strong method of joinery, and biscuits and pocket-hole screws are as good as it gets right now in combining strength and speed.

This Amera hutch is face framed. *(Merillat)*

Designing the Face Frame

There are several important jobs done by the face frame. First, maybe foremost, is the addition it makes to the overall appearance of a traditional job of cabinetry. It covers the raw plywood edges of the case and provides a finished look. The face frame also is the base for hardware (hinge) installation, which means it must be sturdy and well connected to the case. When designing the face frame, we need to constantly keep in mind that finishes go on more evenly when end grain is not exposed (end grain soaks up extra finish, so it looks darker than the rest of the frame). Thus, face frames are best designed to expose as little end grain as possible.

Designing to keep from exposing end grain isn't particularly difficult. Make the bottom rail of a single piece of wood, fitted inside the end stiles. The top rail is made the same way, with any intermediate stiles butting on the rail, keeping the end grain

> There are two basic methods of forming face frames. One uses biscuits, or dowels, and glue; and the other uses pocket screws, with or without glue.

A face frame kitchen from Wellborn. *(Wellborn Cabinet, Inc.)*

Wellborn's Rosebud face frame cabinets. *(Wellborn Cabinet, Inc.)*

Wellborn's face frame cabinets also look fine in a bathroom application. *(Wellborn Cabinet, Inc.)*

from being visible. Drawer rails are butted against stiles, too. In such construction, each cabinet has only two pieces of end grain showing— at the top edges of the frame (on a base cabinet) where that frame is nearly hidden by the overhang of the countertop. The wall cabinet bottom rail is also below the sight line for most people.

Design includes figuring stile and rail widths and thicknesses. If your supplier is coming through with S4S wood, then you may already have it planed to thickness. Other workers prefer to finish-plane the wood after it is ripped to width, but personally, I like the consistency found when all wood being ripped from, for example, a single 8-inch-wide board is finish-planed before the ripping is done. For oversized

> The face frame does important jobs. It adds to the appearance of a cabinet. It covers the raw plywood edges of the case, for a finished look. The face frame also is the base for hardware (hinge) installation, so it must be sturdy and well connected to the case. Face frames are best designed to expose as little end grain as possible so that finishes go on more evenly.

American Clamping Company's new edge clamps work exceptionally well when it is time to cover edges

stock, plane to $3/4$ inch thick. Plane all boards to thickness at the same time, stepping each one up through the planer successively. In other words, if you have to plane 10 boards from 1 inch down to $3/4$ inch, start with board 1, and go on to board 2, and on through the list before you change the setting to take the final cut. Make the final cut $1/16$ inch or less for greatest smoothness. The overall face frame is best made $1/2$ inch wider than the case width, regardless of the case width. This is an adjustment feature that may or may not be needed; if it is not needed, it is simply left in place.

When that's done, cut boards to rough lengths, and only then rip to specific widths.

Working in the above manner gives the most consistent thicknesses and also means that any snipe problems can be cut away. With a well-adjusted planer, there should be no snipe problems.

Stile and Rail Widths

Rail widths are affected in part by the type of fasteners to be used as well as by the personal taste of your customer. Biscuits cannot do a good job of fastening rails, and stiles that are less than 2 inches wide do not cover biscuit edges: the smallest biscuit, the 0, is $1^3/4$ inches wide. Ryobi makes some smaller R biscuits that work with their detail biscuit joiner (DBJ50), a 3.5-ampere tool that may or may not be tough enough for heavy-duty use: if you get one, immediately buy the accessory carbide-tipped blade. Ryobi's special biscuit sizes start at R1, which is $5/8$ inch wide, and go to R2, which is $3/4$ inch wide, with the largest, the R3, being 1 inch wide. These reduce the needed width of stiles and rails considerably, opening up design considerations that are otherwise limited to screws and dowels.

Porter-Cable a year or so ago announced its face frame biscuit for use with its 557 plate joiner, and the included accessory 2-inch blade.

Wellborn Euro Esprit cabinets.

Wellborn Cambria frameless cabinets.

Wellborn Eurostyle bath, Ashley model.

The kerf here is standard 0.155 to 0.160 (the Ryobi kerf is 0.100, handier for thinner stock), and the face frame (FF) biscuit is designed to work in stock as narrow as $1^1/2$ inches.

The top rail is the widest: On larger wall cabinets, the top rail may exceed 4 inches where that's needed, but on base cabinets the top rail is seldom more than $2^3/4$ inches wide because any more than that robs space from drawers. Rails as narrow as 1 inch are often used; when using biscuit joinery, such rails meet the stile ends, so are not a problem when it comes to accepting biscuits of almost any width. Too, pocket hole screws are available in lengths as short as $1^{1}/_{4}$ inches, which should work even with $^{3}/_{4}$-inch stock: I suggest, though, before committing to use of either very narrow rails or any specific type of joinery, that you make your own check. I know a lot of my tools don't work the same for me as they do for others, and that includes depths and lengths of screws. Most do, but there is always the chance of a kink, so check first and save redesign misery—a misery that includes having to make excuses to the customer.

Biscuit Joining Needs

There are a few specific needs when using biscuits for joinery. At the top of the list is a work surface that holds stock so the stock cannot slide as the joiner is pressed

Amera's Euro entertainment center. The use of taller cabinets stacked on smaller base cabinets on either side of the television creates a unique piece that more closely resembles fine furniture. *(Merillat)*

to the wood and the blade extended to cut the slot. A simple jig or two work well here, and some specific clamp types keep short or long pieces of face frame from sliding. Hold-down clamps from a company like De-Sta-Co easily make up into locks for workpieces. For this use, vertical handle models such as the 247 or 267 series, with flanged bases, work best. These clamps are available in many stores and are useful in many kinds of jigs.

Place the stock in the jig, place the adjusted joiner against the stock edge, start and move the saw into the wood. The next step is to dry-fit, after which you go to glue, no nails.

Marking

Biscuits are always used with boards butted together (for flat panels) and marked across all boards being glued up. The location of each biscuit does need to be marked, but if the board itself is not marked to coincide with the next, or preceding, board, then alignment becomes impossible. Simply make a W shape over the faces of however many boards you'll be joining, and mark the boards (1, 2, 3, 4, 5, etc.) as needed for larger flat panels.

For flat panel work, use a biscuit every 8 to 10 inches. Mark the panels with the W, and then mark across both boards for each biscuit. Make sure the biscuit joiner is not going to place the top side, or bottom side, of the biscuit closer than $1/4$ inch to the surface of the wood being joined, and make the cuts needed. The $1/4$-inch depth is needed to prevent print-through from the biscuit; over time, biscuits may swell a bit more than at the outset, and if the wood around that biscuit is too thin, the print of the biscuit may show at the surface. This is obviously less important at the back of a frame, or the underside of a panel, but is to be avoided if at all possible. If the choice of $1/4$ inch to the visible side and $3/16$ inch to the nonvisible side presents itself, that's fine. The opposite isn't okay.

For frames, assembly is easier. Use a pencil to mark across the joint. Use the joiner to cut the slot, with depth set as above, so there is a minimum of $1/4$-inch clearance above and below the biscuit. Do this at every subassembly joint, using the appropriate size biscuit. You can also do the slots in the ends of the stiles. Do not slot the top and bottom rails at this point.

Next, dry-fit the assembly, making sure all measurements and placements are exact once you've squared the face frame. Most of the time, this works best with the frame lying back down on the assembly bench. Set clamps to within about 1 inch of final adjustment, pull things apart, and start at one side to assemble, slowly. You have not yet cut the slots for the splines to fasten to the top and bottom rails, and this must be kept in mind.

Glue up the unit in a manner similar to gluing up the case: For rails and splines to fit around drawers, glue up that subassembly first. Make sure the glue gets down in the slot for the biscuits. Square and clamp carefully to maintain square. Wipe off excess glues. Now, check final widths so you can be sure of a fit on top and bottom rails. After about 45 minutes of glue setup time, mark and cut the slots for the biscuits on the top and bottom rails, fitting them as needed to the stiles in the subassembly or subassemblies. This allows for greatest accuracy of placement; it is faster to go ahead and mark and cut all biscuit slots with the total loose assembly, if everything seems to check out. Should something shift, for whatever reason, after marks are made and slots are cut, not only does the time savings disappear, but also time from other jobs is eaten up. It seems to me that here slow and simple is best.

Continually check alignment during the final dry fit with glued-up subassemblies. Now, fit the top and bottom rails and the end and intermediate stiles that are not glued up as subassemblies. Square, clamp, and set aside after rechecking squareness.

This is possibly the easiest, though probably not quite the fastest, method of assembling and fastening face frames. The use of biscuits gives a range of possible adjustments during dry fit that are not at all possible with dowels. Dowels make dry fit itself exceptionally difficult, as tight-fitting dowels may force you to use a soft-faced hammer to knock the pieces apart, which sometimes dimples and damages the pieces of the frame. Biscuits don't cause a tight fit until the glue is in place and penetrating them, at which point they

A top-notch square is essential to any cabinetmaking shop. This Woodcraft engineer's square is used for everything from checking the square of small cabinets to checking the square of squares.

swell to fill the slots completely. Thus, back-and-forth movement to align pieces is possible, even after glue is placed.

Pocket Screwed Face Frames

Pocket screws fit into specially countersunk holes to hold screws at an angle that pulls the frame together (and can pull one piece off a matching fit with another if care isn't taken). Pocket screw countersink bits are special, using a two-steps-in-one bit. There are a number of simple guides for this purpose and one not so simple that speed finish work

Pocket screws fit into specially counter-sunk holes, holding screws at an angle that pulls the frame together. Pocket screw countersink bits are special, using a two-step-in-one bit. There are a number of simple guides for this job and one not so simple that speeds finish work considerably, without increasing production costs too much.

considerably, without greatly increasing production costs. For any shop considering doing primarily face frame work, I'd suggest at the very least taking a good look at the Porter-Cable 552 production pocket cutter. This 13-ampere 115-volt machine accepts wood from $1/2$ to $15/16$ inch thick, cuts a $3/8$-inch pocket and a $9/64$-inch pilot hole, with each hole taking a shade under 2 seconds to complete. The screw angle is $16°$ which eliminates some of the material shifting problems that other pocket hole guides present. Use is about as easy as it gets: Mark the locations on both pieces for the pocket hole centers. Clamp material in the pocket hole cutter. Push the handle forward, and cut the $3/8$-inch pocket hole. Pull the handle back, and the pilot hole is drilled from the edge into the pocket. Move to the next mark and do that. Change ends and continue.

The work clamp adjust easily and quickly: Simply slide the workpiece under the open clamp, and close the clamp. If adjustment is needed, loosen the jam nut and adjust to loosen or tighten.

Size Limits

The recommended workpiece size limits are $1/2$ inch thick minimum, $1 1/2$ inches wide, and $1 1/2$ inches long. Make sure the workpiece is face side up in the machine, the mark on the center notch, and the piece firmly seated against the fence. For use of two nails per member, three marking notches $3/8$ inch to each side of the center notch are used.

Close the clamp and check tightness. Turn on, and pull the black knob away—forward—until it hits its stop. Move the knob back until it contacts the rear stop. Make the movements as smooth as possible. When the knob is released, it returns to the center, or neutral, spot.

A production pocket cutter helps a great deal to speed face frame assembly. *(Porter-Cable)*

You then simply repeat as often as you need to, making sure you keep dust and chips cleared from the operating and cutting areas, and from the area between the workpiece and the fence.

Adjustments of the pocket cut are possible and are thoroughly covered in the manual.

You can find some people who do not like screws in cabinet assembly, though large-scale cabinetmaking factories use plenty. I'm not overly fond of anything that doesn't use glue in cabinet assembly, and I feel that over the years, screws might create a problem by backing out slightly—probably never all the way—because of wood movement. Things could loosen up and get sloppy 5 or 10 years down the road. Thus, I strongly suggest you use glue during assembly, even when using pocket holes and screws.

Shop-Built Jig for Pocket Holes

There is a simple jig for use with a drill press that allows you to try pocket holes without investing in any expensive equipment (and the Kreg pocket hole jig is fairly low-cost—currently, a total package, including 400 screws, is about $210—speedy and easy to use). It works with material from $5/8$ inch to $1^1/4$ inches thick and uses a two-step drill bit to reduce setup time. It is not as fast as the Porter-Cable and gives a steeper angle of attack, so lifting of parts may be a problem. (Generally, when parts lift, it is because the assembler is not paying attention. Simply keeping things aligned with one's hand, though tightly, tends to prevent such lifting and the resulting face mismatch of levels. If hand holding doesn't work, then use a clamp to prevent the parts sliding against one another.)

For the shop-built jig, all you need do is to make a steeply tilted ramp (22° or so—because of the way our saws and miter saws are set up, $22^1/2$° works well). I used a flat plywood board about 14 inches on the long face by 12 inches on the short face. Cut the ramp board with a $22^1/2$° angle on one edge, and match and attach that edge about the center of the base board. I suggest using two strengthening triangles of wood behind the ramp board. Ramp board size should be about 12 inches high by about 12 inches wide. In front of the ramp board, attach a stop at $3/4$ inch (assuming here the face frame is to be of $3/4$-inch-thick stock: you'll need to either work out a way to make this adjustable or make a separate board for each thickness if you use more than $3/4$-inch stock). The jig is clamped to the drill press table, and the two-step pocket hole drill bit is used to drill the holes (the Kreg bit works very well here). Always check on scrap stock before drilling good material: If the angle is off, the drill bit may break through the front of the stock, which is about the last thing you want to happen with your good, expensive hardwood face frame stock. Make any changes needed to eliminate breakout.

Obviously, pocket holes are drilled in the backs of the stock. Before starting drilling, assemble all parts face down and square and mark the screw locations. Use two screws per joint wherever possible ($3/4$- to 1-inch-wide material will accept only a single screw).

Now, drill all pocket holes.

Assembly Using Pocket Hole Joinery

Start the final assembly with any subassemblies, such as the drawer face frame layout. Slide parts to the point where you can clamp each joint to prevent lift-up of one part over the other. Get this glued and assembled, and the screws driven. Wipe off excess glue. Check square and set aside for the glue to set up. Dry-assemble the other parts, in readiness.

When the glue has set up on any subassemblies, place them between stiles at their correct locations, add glue, and screw together. Again, wipe off excess glue.

This is another quick method of assembly, with great accuracy possible. Strength is more than enough, especially with glue added, for any face frame requirements. Assembly requires only a clutched drill-driver, with a long screwdriver bit (the longer, the better: I like a 4- to 6-inch-long bit for this work). The drill may be corded or cordless. A good hand screw or other clamp to prevent parts from shifting is also a large help.

Even without subassemblies, progress from the inside of the face frame outward during assembly, and keep as careful a check on square as you can. With proper dry assembly, getting things together within the open time of your glue shouldn't be a problem. That leaves you enough time, with glue still unset, to clamp and use some wracking force of your own, to bring things into alignment, if such alignment is needed. Once set, lock the clamps to medium pressure and set the entire assembly aside for a day or more.

Cleaning Up the Joint(s)

There will always be some slight clamp marks, glue dribbles, and other problems at the joints of any face frame. There are a number of ways to take care of these small messes (let's hope there are no deep dents, large globules of glue, massive scratches, or similar problems). Many people use a simple cabinet scraper to get off glue and to remove dents and bumps. Large dents may need to be steamed first (a damp cloth over the dented spot, then a hot—I use the wool setting—steam iron). That lifts most of the dent, and scraping or sanding will get rid of any remaining problem.

Some cabinetmakers still use belt sanders to do directional sanding, taking a chance on messing up the adjoining parts. I prefer a random orbit sander to clean up at joints. That way, there is no worry about grain direction scrapes. Start with an 80-grit paper, if necessary, and go through 220-grit. If marks are very mild, start at 120 grit, step to 150, then 180, and finally 220.

While you're cleaning up, use a router to round over the inside edges of the face frames. You may prefer a chamfer; if so, do that. Do the relieving job on outside edges that will be exposed and that will not mate with other edges. You can save router setup time if you learn to use the cornering tools that Veritas makes. The kit consists of two double-ended cornering tools that give radii of $1/16$, $1/8$, $3/16$, and $1/4$ inch. These tools are remarkably easy to use and take no time to set up: Decide on your radius, select the right end, and go. The kit, including the sharpening tool, costs less than $20 right now, which is less than you're going to pay for a chamfering or round-over router bit.

You've got the carcasses done and the face frames done. The time has come to mount the face frames to the carcasses, and then you can install the doors.

Installing Face Frames on Carcasses

There are a number of ways to install face frames on their carcasses. Because overall face frame width is usually $1/2$ inch wider than the cabinet cases, that $1/4$-inch power side must be maintained. (It is for adjustment side to side in the case of out-of-plumb walls, or out-of-level floors. A couple of quick shots with a plane plus shims under the toeboard assembly will bring everything into line with no indication of changes—in most cases.) The simplest way to maintain the 1/4 inch per side is to make up a couple of (permanent) 1/4-inch pieces of stock that can be held under the face frame as it is laid on the case. Place the 1/4-inch stock against the case side, and bring the face frame out flush to the 1/4-inch stock. Bingo. Both sides, assuming correct measurement, are 1/4 inch away from the case sides.

The design of the cabinets can readily incorporate one form of attachment: screws. Screws are ugly, but easily covered, and using a contrasting wood plug on a counterbore over the screw head can add a detail. Walnut, for example, looks great on cherry, oak, maple, or almost any light-colored wood (as do more exotic woods such as ebony, purpleheart, redheart, cocobolo, and others). Place screws close enough to hold the frame on the case (every 8 to 10 inches), and use screws of a length that is about 1 1/2 times the thickness of the face frame material (usually 2 inches does fine). Normally, shorter screws would work fine, but you're screwing into plywood here, the equivalent of screwing into end grain where strength is lower. Use glue if desired. Check for square as you screw the face frame in place.

Biscuits work well as attachments for face frames to carcasses. Slot for biscuits every 8 inches on each carcass surface, and install biscuits as normally done. Clamp and let dry, after checking for square. As always when you use biscuits, make sure you get glue in the biscuit slots on both sides.

Some workers like to nail and glue the face frames in place and then use a putty stick to fill the nail holes. Nail every 10 inches, with as long a finishing nail as possible. Check square and clamp.

You may also use dowels to attach the face frame to the carcass, which provides a good, strong joint, but is a lot of work. Use dowel centers to position dowels on the face frame; that is, drill dowel holes in the carcass, then insert centers, place the face frame on the carcass, and tap at each dowel location. The dowel centers mark the locations you need for the holes that accept the dowels. Drill those holes with absolute accuracy, using a brad point or Forstner bit. Glue dowels into the case, and then dry-assemble the face frame on the case to check fit. Remove the face frame. Make any adjustments needed. Recheck. Remove the frame and place glue on dowels and along the carcass edges. Place the face frame, force down on dowels, clamp, and wipe off excess glue.

At this point, your carcass is ready to have doors made. You have the final dimensions of each opening right at hand once the clamps are removed from attaching the face frame to the case.

Doing Doors

Doors provide entry into much of a kitchen's or bathroom's storage space and provide a large part of the decor as well. Doors may be in any of a multitude of styles, including flat-panel flush-fit, flat-panel overlay (both more common with Eurostyle), flat panel inside stile and rail, raised panel inside stile and rail, arched pattern stile and rail, and batten styles.

The simplest two are the flat panel of laminate-covered MDF or other substrate, which are straightforward and easy to do, so I'll present no instructions for those.

Frame and Panel Doors

Frame and panel doors make up the major part of traditional kitchen and bath cabinetry. Simply put, a frame made up of rails—horizontal members—and stiles—vertical members—is used to hold a panel. There are a multitude of special cutters designed to produce the frames, so that in a couple of passes on a router table or shaper (your choice, but the shaper tends to last longer), your stiles and rails are ready to fit together. They are also grooved to accept a panel, as the corner joints are cut.

There are so many ways of forming frame and panel doors, there is sure to be a way that I missed, but generally you use a bit that will cut an inside profile to match or complement the one you use to cut the raised panel, assuming a raised panel is used. A flat panel may be used with any style of frame, of course.

Frame and panel doors offer the opportunity of having wide, wide door panels of solid wood without worrying about expansion and contraction, because the panel fits loosely in the inside slots of the frame.

> Doors provide entry into much of a kitchen's or bathroom's storage space and provide a large part of the decor as well. Doors are made in a multitude of styles, including flat-panel flush-fit, flat-panel overlay (both more common with Eurostyle), flat panel inside stile and rail, raised panel inside stile and rail, arched pattern stile and rail, and batten styles. The simplest two are the flat panel of laminate-covered MDF or other substrate, which are straightforward and easy to do.

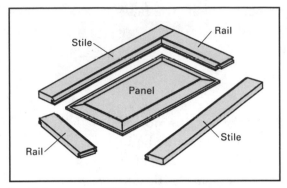

Panel door parts. *(CMT)*

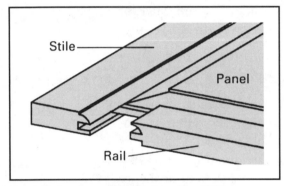

Panel fitting into rail and into stile. *(CMT)*

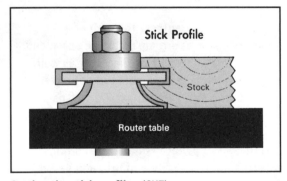

Cutting the stick profile. *(CMT)*

The frame itself is of much smaller components so it shrinks and swells much less, making the door exceptionally stable, even compared to solid plywood doors.

Good joints, though, are essential to a strong cabinet door. And cabinet doors are often opened and closed more than doors in a home, and so need really strong construction methods, tight joints, good gluing, extreme care with squareness, and good clamping. Stock must be perfectly flat and straight and correctly sized before you start to cut the joints.

Companies such as Jesada and CMT produce router bit sets that provide all the bits needed to cut all the parts in a cabinet, including drawer joints. Included in those sets are stile and rail sets that produce proper joints for the door frame parts, with far less work than is needed with most other methods of producing such joinery. Generally, the bits are available in an array of styles; that is, the panel-raising bit will cut an ogee, so the stile and rail bits cut surfaces to match (on the inside edges of the door frames). Also available are standard, or straight, sets; bevel and radius sets; and cove sets. For the nervous among us, there is a vertical raised-panel bit that doesn't get that 3$\frac{1}{8}$-inch-diameter horizontal bit whistling around at 10,000-plus rpm. Personally, I don't like the combination rail and stile bits as much as I do the two-bit sets, but they are effective for those who do like them. Similar cutters are available for shapers, with bit diameters up to 6 inches for the raised-panel cutters.

A few test cuts are usually in order for the first starts on door frames: I'd rather waste 10 feet of scrap stock than a couple feet of high-priced good stuff.

One note: It makes sense to get ready and cut all the stiles and rails for all the cabinets at the same time, if at all possible. (You may not have shop room to assemble all carcasses and face frames and set them around for door work.) Doing them all at once, with the exact same router table or shaper settings, ensures a greater degree of uniformity. Of course, if you're sure the shaper or router table won't have its settings changed between jobs, you're also okay; but since you need two settings, one for each bit or cutter, this is unlikely to happen. This is also the type of work that can benefit from the use of a powered stock feeder on a shaper.

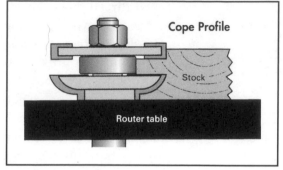

Cutting the cope profile. *(CMT)*

A few test cuts are in order for the first starts on door frames: Waste 10 feet of scrap stock rather than a couple of feet of high-priced, good stuff.

My preference for clamping setups for frame and panel doors is the Bessey K body KP framing system. It is designed for exactly this use, and nothing I've found does it as well, never mind better. But the system is not cheap, so you'll probably start, and maybe stay, with regular bar clamps.

Panels

Forming panels for frame and panel construction can be very simple or very complex. The simple types are flat panels, 1/4 inch thick, of hardwood plywood. When carefully cut (about 1/4 to 5/16 inch oversized in

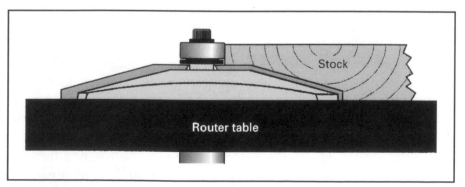

Cutting raised-panel edges (router-table style). *(CMT)*

each direction—horizontally and vertically), these panels just slip in place, and you glue the frame around them, making sure no glue attaches to the panels. The biggest argument here is over whether panels should be finished before being inserted in the slots. Expansion and contraction may show unfinished lines on the panels if overall finishing isn't done before the panels are inserted into the frames. That means the frames cannot be glued until after the panels are finished, and it also means some extra care in finishing the door to prevent too great a buildup of finish, but this generally results in a more professional-looking job. Finish both the frame and the panel, making certain not to get any finish on the areas that must be glued. Even stain is best kept off such places, because wood glues are designed to work with bare wood.

Frame and panel construction is most often used with overlay doors, that is, the door frame is cut with a lip that fits over the face frame edges. This is a very simple method, done quickly on a shaper or router table with a straight bit (if the outside edges of the frame are to rounded over, the bit may be of a type to do that in the same pass). Check the needs of the hinges being used for the size of the overlay. That also affects the side of the doors, and you must remember when double doors meet, there is no overlay on the meeting side (center).

For more complicated panels, the same finishing argument holds: Finish the entire assembly, keeping the finish off the glued areas.

Solid-wood panel doors are used when raised paneling is desired. Usually, this wood is glued up from narrow (about 3- to 4-inch-wide) strips to reduce problems with warping and cupping over the years. The gluing is done carefully to reduce planing needs. Or you may start with oversized stock (7/8- or 1-inch, for example, when 3/4-inch is needed) and plane to finished size. Lay out the boards to be glued into a panel. Check grain and edge fit and joint edges on any that need it. Lay out clamps—this is another area where the blocky head of the K body clamps really shines—with open jaws up, a clamp every 18 to 20 inches, with the array opened to about 2 inches (at most) more than you need to fit the boards inside. Test-fit.

Apply glue, reassemble, and draw up clamps snugly. Use a framing square to check surface flatness and a soft-faced hammer or mallet to tap raised-board edges down flat. Some people use biscuits for such joinery: I've never found it necessary, though it can help alignment.

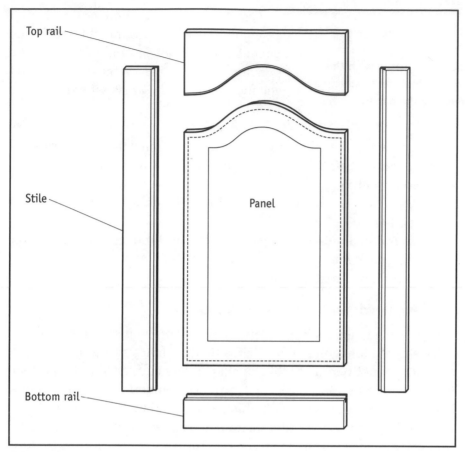

Raised-panel parts, curbed panels. *(Jesada)*

Because of the stresses that machining a raised panel creates (you're taking off one devil of a lot of wood in a single pass, or even for two passes), let the glue dry at least 24 hours, and 36 hours in humid weather.

Obviously, you want to test the results again. Use a test board to make sure you're getting a piece that will fit into your frame slots. Check the size and width overall before machining, of course, but also check the thickness of tongue and depth of penetration into the grooves in the frame before you make cuts in the good material. Adjust the shaper or router until you have the fit you need.

Machine cross-grain ends first, then machine with the grain. Cross-grain is most likely to splinter. Cutting with the grain last removes the splinters, reducing the chances of wasting the panel.

Machine cross-grain ends first, then machine with the grain. Cross-grain is most likely to splinter. Cutting with the grain last removes the splinters, reducing the chances of wasting the panel.

Check the fit in grooves, and dry-assemble the frame around the machined panel. If all is well, go on and finish the panel, rails, and stiles, being careful to not get any finish on the areas where you will be gluing the frame.

Recheck the fit in the grooves. You may have to clean up the grooves to get a perfect fit once the finish is added to the panels. Assemble as with a flat panel now, gluing carefully to keep from spotting glue on the panel ends.

At this point, if you're cutting a lip on the doors for an overlay, go ahead and do so.

Measure for holes for hinges and catches and knobs or handles, drill, and check fits.

You can now finish the cabinet.

Fancier Work on Frame and Panel Doors

It is possible to emulate the crowned-top doors you see on big factory cabinetry without going in for CNC and other fancy machinery. You do need a top-quality router table, or shaper, and enough skill to use templates. You can either make the templates yourself or use those made by a company such as Jesada (which also has plans for templates).

Working from Jesada's template patterns, we find Classical, Colonial and Crown patterns for the panel top and for the top rail. These are offered in a variety of sizes, so you can fit just about any situation. The first step is to select the material for the templates. If you use plywood, make sure it is at least $3/8$ inch thick and free of voids, loose patches, and such. Phenol also works well, as do most acrylics. Lexan (polycarbonate) works well, but is very costly.

Select your size, and cut your template material at least 4 inches longer and 4 inches wider than the maximum finished door width is to be. Put the pattern on template material with spray adhesive. Align center marks for the material and for the pattern. Use a band saw or scroll saw to cut the template (if you're good with a jigsaw, that can work, too). Use a sander to smooth any rough edges. Jesada's patterns come overlaid on one another on a large sheet of paper. If you're using more than a few from a sheet, and those are well separated, you'll want to get copies made so you can have separate patterns for each template.

These are set up so that a single template can cover a range of widths, but not all, because the radius of the arch changes too much as the width of the door changes. For myself, I'd want three to four copies just to allow for the possibility that my scissors might slip.

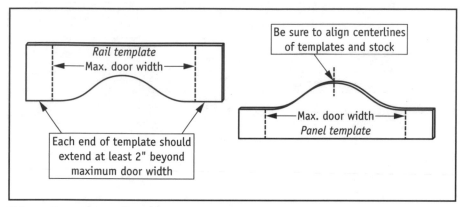

Using rail templates. *(Jesada)*

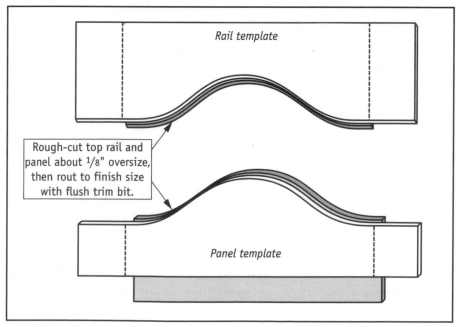

Making rough cuts first works best. *(Jesada)*

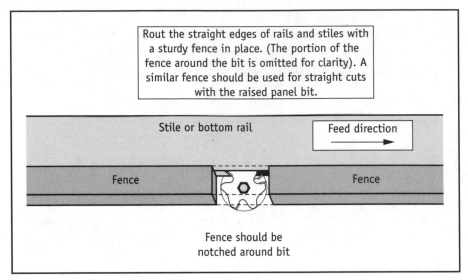

Rout the straight edges of rails and stiles with a sturdy fence in place. (The portion of the fence around the bit is omitted for clarity). A similar fence should be used for straight cuts with the raised panel bit.

Stile or bottom rail

Feed direction

Fence

Fence

Fence should be notched around bit

Use a good, solid fence. *(Jesada)*

When cutting stock for the frames, cut the top rail $4^{1}/_{2}$ inches wide to leave room for the arched template cut. Make sure your panels are cut allowing for a $^{5}/_{16}$-inch insertion into the rail and stile grooves. Fasten the template to the top of the door frame, with the arch facing down. The way to get the best finished cut is to rough-cut the material—preferably with a band saw—to within about $^{1}/_{8}$ inch of the template edge. Then rout or shape to the final edge with a flush trim bit.

Do the same for the top of the door.

Now, it is time to rout the straight edges of the rails and stiles. These go as you would expect, and just as if the entire door were to be straight-edged.

For the curved edge of the top rail, leave the template on the rail, and feed the part into the bit—using push blocks. Do not get your hands involved in this part, as a slip will cost a finger, or worse—with the template riding on the bearing on top of the bit.

Once the stiles and rails are done, rout the door. Set up the panel-raising bit, as you would normally, testing it with the correct thickness of scrap stock. Rout the straight edges and bottom (rout the cross-grain bottom first) with the normal fence set up. For routing the curved top of the panel, remove the outfeed portion of the fence. Start the feed

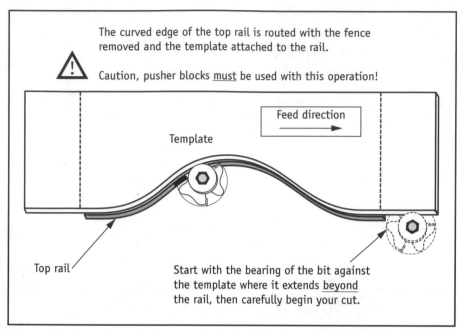

The curved edge of the top rail is routed with the fence removed and the template attached to the rail.

⚠ Caution, pusher blocks <u>must</u> be used with this operation!

Feed direction

Template

Top rail

Start with the bearing of the bit against the template where it extends <u>beyond</u> the rail, then carefully begin your cut.

Remove the fence to route the curved edges of the frame, making sure to use pusher blocks. *(Jesada)*

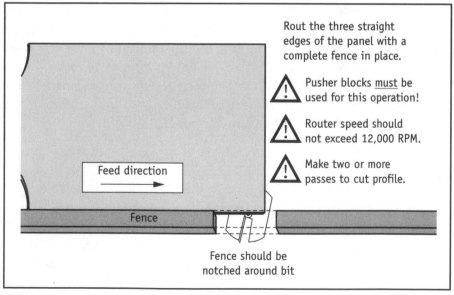

Rout the three straight edges of the panel with a complete fence in place.

⚠ Pusher blocks <u>must</u> be used for this operation!

⚠ Router speed should not exceed 12,000 RPM.

⚠ Make two or more passes to cut profile.

Feed direction

Fence

Fence should be notched around bit

Rout straight edges with the fence in place. *(Jesada)*

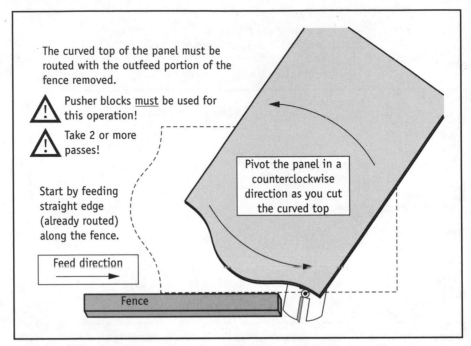

The curved top of the panel must be routed with the outfeed portion of the fence removed.

⚠ Pusher blocks <u>must</u> be used for this operation!

⚠ Take 2 or more passes!

Start by feeding straight edge (already routed) along the fence.

Feed direction →

Pivot the panel in a counterclockwise direction as you cut the curved top

Fence

Remove the fence and use pusher blocks with the curved panel. *(Jesada)*

with the already routed straight edge, and use pusher blocks. Pivot the panel counterclockwise as you feed. With this operation, you should take more than one pass. Depending on the amount of raise you want on the panel, two, or even three, passes are in order. And make sure to keep the bit speed down to about 10,000 rpm, or even somewhat less.

Test all the parts for fit before assembly.

Finish as you would with a straight panel.

Building and Attaching Drawers

Drawers are the most intimidating job for many beginning cabinet-makers, though they need not be. They are simply another box, one that usually needs greater strength because of the use to which it is put, but otherwise is no different from hundreds of other boxes.

Boxes may be made with butt joints, lock joints, finger joints, dove-tails, and dowel joints, including biscuits. The butt joint is obviously the easiest. Adding biscuits makes it hardly less easy. Lock joints are next on ease of cutting scale, with finger joints being more complex and time-consuming. Dovetails take the longest, but can produce the greatest-looking joints—there is some controversy over whether glued dovetails are stronger than glued finger joints, because finger joints tend to have more glue surface area, but it really isn't worth dealing with. Drawer doweling is the easiest kind of doweling, or at least the way I do it. Simply make butt joints, with the back and front of the drawer inside the sides. Then drill holes for the dowels through the sides into the back and front material, placing a dowel every 2 inches (approximately: space these to be aesthetically pleasing as well as sturdy) and tapping

> Drawer boxes may be made with butt joints, lock joints, finger joints, dove-tails, and dowel joints, plus biscuit joints. The butt joint is obviously the easiest. Adding biscuits makes it hardly less easy. Lock joints are next on the ease-of-cutting scale, with finger joints being more complex and time-consuming. Dovetails take the longest, but make the best-looking joints.

each in flush with the sides, with glue. Because most drawer forces are pull forces, this makes an amazingly sturdy joint.

For the most part, good- to top-quality kitchen and bathroom cabinet drawers work with only one of two options: lock joints or dovetails. Public perception of the finger joint is that it is a lesser joint than a dovetailed unit—it is not—so it's usually not worth setting up for and getting through with its repetitive cuts on a table saw, or with a router jig. The above-mentioned technique of doweling is also one that seems well accepted, though the dowel joint is probably not nearly as strong as the finger joint.

Sides and Backs

You always have a choice of materials, of course, but a decent hardwood drawer front at the very least should have a decent hardwood backup construction. I like to use yellow poplar for drawer sides and backs. This is a stable, easy to work, and widely available hardwood that is not really attractive enough to use as a drawer front, or for any cabinetry that is exposed, but that works well in interior locations where moderate to high strength is needed. It is also low-cost and straight-grained and accepts most finishes well. It does fuzz up a bit when not sanded correctly, but that's not a big problem. Poplar takes clear finishes well, but blotches with many stains.

Drawer Bottoms

Drawer bottoms are a different case. Because most drawer bottoms are more than 12 inches wide, and some may be as much as 30 inches, a plywood is a more suitable material as it retains stability in both directions better. (In much older case building, and in furniture, solid poplar drawer bottoms up to $3/4$ inch thick can be found, with planed beveled edges to fit into approximate $1/4$-inch dadoes around the drawer sides.) For small drawers (no more than 16 inches), a $1/4$-inch hardwood plywood, or an MDF with a laminate face, works well. I've used $1/4$-inch plywood in shallow desk drawers that were 30 inches wide, but I wouldn't do the same in kitchen or bathroom drawers. The weight load is entirely different. Use $3/8$-inch plywood in drawers to 26 or 28 inches wide. Give some thought to $7/16$-inch materials in wider drawers. (And, when designing, remember that the wider the drawer, the easier it is to jam it up when pulling it open or pushing it shut.) Always make your dadoes deep enough (depth is equal to one-half the thickness of the drawer sides and back) to allow you to cut drawer bottoms enough undersized to allow some movement for humidity without adding the worry of the bottom dropping out. Minimum dado depth is $5/16$ inch, and $3/8$ inch is better, which means sides and back need to be $5/8$ to $3/4$ inch thick, with the thicker part preferred.

> Because most drawer bottoms are usually more than 12 inches wide, and some may be as much as 30 inches, plywood is a more suitable material as it retains stability in both directions better.

Joints

As noted, the simplest joints speed production—butt joints, pinned with brads, for example, are about as easy as it gets. But they are less durable than almost any other form of joint, especially in a drawer, and are going to create a perception of lower quality. It is almost imperative to use dovetail or doweled butt joints to create a true impression of quality. If you are in a market with some understanding of wood, then finger joints are also good. With new jigs, and new uses of old jigs, such joints are as quickly made, as are dovetail joints. Butt joints with dowels can work in your favor: Make the butt joints, with pinning but with clamping, and then dowel with contrasting-color dowels. Dowels are cut long, tapped in, and trimmed with a flexible dowel saw, doing the entire drawer (in most cases deep drawers take two or three cuts) in a single pass with the handsaw.

The joinery section already has discussed the use of the most easily set up dovetail jig, the Keller. For many furniture uses, the Leigh and Porter-Cable Omni-Jig are preferable, as either one can give you more and different appearances, with varying-width pins and tails and varying placement. The Keller jig is a single-placement type, with two jigs used to produce the most common sizes. Once set up, though, it never needs to be reset and is as easy to set up for use as is a handsaw, or very nearly so. We'll quickly look it over again.

Once the drawer sides are cut, check assembly; dovetails need to fit together firmly, without major forcing (though it often takes a soft-faced hammer to get them together, and apart). Check for length while

Setting up the router for a joint jig.

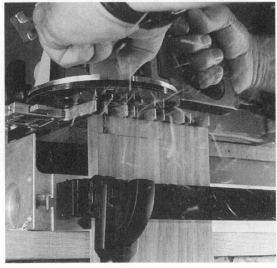

Leigh jig in use.

doing the dry-fitting. Set clamps to within 1 inch of final length. Check for ease of bringing into square. Also check final dimensions.

If the fit is okay, apply glue, assemble, clamp and check for square, and wipe off excess glue. Leave clamps in place, with yellow woodworker's glue, at least 1 hour.

For doweled butt joints, assemble for fit and to check clamp openings. Check square and overall size. Disassemble. Next, assemble with glue, clamp, check for square, and set aside for at least 1 hour. Clamps will interfere with drilling for dowels, so do that after clamps are removed; drill to size and depth for dowels, insert the dowels with glue, and trim. Dowels may be inserted slightly over-length—say $1/4$ inch—and may then be trimmed to final length; or you may use a long dowel for each hole, and trim, saving some precutting of longer dowels.

Drawer Fronts

Drawer fronts may be the simple front piece of the drawer or may be an added piece. In overlay construction, which most face frame work is, the drawer is going to be an extra, larger piece. Edges of plywood or veneered medium-density fiberboard (MDF) must be finished. The piece can then be screwed in place (from the inside of the drawer, into the back of the front, out; also use glue). Make absolutely sure all overhangs are correct. For example, if the drawer is to be $1/2$ inch oversized, make sure you have $1/4$ inch on each surface. Check with a $1/4$-inch piece of stock, getting one side, and the top or bottom, flush with the extra stock surface. That ensures that the other side is also set correctly—if the overhang is to be the same on all sides. Of course, the "feeler gauge" stock must be one-half the total correct overhang, whatever that might be, or the exact overhang for that particular part. As an example, if a drawer is to overhang $3/8$ inch at its bottom, $1/2$ inch at its top, and $1/2$ inch at each side, you make one $1/2$-inch "feeler gauge." The $3/8$-inch bottom overhang will be there for you when you check the top overhang—just don't forget which is which.

I suggest you mark and save each of these gauges. Overhangs don't come in an infinite variety, so being able to reach up and grab a piece of 1-inch by 12-inch by $1/4$-inch wood is faster than having to make one up. Much of cabinetmaking lends itself to such savings of jigs and measuring tools. An open-grid set of shelves can aid in storing such things where they are easily found, with variably sized openings to accept different jig types and styles.

> The drawer front may be the simple front piece of the drawer or an added piece. In overlay construction, which most face frame work is, the drawer is going to be an extra, larger piece.

> Make jigs to get the right overhang. Mark and save each of these gauges. Being able to reach up and grab a piece of 1-inch by 12-inch by $1/4$-inch wood is faster than having to make one up. Much of cabinetmaking lends itself to such savings of jigs and measuring tools. An open-grid set of shelves can aid in storing such things where they are easily found, with variably sized openings to accept different jig types and styles.

Material Selection

Drawer fronts are normally made from wood matching the material used to form the panels or the frames. They may be solid or veneer. In most cases, the grain, or figure, runs horizontally, to match the rails, even though door grain runs vertically. This is a point you may wish to talk over with your client. There are no absolutes, but generally, drawer fronts that are less than 8 to 10 inches high look better with horizontal grain, and there is seldom any real reason to deviate from that. Too, once you've decided on the horizontal direction, you don't want to vary it for taller drawer fronts; that simply looks odd,

> Drawer fronts are normally made from wood matching the material used to form the panels or the frames. They may be solid or veneer. In most cases, the grain, or figure, runs horizontally, to match the rails, even though door grain runs vertically.

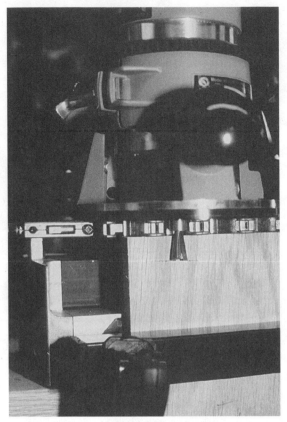

Checking bit depth on work face.

out of place, and incomplete, with the feeling being that someone was unable to decide what to do.

Most of the time, drawer fronts are made from material that is not quite what I call "full thick"; that is, it is less than $3/4$ inch thick. There is nothing that says it can't be $3/4$ inch thick, and the red oak on the desk drawer to my right indicates this looks just fine. It is backed by $3/4$-inch Baltic birch plywood for the drawer front, and overall drawer height is 11 inches. The red oak looks fine to me, with grain running horizontally, and I built this desk about 5 years ago. Basically, though, full-thick lumber is seldom necessary on what is totally an appearance addition.

Hardware Installation

Drawer hardware consists of one thing, the handle. Sometimes a catch is added, but drawer stops are normally part of the drawer slide assembly, which may or may not be hardware. (My desk uses maple guides; generally, kitchen and bath drawers get enough heavier use than desks to deserve actual hardware from Accuride, Knape & Vogt, or another top maker.) These slides allow full extension of the drawer and are available in styles that accept as much as 150 pounds of weight per drawer. They are easy to install and end up costing less than most lower-cost alternatives that take time to put in place and that also wear out more quickly.

Handle installation is done with a jig that centers the handle holes on the drawer front, at a distance down from the top—or up from the bottom—that you and your client prefer. Make sure you have a jig for each handle size you use and for each drawer size. Drill the appropriate size holes for the screws that come from the inside of the drawer front into the handles. Place the screws, set the handle, screw down. You're done checking. Remove the hardware for finishing.

Slides

Not all slides install exactly the same measurements or technique, with lots depending on model and size. Similarities, though, are very strong. At the moment, I'm getting ready to install a set of Accuride C2006 slides I've had on hand for a decade. These are all ball-bearing slides that do not have a dismountable moving member. Moving-member slides are generally easier to install, especially when the design has sidewalls, which this one doesn't (under a desk). This unit screws into the underside of the top, after the drawer is mounted in place. Pull out the slide, and mark and drill holes 1 and 2 at the back of the cabinet or

Cutting joints with the Leigh jig.

desk. Push in the slide, holding it in the same position, and do the same with the front holes. Do the same on the second slide. Mark and drill for each slide's desk underside attachment after attaching the slides to the drawer sides.

Dismountable slides work differently. I'm getting ready to install a pair of Knape & Vogt 8400 full-extension slides. These are 100-pound-limit, full ball-bearing units. You need $1/2$-inch side clearance per drawer to use them, and the drawer itself must be 1 inch to $1^{1}/16$ inches less than the drawer opening width. Screws are provided, supposedly in a Euroscrew, whatever that is. Regardless, should you lose some, they're #7 × $1/2$-inch pan head screws. You begin by fastening two 1 by 2 (actual size is $3/4$ inch by $1^{5}/8$ inches) to the rear inside wall of the cabinet; this is a vertical mount. The strips are aligned so that there is $1^{1}/4$ inch minimum of each strip clear of the face frame. (This requires some checking, as the strips are about 22 inches behind the face frame, at the back of the carcass, but simply make a $1^{1}/4$-inch-wide piece of wood the exact length—give or take $1/8$ to $1/4$ inch—of the cabinet inside depth. This is another keeper, as over time you'll use the same slides, cabinet construction, depth, etc. Just mark use and length on the piece.)

Rear brackets go into the uprights, using two holes per bracket (brackets have the open sides facing to exterior of the drawer side).

Pop the slides apart by depressing the disconnect levers.

Install the in-cabinet sections of the slides, adjusting as needed to make sure the front screw hole center for the slide is $7/16$ inch in from the front edge of the face frame, and the screw hole is centered $11/4$ inches up from the top of the bottom (underdrawer rail) horizontal member.

Now, take the drawer and mount the disconnected piece of the slide $1/16$ inch from the inside edge of the drawer front, with the center of the two fastening holes 1 inch up from the drawer bottom. Use two screws to hold the piece in place, and slide the drawer into position. Final adjustment may be needed, so open and close the drawer several times to check.

Eurostyle Fit

This system is designed to be a nonadjustable fit with 32-mm systems. The holes are drilled and ready to go. Begin by installing the disconnected section on the drawer (one to each side, of course), with $1/8$ inch (3 mm) from the front of the drawer, and the same 1-inch (or 25-mm) spacing to hole centers for attachment screws. The cabinet piece then installs with three screws into already drilled holes in the cabinet.

Insert the drawer and close. Presumably, there is no adjustment needed if the 32-mm system was used. Again, open and close drawer to check for exact fit, and make any adjustments that are needed.

At this point, all the cabinets are assembled and ready to go in the finish room. Naturally, any hardware that has been installed, in areas to be finished, must be removed. Taping is not acceptable.

Final sanding should come down to at least a 150 grit, and maybe as much as a 220 grit. Further sanding to finer finishes is a waste of time and may interfere with finish adherence. We'll go over that more in the finish preparation chapter coming up next.

Finishing: Put a Glow on It

There are almost as many ways to finish a cabinet as there are to build it, but for commercial purposes today, one method really stands out over the others: HVLP. For speed, economy of material, and great results, high-volume low-pressure (HVLP) spray systems are the choice in most situations where you are not hand finishing. In fact, I cannot think of a really good reason to ever use a standard high-pressure spray gun at all. A few times, you may want a hand-rubbed lacquer for your clients—or the client may demand hand-rubbed lacquers—but for the most part, spraying lacquer with an HVLP system or conversion gun is the way to go.

The process of finishing most cabinets for the bathroom and kitchen is fairly simple. Take the materials in the white (without hardware or accessories) to at least a 180-grit sanding. For the most part, random orbit sanders work best here, though regular finish sanders (oscillating) and belt sanders may also be useful from time to time. For superfine finishes, go to 220 grit; anything finer than a 220-grit sanding is a waste of time, unless you're smoothing a coat of finish.

Easy Steps to a Great Finish

The first step in any finish job is to get the surfaces to be finished clean and smooth. The work starts with removal of excessive glue and goes

Finished pillar and cabinets at J. R. Burnette's shop. The pillar is for a mantelpiece.

on to sanding, or scraping, the wood surface to a smoothness that will accept your selected finish (we'll cover the actual finishes shortly).

Glue is best removed by scraping, though in a pinch sanding will get most of it. The problem with excessive glue spillage and general sloppiness—and squeeze-out from joints—is that the glue tends to fill the pores of the wood, which means stains don't penetrate properly. The result is a splotchy-looking finish. Wiping off excess glue with a damp rag is a time-honored solution that doesn't work too well. Most woodworking glues are water-soluble, so as much seeps even deeper into the wood pores as is wiped off when water is added. Thus, scraping with a knife or scraper is almost always the best method of glue removal. Of course, that means glue cannot be removed until the glued joints are thoroughly dried, so that all squeeze-out and spillage are hard enough to allow scraping.

Once as much glue as possible is removed, you start the smoothing process.

Depending on the quality of the wood, start sanding with 80- or 100-grit paper. In

> The first step in any finish job is to get the surfaces to be finished clean and smooth. The work starts with removal of excessive glue and goes on to sanding, or scraping, the wood surface to a smoothness that will accept your selected finish. If this step is shirked, the finish won't be good.

most cases, 100 grit is a fine starting point, and leaves you with only two or three paper changes to reach the optimum smoothness for clear finishes over stains (and even over plain wood). Save those 320- and finer-grit sandpapers for smoothing between coats of finish, not for slicking wood down for the first coat of finish.

> The 100-grit is a fine starting point, and leaves you with only two or three paper changes to reach the optimum smoothness for clear finishes over stains (and even over plain wood). Save those 320- and finer-grit sandpapers for smoothing between coats of finish, not for slicking wood down for the first coat of finish.

If a water-based stain has been used, you need to finish smoothing with a 180-grit paper, as water raises the wood grain. If an oil-based stain has been used, you may not need to do any further smoothing. I like to go over most stains with a 0000 steel wool to make sure there is no buildup of the stain in corners and in decorations. The steel wool is so fine it doesn't reduce the coloring effects of the stain.

For final cleanup, before staining or coating with finish, always start with a wipe-down with a dry rag, and go on to a vacuuming or blowing off (most experts now say this can cause lung damage, so you

Finishing sanders are ideal in tight corners. (Makita)

Before finishing, hardware is installed to check fit and then removed.

may want to do only the vacuuming). Use a tack cloth for the final wipe-down before coating with stain or finish.

Staining

The cabinets are then stained, using your selection of stains. In its simplest form, staining is nothing more than applying a colored liquid to wood. For the most part, that's what cabinetmakers' stains are, and specialized techniques such as glazing, toning, and shading, all done so the wood remains visible through the color, are special-order items offered by only a few shops. I'd suggest, for those who really wish to do such work, that you get in contact with Reader's Digest Books and order a copy of Bob Flexner's *Understanding Wood Finishing.* (The book was originally published by Rodale Books, which recently sold its woodworking operations to Reader's Digest.) There is no more succinct and accurate overall coverage of wood finishing in existence.

We'll stick to simple stains here, with a quick look at applications. There are numerous ways to go, though for most, pigment in an oil-based binder works well and gives leeway in removing excess. For greater coloration, pigment and dye in an oil-based solvent work because the dye penetrates more deeply than plain pigment, particularly in hardwoods such as maple. Both these dyes need to cure overnight before a finish is applied.

Pigment in a water-based solvent cures rapidly, and you can apply finish to the piece within 4 hours most of the time. Pigment and dye in a water-based solvent color wood more deeply than a pigment-only stain; with both of these, the water base has at least a slight tendency to raise the grain in some

> Use a tack cloth for the final wipe-down before coating with stain or finish.

Jig for setting hardware.

woods.

Stains are used to give wood a color that matches a desire or need of the customer's, and, while doing so, it is also used to match up differences in color between sapwoods and heartwoods. Whatever type of stain you elect to use, I suggest at the start you begin by using no more than two types, and make up plenty of sample pieces about 12 inches by 18 inches to spray or wipe stain on. Use a number of different species of wood to get the feel of the stain across as many of the materials you'll be using as possible.

Notes on Water-Based Finishes

As a commercial shop, your cabinetmaking enterprise will be required to meet volatile emissions standards, whether by the use of extraction machines (including fans) or by the use of solvents that do not emit many, if any, volatile compounds. While nitrocellulose lacquer remains important, spraying that, and many other finishes, can create problems in the environment, so requires specialized equipment. Most water-based finishes do not require such special equipment. Note that regardless of the type of finish being applied, lung and eye protection

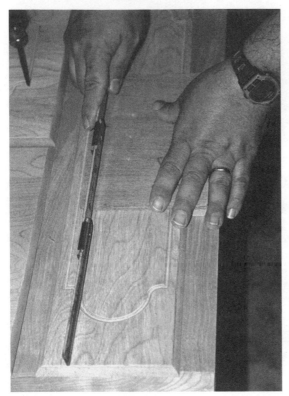

Set jig at the correct distance.

is essential.

Generally, the best commercial finish is durable, easy to apply, with an easy-to-reach final sheen of a level that matches your customer's desires. Various lacquers fit this bill better than anything else.

Open-Pore Woods

For oak and other open-grained woods, you may want to use a filler or use a sanding sealer or a filler coat of the finish, sanding it thoroughly again. At this point you might wish to use a finer grit than 220. Several coats are needed to fill the pores, with each coat being cut back to level the surface. You can also spray on several coats, allowing minimal time between each coat, and then sand them back. When sanding, be careful not to go through to the wood.

This is another of those skills that is gained through practice, which means a few more 12-inch by 18-inch boards are going to be very handy. This kind of filling offers options, ranging from an almost unfilled look, right on up to completely filled. It is also by far the quickest way to fill cherry, maple, and similar small-pore woods.

Paste wood fillers such as Behlen's Por-O-Lac wipe on and then are forced into the wood grain with a spreader, usually plastic (auto body finish spreaders work very well). Such fillers come natural and colored. Colored fillers also serve as the stain, saving one step in the finishing process. Drying time varies according to what the maker uses in the filler, and depending on whether it is a water- or oil-based solvent type.

The biggest warning with paste wood fillers concerns application: Don't cover too great an area at one time, because once the stuff starts to harden, you can't wipe it off properly, which means scraping, or, in the best case, a lot of work wiping it off with a hard scrubbing motion. Wiping is not easy work even when all goes well, so it's best not to make this particular mistake.

Lacquer Finishing

For many types of cabinetry, a lacquer finish is the way to go. Lacquers resist water, heat, alcohol, many acids and alkalies, and the general run of kitchen and bathroom chemicals, at least in mild applications. Nitrocellulose is usually the lacquer of choice, but cellulose acetate butyrate (CAB, water white) yellows less over time and is less amber on original application, so allows more of the wood's natural color—or stain—to show through. CAB is considerably more expensive than nitrocellulose, and the new water-soluble finishes do not yellow at all, so CAB has little reason for its price premium these days.

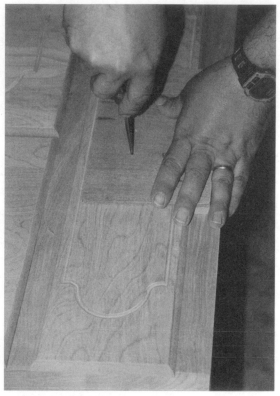

Mark holes with an awl.

Lacquer applies easily with spray equipment, with a rapid drying time. Fast drying time cuts production costs because you can get three or four coats on in a single day, but it also saves problems because of airborne dust: the faster the finish dries, the less chance there is of its being contaminated by dust. Lacquer is easy to repair, has good film clarity (it looks very deep, as if many, many coats have been applied, when in fact only four or five may have been), and comes in many formulations of colors and styles, including types that crackle. Cost is also relatively low, and screwy weather is not really a problem, because you can change the formulation to fit the hot, dry weather that creates finishing problems (the lacquer dries before it hits the surface, thus not flowing out properly). Lacquer retarder works here, and it also works where the evaporating solvents draw air into the finish in humid weather, creating white blushes. In colder weather, regular lacquer thinner is too slow to evaporate, so a fast type is used.

Where Lacquer Is Going

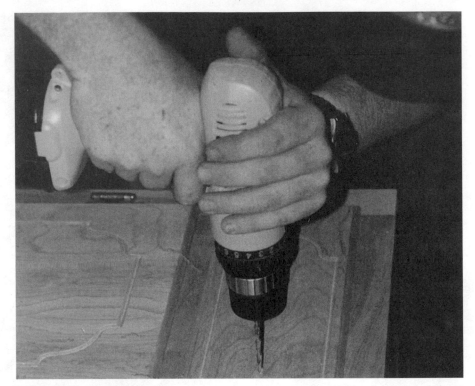

Drill holes for hardware.

Install hardware, here hinges, to check fit.

Roll-around glue and finishing station.

There are problems that today's world has brought to lacquers, and that actually make them less sensible than other finishes for the small shop. The solvents are polluting, toxic, and flammable to a high degree, and you need lots of them. Heat, acid, and solvent resistance are low in comparison to modern finishes. Lacquer also scratches easily.

All these features make lacquer less than ideal for kitchen and bath use, though for many years it was the primary way to go.

Varnish

The greatest finish durability is found with varnish. Varnish includes many polyurethane formulations; it resists wear, solvents, and acids and provides an excellent water barrier. Varnish is also cheap and builds fast to give a fine finish in a rush. At one time, linseed oil was the base (oil is cooked with a resin and then driers are added to speed curing to make varnish). Other oils have since taken over because of lower cost, and because they yellow less than linseed oil.

Synthetic resins now replace the natural resins once used. Early on, phenolic resin was developed, after which alkyd resins came into being. Alkyd resins are cheap and are now the industry standards. Polyurethane resins were created next, and they are exceptionally tough—the pure polyurethane finishes come in two parts, like epoxies.

Driers are added to all the varnishes to speed drying time, which they do by speeding up oxidation.

Varnish, compared to lacquer, is slow-curing, taking at least 1 hour to dry dust-free. Only a single coat can be applied in a day. It also does not spray well.

The true modern standard for the small commercial shop is found in the water-based finish.

Water-Based Finishes

It must sometimes seem as if any finish works well, except for two items, at least, that make it partly unsuitable for its chosen job. Water-based finishes use acrylics and polyurethanes that are manufactured in tiny drops that disperse in water. Glycol ether is added as a solvent, because it dries more slowly than water. As the water evaporates, the droplets come together, at which point the glycol ether softens the outer molecules of the droplets, letting them join. The glycol ether solvent then evaporates, and a film is formed.

The result is a very tough finish that is scratch-resistant. Resistance to heat, solvents, acids, and alkalies is lower than that for polyurethane varnish, and the barrier against water is not as good. In other words, the surface tends to breathe, so wood is still exchanging moisture, though at a much slower rate. With the reduced solvent content, water-based finishes produce few volatile organic compounds (VOCs). They also clean up with water. The finish produced is clear and colorless, with no yellowing, which can be startling when you first see wood finished with water-based materials.

Kicking perfection in the tail again, we find that water-based finishes are not as easy to spray as lacquers (they also don't brush as easily as varnishes). Too, the first coat will raise some fuzz because of its water content. The first sanding eliminates this problem, and finish coats build exceptionally well because water-based finishes have very high solids contents. You may also use a solvent-based stain, which will seal the wood and reduce grain raising. Shellac may also be used to seal the wood before applying the water-based finish, or, as above, you may simply apply a wash coat and sand the finish. You may need to apply two coats to reduce all the raised grain.

Water-based finishes can also create rust problems in spray guns and, if steel fasteners are used, in the cabinets themselves. Rust is a possibility even on the cans the finish comes in. Make sure all chances for rust are removed, because rust causes black stains on the finish.

Because water has high surface tension, it doesn't flow out well. There are numerous flow enhancers out there, including a drop or two of detergent. Naturally, such surfactants foam when stirred, so defoamers are then added. Weather can also affect water-based finishes more than it does solvent-based types. Hot, dry conditions can create too fast drying times and give poor flow-out. Add 10 to 20 percent distilled water to extend the evaporation time. Manufacturers also make solvents for the same purpose; distilled water is much cheaper. Do not finish with water-based materials when room temperature is below 65 degrees Fahrenheit. High humidity may create slow curing. Place the finished work so that there is a distinct airflow over its surface, to speed drying.

Spraying

Spray guns are the way to go for commercial shops. As noted earlier, though, there is no real reason to go with a standard high-pressure spray setup unless you already have the equipment. Where HVLP units shine is in the application of finishes such as polyurethanes, stains, and most water-based polyurethanes. In fact, water-based finishes and HVLP units seem almost born for each other, and each adds a great deal to the ease of getting a top-quality finish with only a rational amount of practice. I often thin UGL's ZAR polyurethane and Minwax's polyurethane about 40 percent. These particular water-based polyurethanes are formulated for brushing, so are thicker than most. When thinned, they go on as well as any other. Some time ago, I finished a project with Behlen's satin polyurethane that needed no thinning at all. Hydrocote also goes on nicely without thinning, and there are a wide array of

> Spray guns are the way to go for commercial shops. HVLP spray units excel in the application of polyurethanes, stains, and most water-based polyurethanes. In fact, water-based finishes and HVLP spray units seem almost born for each other, and each adds a great deal to the ease of getting a top-quality finish.

brands out there that some people swear by (and others swear at).

Where to Work

Spraying is not something you can do in an enclosed, unventilated room, but care must be used with the type of ventilation. An indoors spray booth that is properly set up reduces problems with VOCs and reduces contamination of finishes by shop dust. Any ventilation, and there must be some, has to be achieved with an explosionproof fan. Even with water-based finishes, regular window fans are a fire hazard: Over time, the buildup of overspray around the motor parts creates an insulation source that raises heat until the fan will catch fire.

The physical size of the spray booth need not be much greater than the largest piece you expect to finish in there. Thus, if you don't build anything larger than 8 feet long, a 12-foot-long room about 5 feet wide offers sufficient, if minimal, finishing space. If you can spare more room, then do so, for then you can finish several pieces at one time.

> Spraying is not something you can do in an enclosed, unventilated room.

Good dust masks are essential during sanding.

HVLP Facts

The particular greatest point of HVLP spray systems is the reversal of the amount of spray staying on the surface versus the amount that is wasted. High-pressure spray guns seldom get more than 15 percent of the material on the surface being coated; HVLP spray guns get 80 to 85 percent of the material on the cabinetry. This means a great savings in finishing materials cost, less gunk to get in your lungs, and less on you, your clothing, and your glasses (or goggles).

Use a drip stick, or a viscosity cup, to test the material: For drip stick testing, you need a single second between drips. If the liquid drips faster, it's too thin. If it drips more slowly, it's too thick.

Save shirt cardboards and old boxes so you'll have plenty of cardboard on hand

A light-duty HVLP gun works well as a start. *(Campbell-Hausfeld)*

for test spraying.

Naturally enough, proper preparation remains the big secret in getting a good finish on any project. Rough wood gives a rough finish. Dirty wood gives a rough finish. Sloppy habits make for all kinds of finish problems. Thus, sand to at least 200 grit, and sand carefully, with the grain whenever possible. Clean up with a tack cloth. If you use air pressure to clean up, wear a dust mask.

Make sure your equipment is clean; and especially make sure all mixing sticks are clean. Mix the finish well, stirring carefully. (Do not shake clear finishes. Shaking causes air bubbles, which add another dimension to your finish, a dimension you don't want. Too rapid and rough stirring does the same thing.) Determine just about how much finish you're going to need, and pour that amount into a separate container. Seal the remains in the original container and store it.

Check the surfaces to make sure they're dust-free.

Check the gun's inside and outside to make sure it is free of dust and other contaminants. Anything that isn't clean must be cleaned.

Check the viscosity of the finish, and thin until it's right.

Fill the gun, but no more than three-quarters full (this leaves room for further thinning, if needed). Seal, and hook up the air hose.

Begin with a few practice strokes (what you should really be starting right now is a whole series of practice strokes on scrap wood). For most work, you'll want to be fairly close to the surface being finished, starting at 9 or 10 inches and moving in to 5 inches to increase the amount of finish laid on. Move the gun steadily, with a slow, even (that is, maintain the same distance from the surface, from the beginning of your pass to the end of your pass) sweeping motion. If you start the sweep off the surface being coated and squeeze the trigger after the gun is moving, you'll get a better job. Overlap strokes by about 40 percent.

You may adjust the spray gun to offer a round, horizontal, or vertical pattern; use whichever pattern you feel will logically cover the best for the work. The gun also has a material control knob, which is turned to increase or decrease the flow of material. Airflow may also be controlled with an airflow control knob.

Material comes out of the HVLP spray gun at a slower pace than it does from most high-pressure spray guns. It is thus simpler and easier to control, and there is less tendency to drip or run.

Practice. And practice some more. Use different weights of finish material so you'll get used to different application rates. The odd thing about HVLP systems is that they're easy to use for the novice, but respond nicely to increases in experience and skill. The more experience you get, the better the results are going to be.

Keep running tests and getting the hang of things. You'll use up a couple quarts of finish and mess up a lot of cardboard and spare wood that way, but when the time comes, you'll be ready to spray almost any project, with great results.

Do not do a half-and-half. That is, don't finish one-half or three-quarters of what's in the gun and then refill, trying to adjust from that point. You run the risk of uneven finish application if you work in this manner. Plan your finishing patterns so that you reduce waste and overspray.

Arrange an overhead light source that lets you clearly see how the finish is going on: Above and in front works best.

- Use the largest spray pattern you can to reduce the number of strokes.

- Start each stroke off the surface being finished, and end off the surface, too.

- Release the trigger at the end of each stroke.

- As much as possible, keep the gun perpendicular to the work.

- Keep the gun the same distance from the wood, from one side to the other—about 6 to 10 inches is normally the best.

- Maintain the same moderate speed from side to side.

- Overlap strokes by one-half to get an even thickness.

- Spray edges of flat panels first.

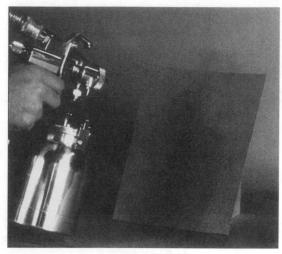

Test spray is essential to good work.

And cleanup is easy. Start by spraying solvent through the gun. The guns disassemble quickly and are then sloshed in suitable cleaning solvent. You can just set the whole disassembled gun in a bucket of water when you are using waterborne materials. Slosh, remove, and shake partly dry. Let air-dry and reassemble. If you're using water-based materials, start by making sure your gun has only brass and aluminum parts.

One of the worst jobs of any spraying work is over and done with in 5 minutes or less. That's very hard to beat. As is the HVLP spray system overall.

Installing Base and Wall Cabinets

The hard work is done, and you've got a gorgeous set of cabinets, or a single wall or base cabinet, ready to take to your client's home for installation. Of course, during transport, for which you have a van large enough to carry the base and upper case cabinetry, you protect the cabinets and their finishes as if they were precious and fragile, for they are. Any scratches or blotches or scrapes added to the job now detract from the appearance of the final installation, and from the value your customer is going to put on your work.

The best protection during travel is mover's blankets. These are available from many sources. Enough to protect a kitchen's worth of cabinetry is fairly expensive, so you may wish to start out with old blankets and quilts from your and your friends' homes. The setting, though, sometimes affects the pleasure of the receiver: Those raggedy, old blankets may detract enough from the effect of your cabinet wizardry to make you sorry you didn't make the investment in a more professional-looking kind of protection. Do so as quickly as profits allow. Generally, too, it is easier and quicker to arrange the professional-style blankets on and around the surfaces to be protected, and you don't have to worry about blankets falling apart and leaving pieces unprotected.

Checking the Area of Installation

By the time you arrive to install the cabinets, appliances should already be in place. If they're not, then you may face a few problems with final setup. If the actual appliances are not yet in place, the various hookups *must* be. You cannot install, for example, a kitchen sink cabinet, or a bathroom lavatory cabinet, when no plumbing has been installed for the sink or lavatory. It is usually possible to install cabinetry around the spots for ranges and refrigerators, but the refrigerator may need plumbing for its icemaker, while the range may be a gas style that requires a through-the-wall gas connection, as well as an outlet. (And if it's a 220-volt range, that outlet must already be in place, for best results, though ranges are relatively easy to move through a few feet to space things out properly.) Refrigerators may also be moved a bit for proper spacing, but if a water line is needed, that water line is of copper, and not enough slack is allowed, then you may have a problem. No appliance but the kitchen sink, usually, is totally immovable, though dishwashers come close.

> By the time you arrive to install the cabinets, appliances should already be in place.

This is the time to recheck all measurements, and to check your filler strips that you should have ready to go. Things always change at least a little. You expect that the builder or renovator who is doing the work has not made major kitchen changes without letting you know. It happens. It is the builder's fault. But that is insignificant when compared to customer disappointment. One note, then (and this is especially true on new homes and remodels where kitchen size and layout changes are drastic): Keep in touch with the builder, making sure you know how things are going from day to day. This is a business step that is going to be important on only one job in five, or even fewer, but you never know what any one job may lead to in the way of later jobs, so your best and neatest work is always necessary. Usually, any deviations can be corrected with filler strips, and with sanding down, or planing, the extra allowances on the cabinets themselves.

> Check your filler strips (have those ready to go). Things always change at least a little.

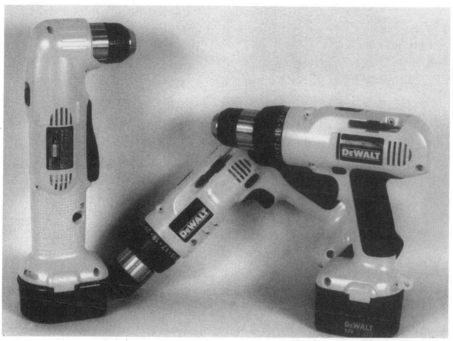

An array of useful cabinetmaker's cordless tools, especially for installation needs.

Toeboards (Base Systems)

Between the cabinet bottom and the floor is the base system, or the toe-board layout. Depending on preferences, you may have a toeboard that is part of a full frame under the base cabinet, or simply a set of parallel rails. For most cabinetry, the full base system is far preferable, as it gives support at cabinet ends. These will already be sized, built, and finished (where needed) in most jobs. Set-in ledger strips are included on the bases, so that the base cabinets can be fixed to the floor. (Not everyone does this. If you have a full ceramic tile floor running up under the cabinets, this step is best eliminated, because replacement cabinets in 15 to 20 years will not fit in the same holes. Too, ceramic tile is a stinker to drill without breaking. To prevent shifting, lay down a couple of $1/8$-inch strips to caulk under the ledger strips; or if there are no ledger strips, lay the caulk down under the toeboard edges, at the floor. The caulk idea works on any floor, and the caulk should not be replaced with Liquid Nails or similar construction cements because

> Pop chalk lines along the distance out from the walls where you'll be placing the toeboard system.

the idea is to lay in a nondestructive holder that can be removed when the cabinets are replaced.) Generally, the nailer strip and some 3-inch screws into studs are the best way to go to fasten almost any base cabinet nearly immovably in place. The toeboards are fastened with 2-inch screws, or 3-inch nails through the flooring, or, as above, caulk that prevents shifting. Screws should not be driven tightly, because the base cabinets must be leveled, and you may need to slip some shim stock under the toeboard bottoms, to make sure all is on the level. Nails can be pried up a bit, of course, but the screws need some beforehand slack.

Pop chalk lines along the distance out from the walls where you'll be placing the toeboard system. The toeboard system should have enough room to actually allow a set of toes under at the front, maybe 2½ to 3 inches, while a 2-inch inset at the rear (from the wall) makes it easier to align the toe rails in areas where walls are rough, slightly out of plumb, or otherwise messed up. Simply find the studs and go, making sure the piece that is designed for an area goes into that area.

Level the toeboards, shimming as needed.

Finding Studs

One of the fun jobs of any installation is locating the studs to which the cabinets must be attached. There are dozens of ways to find the studs, some of which work some of the time, many of which don't work, and a few of which work with some frequency. (You can drive a 16d nail every 1 inch, for example, starting 5 inches to each side of where you expect to find a stud; locate one stud; measure 16 or 24 inches off-center (OC) from there; and go on. Of course, the wall is pretty torn up— but this doesn't really matter except as a point of workmanship, unless there is no back to the cabinet being installed. Measure the next, etc. It works just about every time, but eats time and makes a mess.)

There are several other methods to find studs, two of which come from a California company called Zircon. Zircon recently announced its new line of stud finders, including the contractor line of three models which has one that even locates edges of wood and metal studs. The $10 model can find studs through ¾ inch of drywall

(Sheetrock, etc.) and has a light-emitting diode (LED) display that comes on as a stud edge is found. This one is about the size of a candy bar and has a pocket-clip. For those who need to find the center of a stud, instead of having to work to find both edges, the StudVision Pro finds those centers and has an LCD display that actually pictures the stud. It works to depths of 2 inches and projects a beam of light when the stud is found. This is an easy-handling, quickly calibrated tool that is really worth the $65 it costs, at least to the installing cabinetmaker.

The MultiScanner Pro works to find the edges of wood or metal studs, and then it projects a beam of light to the point. It works through $1^1/2$ inches of drywall, and it can find rebar through 3 inches of concrete (not all that handy for cabinetry, but intriguing, nonetheless). It will also look for hot alternating-current (ac) wiring and can be used as an ac scanner only with other modes shut off. This one is superb for the cabinet installer. It costs about $60.

These tools may not be dead-on essential, but for the price of any of them, they do their jobs well enough to make your job a lot easier, less messy, quicker, and, with the ac scan, safer. The ac scan is of greatest importance in old work, but can also be a lifesaver in new work. I would really like to see that combined with the StudVision's stud center finder, but that hasn't happened yet.

Thus, locate and mark the studs so you can come back to them easily once the cabinet is roughly in place. (Draw a line up over the backs of the cabinets, to where the backsplash on the countertop will cover the mark.) Use a level to get a plumb mark if you don't use a finder to get the stud center or edges.

While shimming toeboards, level the sections to each other. This is an appearance and quality feature: Each toeboard should have its top level with the top of the one before it, even though it is across a gap where an appliance fits. The gaps are simply bridged with a 6- or 8-foot level. That adds to the tool list, but if you can find a perfectly straight 2 by 4 that is 8 feet long, you can set your level on top of that. These days, the simplest way to get a perfectly flat 2 by 4 seems to be to start with a 2 by 8 and cut it down in the shop, jointing it to a perfect flatness after ripping it to almost-perfect flatness (as always, joint one side first, then rip the parallel side).

> **While shimming toeboards, level the sections to one another.**

You can also build on-site toeboard sets if you have elected not to use 1-inch material in constructing your toeboards. Usually, such toeboards are made of 2-by-4 stock, cut to fit on site, and then covered with $1/4$- to $1/2$-inch hardwood plywood finished to match the cabinets (it may also be a flat colored panel of whatever board or laminate serves best). In such cases, the toeboard frames are built and set in place, and the screws are driven through the front of the 2-by-4s into the subfloor and floor to hold them in place, after the toeboards are leveled both individually and to each other, and shimmed. Drywall screws 3 inches long are usually used for the angled fasteners, and they may be backed up by drywall screws run into the studs at the back of the wall, though this is probably overkill.

Positioning Base Cabinets

Set the base cabinets loosely on their toeboards. You should have plans that show where each cabinet goes, and those plans should relate to the toeboard sets you've built and brought along, so the toeboards are positioned exactly where their cabinets will be (within a fraction of an inch). Check toeboards before you position the base cabinets. Do the appliances fit in their respective areas, according to toeboard sizes? If the appliances are not on hand, you have to accept the dimensions supplied by the renovator or builder and hope they're not mistaken. Most dishwashers are of similar sizes, but ranges and refrigerators vary so much that it is possible to run into major problems converting cabinetry to fit if models are changed after you build the cabinets. Something of the same thing holds true for sink base cabinets in kitchens and lavatory bases in bathrooms. The variations available in kitchen sinks today have to be seen to be believed, and not all fit the old standards (though a good 6- or 8-foot-long cabinet will accept everything I know of on the current market). The bigger problem here comes with the plumbing that must come up through the bottom of the cabinet, with two supply lines and a drain, plus possible extra supply lines and drain for the dish-

> Set the base cabinets loosely on their toeboards. Have plans that show where each cabinet goes. Those plans should relate to the toeboard sets you've built and brought along, so the toeboards are positioned exactly where their cabinets will be (within a fraction of an inch).

washer. A note here is in order: To date, some dishwashers have been installed with the drain fitting cut into the sink drain fitting, and below the trap. This may or may not be code-legal in your area, but it is a bad idea, because it gives an outlet for sewer gas (methane) into the house. If the plumber has cut the dishwasher drain below the trap for the kitchen sink and hasn't used a trap on the drain line, you should ask that it be changed. This probably isn't one of a cabinetmaker's main interests, but it is a strong safety factor for the cabinetmaker's customer.

Hold Countertop Height

Whatever leveling system you use, try to hold onto your base cabinet countertop height: 36 inches total is most common and fits most appliance heights. To stay at the cabinet height you need for this final height, you must know the finished height, on the cabinet, of your countertop. Reduce the top height by that much, and maintain that height around the wall as the base cabinets are placed.

Set the first cabinet against a wall, and use a scribe to trim down the cabinet face frame stile to match any unevenness in the wall. Use a jigsaw (bayonet saw and saber saw are other names) to trim to the line, the plane to a slight underbevel so only a clean line shows. If the jigsaw cut is too wide, a rasp such as Stanley's Surform can be used, as can a belt sander. Simply scribe the line and sand to it. Again, use the hand plane to cut an underbevel.

The cabinets were designed to leave enough room to position a replacement appliance when the time comes. (Today's appliances are all supposedly engineered for a minimum 7-year life, but most will outlast that by years, though people may like a change with greater frequency.) A couple of inches around a range is plenty, while a bit more is essential around the bulk and weight of a refrigerator. Drop-in ranges and built-in refrigerators, of course, cannot be replaced by slip-in units without major cabinetry changes. The same holds true for dishwashers.

> If everything is level and plumb, the next cabinet should butt cleanly against the first cabinet, and on down the line and around the kitchen. Everything may not be level and plumb, but the leveling of the toeboards should set the cabinets in place, with only minor needs for filler boards.

> Once a run of base cabinets is finished on the wall, before fastening, use a long straightedge to make sure the faces are all even. Shift if needed, and use clamps to hold the cases together in a unit while fastening with screws.

If everything is level and plumb, the next cabinet should butt cleanly against the first cabinet, and on down the line and around the kitchen. Of course, everything won't be level and plumb, but the leveling of the toeboards should keep the cabinets pretty much in place, with only minor needs for filler boards.

Once a run of base cabinets is finished on the wall, before fastening, use a long straightedge to make sure the faces are all even. Move as needed, and use clamps to hold the cases together in a unit while fastening with screws. Keep a running check of the level of the top of the cabinets, and shim at the toeboard as needed.

The cabinets are now fastened to one another with 1¼-inch screws, from the inside. Next, use 1⅝-inch screws to fasten the cabinets to the toeboard frame. The cabinets are finally attached to the wall by 2½-inch drywall screws through the nailer strip and into the studs.

Use a story stick to transfer plumbing layouts to the cabinets where holes must be cut to match. Holes may be in the back of the cabinet, or in the bottom, depending on how the plumbing stub-outs are placed. Simply place the story stick plumb alongside the items to be marked. Use a short level (18 inches or so) to mark the lines from the stub-outs (or the actual pipe runs) to the back or base of the cabinet, marking to both sides of all pipes. To get horizontal and vertical alignment, mark this first run with the word *vertical.* Now, take a second story stick and place it so that it lines up with the cabinet edge of the existing cabinet, horizontally. Mark the runs in the same manner, making sure they're plumb as you move to the story stick.

Use the two story sticks to transfer the openings to the back or base of the cabinet, and drill your holes. For such holes, use hole saws, which cut more cleanly with large holes and tend to do less splintering than large drill bits.

Upper Cabinet Installation

Installing upper cabinets is easiest with special lifts that pop the unit up into position, where you level and plumb it without much effort.

Then you can just screw it into the appropriate stud and go on to the next. Such lifts are costly and not usually included in the start-up cabinetry business's tool list. In fact, many smaller shops never get such lifts.

> Because it can be impossible to see marks that locate studs, measure on cabinet backs the distances from mark to mark on the wall, and predrill screw holes in the backs.

The simplest way to work the raise is to use automatic scissors jacks placed on a flat piece of plywood on top of a temporary countertop, with the plywood at least as large as the base of the cabinet being lifted, which is then blocked to almost the needed height. Scissors jacks are the most useful because they are infinitely adjustable, allowing easy leveling of the cabinets.

As always, begin the installation by locating studs (these should be the same studs found for the lower cabinets), then snap or draw a level line for the height of the cabinet bottom. This line goes all around the room.

Again, start in a corner, setting up as above. Get the cabinet up and shimmed into its final position, with the doors removed. Level the cabinet with the jacks, and shim to get the face plumb. Use your scribe to mark the line down the cabinet (if the face frame butts the wall—corner cabinets that have no face frames butting the wall are simply shimmed into plumb and fastened to the studs).

Once the first corner unit is in place, assemble the remaining cabinets on that wall to one another. (Depending on the help available, you may choose to assemble 6-foot runs or 8-foot runs. Anything more than an 8-foot run is pushing your luck.) Use $1^{1}/_{4}$-inch no. 8 screws to hold the runs together.

Because it can be impossible to see marks locating studs, measure on cabinet backs the distances from mark to mark on the wall and predrill screw holes in the backs.

Use $2^{1}/_{2}$- or 3-inch no. 10 or 12 screws to attach wall cabinets: I don't like to use drywall screws on upper cabinets, because the drywall cabinets lack the shear strength needed, in my opinion, to hold the loads sometimes imposed on these units. To run the screws in with a driver-drill, use square drive or Phillips head screws. Do not immediately screw cabinets down tight to the wall because you may need to do some shimming to get the faces even.

Final Steps

Toeboard molding is nearly the final application. Once the final floor is in place, quarter-round molding is cut to fit at the toeboards. Miter at the inside and outside corners as neatly as possible; if an inside joint won't miter correctly, then cope the joint. Coping is simple: Bring the first piece of molding to a butt, or near butt, with the wall. Cut the molding that is to match at a 45° angle. Using a coping saw, cope along the outline of the profile on the 45° molding. Check the fit and use a wood rasp to adjust. Apply. Nail. Use finishing nails about one per foot or so.

Some kinds of cabinetry require molding at the tops of walls, too. Because this will usually have to match the cabinets, it's often the cabinetmaker's job. Usually, this means crown molding, which is a nuisance to do but which can be done relatively well. Outside corners are the easiest. Set the molding in your miter saw at the same angle it will occupy on the wall, and simply cut the angle at 45°. You can build a jig for your miter saw, or you can simply hold the material tightly in place. The inside angles are most easily coped, as above. Bring the flat piece almost to the adjacent wall. Cut the next piece at a 45° angle, and cope along the profile. Check the fit, and use a rasp to fine-tune things. This type of molding requires long finishing nails; because pneumatic finish nailers are limited to a top $2^1/2$-inch-long nail, they're generally unsuited for crown molding, because it needs a minimum 3-inch finish nail. Nails are driven in a cross pattern, with the lower nail driving into the top plate of the wall while the upper nail drives down into the stud.

At this point, you're just about finished. All hardware can be replaced. Doors are rehung. The overall fit and finish get a last check.

And you head for the countertop.

Topping It

Plastic laminates have been around for some time now, and we even find several resin-based solid materials for use as countertops. In today's world, people also use poured concrete and granite, both polished and matte, and other materials. I'm baffled as to the need for concrete, and I have no skills or advice to impart about real stone, so we will here look at countertops of laminated plastic, materials such as Formica and Wilsonart, over a substrate of medium-density fiberboard (MDF) or plywood.

I'm not going to do a long song and dance about plastic laminates, but a short look at what they really are is helpful in determining which you'd care to use. As a start, the type of laminate we're looking at is quite thin and has little innate strength. That's the reason for the substrate, of course, and the reason for much of the contact adhesive business in the world.

There are differences from maker to maker, but the basis of high-pressure decorative laminates is several sheets of brown Kraft paper, saturated with phenolic resin. These sheets are stacked in a press, and a decorative sheet is added. This sandwich is then subjected to a good deal of pressure and heat, so that all forms into a solid sheet of plastic.

Two of the three types do not apply to general cabinetmaking work—that is, work the cabinetmaker is doing in her or his own shop. Vertical laminates are too thin for countertop use. Postforming laminates are made of a resin that bends to conform to tight radii when heated to about 350 degrees and subjected to pressures of about 1000 pounds per square inch (psi). The equipment to do postforming work costs several thousand dollars, making it easier to buy postformed materials than to produce them.

Horizontal grade laminates come in a wide variety of sizes, shapes, shades, and surfaces and are specially designed and produced for use as countertops (horizontal, right?). Of the most popular special-purpose styles, there are solid-color decorative laminates, decorative metallic laminates (not useful for countertops, but can be used for accents in gold, chrome, bronze, copper, and other colors), wood veneer laminates (these can include real wood veneers applied to a laminate backer), and the general various other laminates in semigloss, gloss, and matte finishes that may be marbleized, or otherwise treated to look like just about any material you could want.

Some General Don'ts for Decorative Laminates

Avoid bold, trendy colors. You may be in love with red and dark green right now, but 5 years down the road, is your customer going to thank you for introducing it to his or her home? I remember once doing my own kitchen in bright royal blue and bright yellow. It looked great for a few weeks, but the walls repeated the tile pattern and colors, and after a month, it was tiresome. I stayed in that apartment for another 4 years and never did change the colors, from lack of time and interest—it was, after all, owned by someone else, and I'd already put in lots of work on the place. But we sure ate out a lot.

Avoid dark colors. This is a practical avoidance, for dark colors in a kitchen or bath show every detail of any kind of dust, dirt, or water ring. Scratches that reach into the laminate's backer tend to show up as light, too.

Avoid high-gloss surfaces. Again, this is practical because gloss is gorgeous at the start, especially on a sample chip, but

> In almost all instances, matte finishes in neutral colors are best; not everyone is going to want matte finishes in neutral colors. The point is simply to make sure your customer knows probable problems associated with the selection of gloss, dark, or metallic surfaces for countertops, and understands that trendy colors are not going to be trendy for very long. (Give some thought to kitchen colors of the late 1970s and early 1980s: Coppertone and avocado bring giggles today, but almond is still at work.) If all that is understood at the start, then the customer should be satisfied, even when things turn out as you predicted.

rapidly becomes a problem as it scratches, reflects light back into the owner's eyes, and highlights any imperfections. Too, as with dark colors, minor bits of soil and damage tend to show up as major imperfections.

In other words, in almost all instances, matte finishes in neutral colors are best. Now, that being said, not everyone is going to want matte finishes in neutral colors, so do with the above advice what you will. The point is simply to make sure the customer knows the possible and probable problems associated with the selection of gloss, dark, or metallic surfaces for countertops, and understands that trendy colors are not going to be trendy for very long. If all that is understood at the outset, then the customer should be satisfied even when things turn out as you predicted.

Building Up a Countertop

Today's countertops are usually of the built-up style. (Corian, as a sold material, is not built up in the same sense, but DuPont requires a special 3-day school for people who wish to work on Corian countertops, so we're not going to be redundant here and supply information on a material you cannot buy without taking the course from them.) In its essentials, a base, or substrate, such as MDF is used, and laminate is adhered to the top, with any of a number of types of edging used to finish off the countertop.

Your fastest countertop option is a dealer's postformed countertop that will drop into place, needing only mitering at corners and final trimming. These are relatively low-cost, easy-to-install options, and for the customer pleased with them, they are a great choice. For the customer who wants something different, unusual, or just a color or shape not readily available in postformed types, then the work starts. Today's countertops are generally built-up, cleated, double-layer types that do not use simple edge strips as in the past. This is due in part to use of Eurostyle cabinetry where edged styles might easily interfere with drawers opening, but it also offers general advantages in strength and portability. Take MDF as an example. A $1^{1}/_{4}$-inch-thick type is exceptionally heavy, as is a $1^{1}/_{8}$-inch model, and both still retain enough flexibility to be hard to transport. Drop back to $^{3}/_{4}$-inch MDF, and you lose much of the excess weight, but of course have an even floppier piece of substrate. Plywood, wood, or MDF may be used as

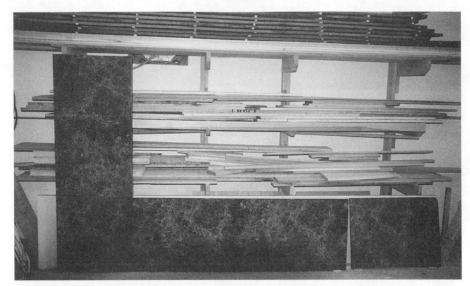

Built-up laminate countertop for kitchen corner.

Heavy-duty laminate roller is great for getting things properly stuck together.

build-up strips for the cleats that strengthen the $3/4$-inch MDF (or particleboard or plywood) substrate. Generally, though, I'd prefer to go with wood or plywood, regardless of the type of substrate, because it is stronger. Plywood is actually the best selection, because of a combination of strength and lack of worry about splitting during fastening.

Cleats are placed exactly at all edges, and seams should be backed by wider cleats (as much as 6 inches wide). The sink cutout area is braced with cleats, front, back, and sides.

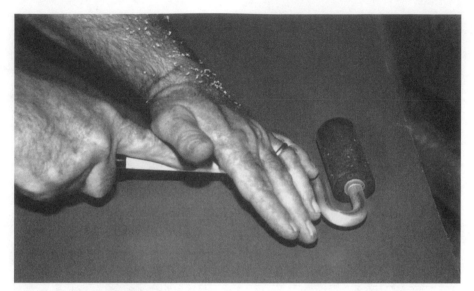

Here I'm rolling out the laminate.

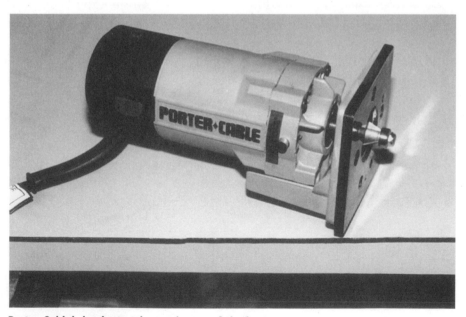

Porter-Cable's laminate trimmer is one of the best.

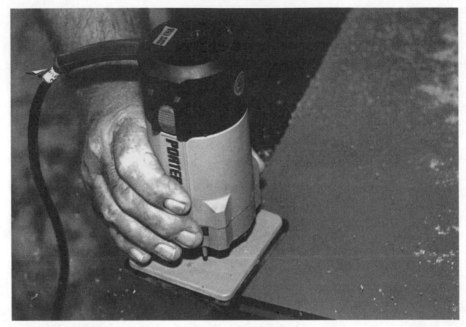

Trimming edges.

The front cleat must be wide enough to rest on the top of the cabinet frame (or front, if Eurostyle is used), with at least a 1/2-inch projection past any drawer or door thicknesses. (This is for two purposes. First, aesthetically, having the top extend past the drawer front edges, or the door edges, looks much better. Second, it serves to feed drips over the door or drawer edge so that contents of the interiors don't get drenched or soaked with a liquid messier than water.) Generally, a 1-inch overhang does the job, but for extra-thick drawer fronts, you may need a 1 1/8-inch overhang.

Backsplashes

There is no written rule as to the height of backsplashes, but over the years 4 inches seems to have become the primary figure used. Four inches looks good and protects a reasonable way up the wall.

On remodels in old work, the height of the electrical receptacles may be the determining factor: If they're less than 4 inches above the new countertop, then the backsplash is best lowered to suit. Of course, it's always possible to build around such obstacles, cutting a notch out of the backsplash, but it looks odd and tends to be a grease and grime catcher. In other areas, windows may fit close to the countertop and prevent a full 4 inches, or may open up a bit and suggest making the backsplash $4^{1}/2$ or even $4^{3}/4$ inches. That eliminates a thin, narrow strip that looks a bit off, and it cuts out the need for painting that same skinny, hard-to-paint section.

There is also no written rule about backsplash thickness. Generally, $3/4$-inch material of the same substrate as is used for the countertop is considered fine. It may be glued and screwed (from underneath) into place. Thinner pieces may also be used, but should be set into rabbets cut into the countertop, before any laminate is added—and, if it isn't obvious, before the substrate is set in place on the base cabinets. You have a choice in application with laminates; they may be added in the shop or in the field. Adding them in the shop is simplest, usually, but where it wouldn't be, adding laminate in the field also works.

Adding the laminate is an essential skill. Once the countertop is shaped and cleated, the laminate is ready to go on.

Working with Sheet Laminate

Laminate is not particularly fragile as generally handled around the shop, which is a good thing, because there is a lot of handling to do. Start by laying out the countertop, face up. Get any measurements needed, and plot the laminate layout to give as few seams as possible. Cut the laminate to rough size, leaving $1/2$-inch or so overhang all around.

Check out the edge sizes and strips, and get those ready to go. Cut the edge strips to size, and lay them out along their specific edge locations, back side up. Edge strips should be cut with about $1/8$-inch overlap on the top (this is trimmed soon). I like to keep the bottom flush, but you must make certain you are able, first, to work with a piece of laminate that has that bottom edge cut straight and, second, to keep the bottom flush. If you have a long edge to do (more than about 4 feet), a helper is handy. But if no helper can be found, roll the laminate, holding it in your left hand as you apply with your right. (Lefties, reverse this procedure if you like—everyone may have to switch when going in directions incompatible with handedness.)

Coat the side with contact cement. I strongly prefer the water-based types for several reasons. First, they are less likely to create fume problems; second, they change color as they dry, so it is easy to know when they're ready to work. Contact cement bonds best when the surfaces have *just* finished drying. If you let it go too long, simply recoat. Apply an even layer of adhesive to both the edge of the countertop and the back of the edging. Use a small roller (2-inch-wide paint rollers are great here).

Start the application as soon as the glue is dry, touching the first end of the edging to the exact start of the countertop edge. Keep the bottom flush, touch the edging down about every 6 to 10 inches, and continue until that strip is in place. Come back with a roller and make sure it is fully in contact with the adhesive. Place all edges at one time.

Use a laminate trimmer to rout the top edge.

Over the Top

The top of the countertop obviously gets a much larger sheet than any edge does. An even coating of contact adhesive goes on just as it does for the edging, both on the countertop and on the back of the laminate.

You should have prepared spacers for the countertop-laminate assembly before hand. These must be at least a few inches wider than the countertop, and they should be numerous enough that the coated laminate can be lowered into place and jiggled until it comes as closely as possible to matching the countertop edges.

Over the years, I've seen Kraft paper used for spacers, and it is often recommended. I've also seen slats from Venetian blinds used, and those work nicely most of the time (newer ones seem too flimsy). Personally, I like $1^3/8$- to 2-inch by about $1^1/2$-inch wood slats, about 36 inches long, as permanent spacers. These keep the laminate from touching any of the countertop until exact placement is ensured.

Start with a spacer each 6 to 8 inches, and see how that works for you. Much depends on the stiffness of the laminate. Get everything lined up after you make sure there is an even coating of contact cement drying on both surfaces.

Slide one spacer out, and press the laminate into place as the contact-cemented surfaces touch. Do the same with the second spacer, the third, and so on down the line. If it is done correctly, you can get an almost exact alignment on one edge, thus needing to do only minor trimming.

In some shops, the contact cement is sprayed on both surfaces, and two people hold the laminate at an angle to the countertop while a third lines up the first edge. That edge is very slowly pressed into place. This requires great care and more expertise than the spacer concept. It also requires three people, which is sometimes an impossibility in the small shop. (I've seen it done with two people on a 10-foot-long countertop, but my brain insisted on adding up the cost of the substrate, the laminate, and the 2 hours of work already in the piece, so I didn't care too much for it, though it worked then and has worked for years for the shop using this method.)

> **Prepare spacers for the countertop-laminate assembly beforehand. Make them at least a few inches wider than the countertop, and numerous enough that the coated laminate can be lowered into place and jiggled until it comes as close as possible to matching the countertop edges, without touching the glue on the countertop. When two dried coats of contact adhesive touch, the bond is instant—on contact—and very strong.**

Edges are routed smooth with a laminate trimmer, which is followed up with a file held flat to the countertop to give a final-finish smoothness. This takes only a few passes with the file, without much pressure.

Other Edging Treatment

There are so many edging options, it is impossible to even mention them all. Almost anything that will attach to the laminate or to its substrate can be used. Some are easier than laminate as edging, and others are not. One particular item that I like is a dual strip of $1/2$-inch-wide by $3/8$-inch-thick wood to match the cabinet wood. This may be done with a flat wood with just relieved edges, or with a molding style; or you may cut a molding that holds a strip of the laminate. In any case, you get about $3/4$ inch of laminate showing between the wood strips, and it looks great. Solid-wood edging to match the cabinet wood also looks fine, and you might want to consider Corian or another solid-resin material as an edging treatment in colors that contrast sharply with those used in the laminate.

This is a spot where imagination helps: I'd suggest taking some different kinds of laminate and wood and other materials and making some samples of a variety of ideas, using your own concepts, or those

This kitchen gets its ambiance from the sunny Harvest finish on Avondale maple cabinets. The detailing in the mitered cabinet doors, and complementing rope and crown moldings, adds an air of classic elegance. *(Merillat)*

listed above, to see what you come up with. You'll need the samples to present the options to your customers, who will almost certainly love both the look and the fact that you can actually promise that no one else will have the same look. (Cut an unusual molding shape for them, add a contrasting color to that, and then nip the corners of the countertops, or otherwise change the shape slightly, and the custom is really custom.) This is another way to please a number of customers without excessive time expenditures that run costs out of sight.

Flexibility

Using Shelves and Other Specialties to Supplement Income

For the first year or two, any business—whether involved with cabinet building or installing or both, in new or old construction—may have problems holding a steady flow of work through the colder months. This unevenness presents problems with keeping your cash flow on a sane and sensible basis. (The consistent bane of small businesses at all times is the cash flow. Getting it to a point where cash flows when you need it to flow is a chore harder than constructing the most complex cabinet or other structure. And it may take several years, which is why it is imperative to always begin working for yourself with a minimum of one year's cash on hand—enough for both business operating expenses and your own personal living expenses. Eighteen months to 2 years is better.) Keeping up some cash flow during cold months is difficult unless you've managed to get to a stage where you can plan indoor work in several structures for the winter. Even after you've been in business for some time, such planning doesn't always succeed.

Thus, there's sometimes a need for a form of business related to cabinetmaking to carry some businesses through the early cold, cold days. The profit margins on individual shelving jobs, regardless of type, may not seem as good as those on full-scale cabinet jobs. But the fact is, if shelving and small cabinetry are to be a source only of

Hutch design. *(KCDw)*

shorter-term income, the profit margins are somewhat less important, as they need only be sufficient to keep things rolling until the business is truly under way. At the same time, it's nice to make a decent yearly income early on, instead of after several years; so keeping profit margins reasonable makes sense. Other specialties, either instead of or in addition to, work well. J. R. Burnette advertises that his shop builds anything if it's made of wood. He does, along with a full load of custom cabinetry, a pretty good line of custom mantelpieces in many designs and patterns for many customers.

When you switch from general cabinetry, no matter how precise, to at least part-time shelf building and shelf installation or other specialized woodworking, defining the variances is far easier than you might expect. You are not going to suddenly need more space for work, in or around your shop area, and you are not going to need a wholly different array of tools, as well as new skills to operate those tools well enough to satisfy many customers. If you do only bookshelves, you'll probably find yourself seriously overtooled; but, as always, that's better than needing tools you don't have, can't afford, and are unable to rent.

Shelf design and building uses basic cabinetmaking skills and can call on the full array of your skills should you decide to make fancier bookshelves.

There is, though, the simple fact that it isn't wise to begin such a business, even as a cold-weather supplement to a cabinetmaking business, without being certain you can sell what can produce. That requires some local checking or knowledge. If you find that almost all bookshelf, box, or custom furniture sales in an area are covered by some low-end unfinished-wood sales, you're not going to make money easily, though it may be done. You must build a reputation for both quality and customer satisfaction; these transfer over from your cabinetry business, and transfer back as well. So, with some good fortune, you may convert customers for one type of cabinet to customers for another type, and vice versa.

At the outset of the addition of some form of cabinetry or custom furniture business, we're looking at a need to define just what is meant by cabinetry—or custom furniture. I have one friend who makes as much as 50 percent of his gross income building simple furniture—usually, though

Stock patterns can help shortcut design work. *(Wellborn Cabinet, Inc.)*

not always, lawn furniture. He has set a pattern for himself in an area. He turns out his own versions, developed over many years, of Adirondack chairs, love seats, tables, and various kinds of picnic tables. (Recently, too, Bobby has built a number of gun cabinets, a solid-walnut corner cabinet with raised panel door construction, heaven only knows how many fancy mailboxes and posts, a cedar bedroom suite, several entertainment center cabinets, and a podium stand for a teacher.)

There are several ways to go about working with shelves and custom furniture, though all involve actual woodworking. The installation of ready-made cabinets, whether from a custom manufacturer or from one who makes only stock models, is a great way to supplement rough periods, too. But you could end up with an installer's reputation, and your reputation might depend on the quality of such cabinetry. Such work does tend to involve extensive kitchen and bathroom remodeling, as does much of cabinetmaking, but the plumbing and wiring work there can be covered by your usual list of subcontractors, who may well be happy to find some extra work during the usual "off" months. Working with furniture does not involve subcontractors, but requires even more complex skills in some areas and even more tools (lathe, shaper, etc.). We look at some of the most commonly needed tool upgrades here, but this isn't a place for a complete treatise on workshop design and construction, beyond a few tips and a general look at how much more complicated the tool situation may make your life. (See also Table 7.)

Working It Out

You must do a quick reexamination of your resources before getting involved in areas other than cabinetry, because even the mildest form is going to take up more space and might create problems with zoning that you have not experienced to this point. If you decide to emphasize bookcases, for example, you may find yourself doing a lot of built-in work with materials no wider than 1 foot (except for the backing boards, which tend to be $1/8$-inch-thick plywood of the same type as the face boards). But, by the same token, you may find you're building barrister's bookcases, and that takes hardware, glass, and more of an assembly-line attitude, though craftsmanship needs are extreme.

TABLE 7 Changes to Live with for Extra Production Work (Shelves or Furniture)

Tool changes	Space needs	Cost range
Table saw: Must rise to a more accurate, durable model that holds accurate settings.	Variable, according to layout, but generally at least 16 feet long and 6 feet wide.	$750 (used) to $3000
Drill press: Floor model is preferable to benchtop model. Ryobi's drilling system is a good one, as are some of the radial head models.	Almost totally depends on work length. Access of 8 feet on each side is handy; may be over any other tool not apt to be in use at the same time.	About $450 up to about $750
Planer: For most light operations, a 12-inch portable planer does nicely, and cost tends to be very reasonable. Larger—15- and 20-inch and up—planers are exceptionally costly, but are essential for cabinet shops.	In-feed of at least 8 feet, with a similar-length out-feed. Overall width from 24 to 36 inches, depending on planer width.	Portable 12-inch planers run from $400 to about $650 and stationary 12-inch and wider planers start at about $1000 and scream up to about $4000.
Jointer: Here, a stationary 8-inch-long bed jointer is almost an essential to accurately planed wood edges. Accurately planed woods edges are essential to good wood joints. Period.	In-feed needs 6 feet, and out-feed the same. Overall width about 2 feet.	About $750 to about $1800, new
Sprayer: A good grade of HVLP finish sprayer is eventually going to be needed. I've recently used the Wagner commercial HVLP unit and am impressed. It is on the low end of the price scale.	Clean area where projects can be protected from dust. Varies according to cabinet or furniture size, or entire shop can be shut down for finishing.	$500 to about $1000 for a single-gun unit
Space: Whew. You figure this and you figure the cost.	Space needs never end and are always larger than what is available.	

For more extensive custom woodwork, it is difficult to determine what kind of shop space someone else is going to need to do a specific job, but any cabinetmaking shop should have the tools on hand.

Consider zoning once more, especially if your workshop is in or around your residence in a semiresidential area. You'll be using loud tools in or near your home, and your neighbors may show a distinct lack of fondness for the money-making whine of the saws, particularly if you're not careful about the hours when you work. If there are possible zoning problems, irritating the neighbors is a great way to bring the problems to a head.

Try to confine your working hours to more or less normal business hours: I've got a friend who cannot understand why people living next to homes he remodels consistently get upset when he starts at, or before, first light. Don didn't seem to understand that the shriek of a circular saw isn't music to the ears of the average citizen at 5 a.m.; but after more than 30 years in the business, he's working in towns more and is beginning to get the drift. Irritating people, in his case, loses customers; but for others it can create hassles that are just as bad.

Floor Space

As noted, floor space is going to be a need, and one that will differ from person to person. I'm something of a klutz, so I like to have wide aisles and plenty of stumbling room elsewhere. Other people work well in a tightly fitted shop and feel uncomfortable with wide-open spaces not filled with tools or materials or storage. Still, any opening between tools, and between tools and benches, needs to be at least 30 inches wide, and a table saw needs a 10-foot lead-in and about 2 feet less for materials run-out. (If you do a lot of long work, feed the table saw out over a workbench with its height adjusted to accept that run-out; then you can use the bench for other things when no one is using the saw.)

Overlap is an important consideration in saving space in shops where only a few people are working. You need to combine tool spaces to see what can profitably—from a standpoint of safety, work efficiency, and space efficiency—be overlapped and what cannot in the selection of tools needed to carry out your basic cabinetmaking functions.

Overlap the feeds for table saw and the planer in one direction, and all may work well; but do it in the other direction, and problems crop

Fully loaded table saw (all accessories) is imperative in a cabinetmaking operation. *(Delta)*

up. Basically, the planer may feed into the front of the table saw as an end result after at least 8 feet of run-out space; but it is probably best not set to run into the table saw from the rear. It can overlap its feed with the table saw from the sides, of course, needing only the blade run down plus rip fence removal to make the saw a handy helper. If you don't envision a need for a planer, then all that is unnecessary; but if you do much cabinetwork, sooner or later you're going to want a planer.

Planers and table saws do best when placed near the main entry door or doors, because they are the machines you're most likely to be feeding long stock into with frequency. Make it easy to get the stock to the tool, and you save time and energy, thus creating a shop where you rush less and don't get extra tired. Taking one's time and being well rested are both foundation stones of workshop safety.

To place a planer so that its outfeed table runs the finished stock over a table saw table, two things are required. First, the tables must be

Big planers need big room but do big work. *(Delta)*

close to the same height, with the table saw lower than the planer, if there is a difference. Second, you must set up so that the chips and dust from the planer do not foul the table saw. The planer turns out more waste in the form of chips and dust than any other tool, a fact that always needs to be considered.

Using planned in-feed and out-feed overlaps, you can design a shop with many tools and relatively small space needs.

A very good friend of mine, Bobby Weaver, has all his table saws feed onto workbenches built at exactly the same height as the saw tables (which are placed on stands Bobby designed and built). He has four table saws at present, an unlikely number for a small shop, but all get used.

No matter what size shop you create, there is a need for some space saving. Even if you could afford to construct an aircraft hanger, you'd eventually fill it. Make a list of the sizes of all your tools, and of the tools you plan to buy. Draw a diagram on ¼-inch graph paper close to scale, and make cardboard cutouts of the various tools and their placements. It's probably best, on major stationary tools such as the planer, table saw, jointer, and band saw, that you make the cutout of a size to include the in-feed and out-feed needs of the tool (run-in, run-out). And mark the height of the tool's table, if known.

Most band saws have considerably higher tables than do table saws, scroll saws, sanders, and so on. So band saws are readily and easily placed close to benches at the rear, and possibly one side, as long as there's sufficient working room for the operator.

Leave yourself enough free space to concentrate on doing your work safely, regardless of which tool you're placing. It pays off in both better work and fewer nicks, cuts, and missing digits. Don't waste space, though, for too much space between tools can be almost as bad as too little. You can run yourself ragged getting wood from one station to another in such cases.

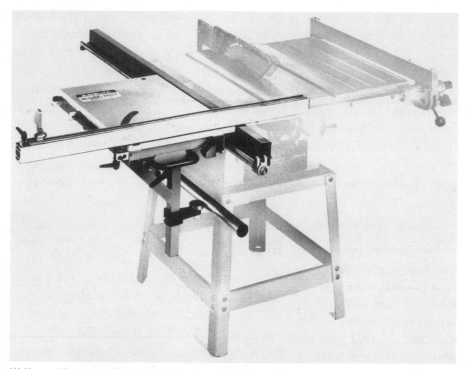

Sliding tables are an immense help in making accurate cuts. *(Delta)*

Try to envision your working habits, based on the projects you now build and those you intend to build. If you start with raw wood, you want a protected area to store the drying wood, and an even better-protected area to store wood that's finished drying. You want easy transfer from the first to the second, or to a vehicle if it is to be taken and kiln-dried after air drying. The seasoned-wood stack needs to be close to the door that leads directly to your planer. From there, you'll want to go to a radial arm saw (possibly), or a power miter saw, to cut the wood to length. After that, you need a jointer and, only then, a table saw. Beyond the table saw, or beside it, you want your drill press and band saw. If you do a lot of lathe work, the lathe will need to be near the table saw, but probably beyond a gluing bench where you might glue sections of countertops.

Finally, there's the finish area. High-volume low-pressure (HVLP) spraying equipment is more compact and, really, easier to use, wastes less material, and so on, but the good basic units start at about

$500 and the cost curve goes sharply upward after that. I find them invaluable, but I know many others who don't use them at all, never have, and don't miss them (or at least they think they don't miss them).

Your finish area may need some enclosure to protect it from dust and to protect other areas from overspray and fumes. If you enclose your area, use a spark-free fan to exhaust air to the outside.

In most small shops, spraying is moderate, and special venting may not be needed. Good shop ventilation is a requirement anyway, with the few windows installed placed so as to give good cross-ventilation.

If you build cabinets from scratch, some form of wood storage is needed; the size depends on the operation's size and style. Starting from raw wood creates greater storage needs and demands more tools (planer, jointer), but provides total control. Plywood also has storage needs, sometimes massive ones, and all wood must be protected from dampness. Evaluate the size and intent of your operation in light of these needs. Plan for wood storage, for the most part, outside your shop. Keeping a few hundred board-feet of lumber inside the shop is a good idea, as is keeping a dozen or so sheets of your most frequently used plywood there, along with a bin of scraps. But storing a large lumber supply indoors can create many problems. Many of the massive cabinet and furniture shops in this area don't store more than a few days' supply indoors. It eats space, for one thing, and space that has a floor and is cooled and heated can be awfully expensive.

Storage

As we've already noted, if you build cabinets from scratch, some form of wood storage is essential, with the size depending on the operation's size and style. Starting from raw wood creates greater storage needs and demands more tools (planer, jointer), but provides total control. Plywood also has storage needs, sometimes massive ones, and all wood must be protected from dampness. Evaluate the size and intent of your operation in light of these additional needs. Plan for wood storage, for the most part, outside your shop. Keeping a few hundred board-feet of lumber inside the shop is a good idea, as is keeping a dozen or so sheets of your most frequently used plywood immediately on hand, along with a bin of scraps. But storing a large lumber supply indoors can create many problems. Even the massive cabinet and furniture shops in this area don't store more than a few days' supply indoors. It eats space, for one thing, and

Accessories can add gusto, and usefulness, to stock cabinetry. This is Merillat's Amera line. *(Merillat)*

space that has a floor and is cooled and heated can be awfully expensive. Wood also tends to carry some passengers as it is brought in, including various insects, plus general woody debris such as grass stems and heads. If you plan to keep a lot of wood on hand, store only your more immediate needs in the shop. Construct a shed to keep the weather off other wood supplies; or use a tarp, or section of metal roof, to cover the pile. And make sure any outdoor pile of boards is stickered. That is, ³⁄₄-inch by ³⁄₄-inch stickers, cut the full width of the stack and *dry,* are placed every 2 feet along each layer. And keep the first layer up at least 6 inches off the ground.

Basically, wood storage for a supplementary woodworking business should be neither larger nor more complex than for your cabinetmaking work.

Starting the Small Business

Starting your own business, no matter the type, is a serious step that entails a lot of consequences, many good and some bad. One of the major reasons that businesses fail is undercapitalization at start-up. Another is the simple fact that the person or persons starting the business don't know enough about what they're doing to successfully set up, run, and operate that particular business.

Cabinetry from the Small Shop

Cabinetry businesses can be run from the home as successfully as any other business, but a decent-size shop usually contains room for an office. Starting a cabinetry business, no matter the emphasis, is a matter of tools and transport more than actual space in the office. So as little as 40 or 50 square feet will serve very well for the base for the business. (But let's not forget the shop, whether owned or rented, because considerable square feet are needed just to get rolling.) In the office, you need room for a desk, a file cabinet, a plans bucket, and a wastebasket. Add a phone line and a chair, and you're just about totally set to begin.

Of course, the list of tools needed to start a cabinetry business is long, and some are quite costly. There are several reasons to learn your business well enough to be working as a foreman cabinetmaker before starting on your own, and using your income to gather needed tools is one solid motive. A cabinetry business can actually use a greater amount of space, and any developing cabinetry business is going to need more space as time passes and the business succeeds. You'll have to hire office personnel eventually, just as you may eventually move on to hiring several shop helpers and cabinetmakers to complete your work. At that point, complexities increase, but naturally any cabinet-making is easier with at least one helper.

You may be the world's greatest cabinet designer and cabinetmaker, but until you get the niggling details of office work, plus the details of the work itself, down pat, you're going to be balanced on the cutting edge—of failure.

Personality

Personality—yours—plays a large role in the success or failure of your small business. The absolute first step in making sure you're going to get that business up and running, and keep it running reasonably happily, is for you to have the right type of personality. If you prefer to lay back and let things ride, if getting there is less than half the fun, and if doing a good job in less time than allotted doesn't seem like much, then forget the whole deal, for you're never going to make it in your own business. If petty details not only drive you to distraction, but also keep you from working, then you won't do the job well. If you are a true self-starter—and don't even think of kidding yourself here, for it's a complete waste of time and money to do so—then you've got a shot at making it, no matter what else is in the offing. If you enjoy starting projects, both simple and complex, and bringing them to a clean close, in good time and on budget, then you've got a great chance of succeeding in setting up a cabinetry business for yourself.

Workweeks and Suppliers

If you don't care who thinks what of how you make a living, you're starting ahead of the game, as you are if you believe the old adage about working for yourself: "I can pick any 60 hours out of the week and work then." It's not totally true, by the way. First, you've got to stay within the 9-to-5 world of the rest of the business world, at least in part. Then you have to be available evenings and at odd hours for estimating and talking to prospective customers. You are going to find yourself working more than 60 hours many weeks, often many weeks in a row, especially in the first 2 years. This cubic increase in working hours should slack off as time passes and success jumps in your lap, but you almost never get a chance to work a 40-hour week when you're self-employed. Either you're short of work and dreaming of working a full week, or you're working two-for-one. If you're going to succeed, you better not be working a short week with any frequency. The loss of spare time is one of the greatest sacrifices you make in working for yourself, if your business is going to have any chance to succeed.

Your Business as a Project

You also need to be organized, detail-oriented, and able to juggle several—often many—projects at the same time. The projects are not always going to be cabinetry projects. Your business itself is a project, and the most important one in many respects. With no business, no other projects get done, and you don't make a living. One problem with success of an intermediate nature in any small business is the temptation to go with what you've got in hand, do the work that's on hand, and let the future worry about itself. I've done it. Others have done it. It cost me many, many months of time and a lot of money. You must balance the marketing with the working. You cannot put in all your time designing, cutting, and gluing wood and then hanging the results. You absolutely must sell the next job, and the job after that, and so on, to be sure of being able to hold onto some of the profit from the job at hand. And you'll have to be an estimator with few peers in order to make a profit on the jobs you bid. The greatest work, safest workers, finest designs, and best craftspeople don't mean a thing unless you make money on your jobs. If you don't make money, you don't stay in business. Money may not be your primary motivation, though the things you can do with money usually are pretty good at getting you going in the morning; but money has to be a major prod, or you'll never work as hard as you're going to need to.

Starter Questions for Opening a Cabinetry Business

Possibly the most important qualifications for working any business are those that are almost totally unrelated to the particular business you start and run. In the case of a cabinetry business, there is a lot of out-of-office contact, but for any small business to be successful, certain qualities are essential to the owner/operator. The following questions will help you determine whether or not you fit the mold.

1. Are you energetic?

2. Are you willing to work longer hours, and work harder, than you ever have before?

3. Are you a self-starter?

4. Do you work well without someone instructing you in every move?

5. Do you work well under pressure?

6. Are you organized?

7. Are you a take-charge kind of person?

8. Are you well disciplined?

9. Are you willing to make the sacrifices you must make to succeed?

10. Do you assume all your business dealings will be with honest people?

11. Will you do the scut work, cleanup, and setup, without feeling angry?

12. Do you work best as a team member?

13. Do you procrastinate?

14. Do you think work must be fun?

15. Are you stern with people who owe you money?

16. Do you consider it necessary to meet or beat schedules?

17. Do you own sufficient cabinetry or cabinetmaking tools and equipment?

18. Do you know as much about cabinetry, or cabinetmaking, as you'll ever need?

19. Do you need a lot of people around?

20. Do you feel that strict follow-up procedures are a waste of time?

21. Do you need peer recognition and approval?

"Yes" answers to questions 1 to 9, 11, and 15 to 17 and "no" answers to questions 10, 12 to 14, and 18 to 21 should make you wonder why you're not already in business for yourself.

Honesty

You'll note our little questionnaire doesn't say anything about how you feel about your own honesty in business dealings. That's because I'm making the assumption that anyone going into business in today's climate realizes that, outside of politics, honesty really is the best policy. As much of a cliché as it is, it's true. Do good work. Do it when you say you will, for the price you name. That's honesty, and it brings repeat business. On the other hand, don't assume the other party in a deal is as honest as you are. Hope that person is, but never assume it. Use good, basic contracts and general credit investigation techniques, just as others do with you. Don't take a client's word that he, or she, has $25,000 to put into a large job unless you see some strong signs of major amounts of cash or an excellent credit rating. Use a local credit investigation outfit, and create clear and concise contracts that bind to payment at specific steps (depending on the job length, cost, and so on).

Where to Begin

The next step in making sure your new cabinetry business is a success is to decide, first, where you're going to begin and, second, how you're going to learn the working procedures that produce a successful list of clients, if you do not already have all the necessary skills. You cannot begin without most of the needed working skills, even with helpers to fill the voids, for it's too difficult to manage another person's work if you don't really know how that person is going to do the job. Small businesses require hands-on knowledge and lots of hands-on work. In the future, you may be able to skim the cream, but first you've got to learn to milk the cow, and that's hard work. As you work a job for someone else to gain knowledge for starting your own business, becoming an entrepreneur, you must also be saving as much as you can, after investing in the tools you're going to need to begin on your own. As a business owner, you will supply more tools, thus will need more tools. You may not be able to save as much as you need to start your own cabinetry business, but you should aim for as much as possible. The less money you borrow, the less interest you must pay on that money as you go through the almost always shaky first year of business.

Three Directions

Cabinetry businesses offer three main directions: You may opt for a basic cabinetry business, doing any kind of installation, with only absolutely needed changes made in your shop. That reduces the need for shop space a great deal, as you can usually order custom or stock cabinets. It also reduces the need for cabinetmaking skills, though you'll need countertop assembly space and knowledge that is really deep. A great many such start-ups then move on to custom cabinetry, changing the emphasis to upper midline and similar styles and doing more custom work in their own shops. That's two: cabinetry, middle-end and middle customization. A third option for a cabinetry business is total custom cabinetry, or producing your own custom cabinets and then installing them in the customers' homes. Here, you can work from a provided plan, or go even further and design your own. The designing of your own cabinets is the top of the line, and you can switch from there to furniture making, or other types of custom woodworking. The basic requirements are no different, but more tools are needed and space needs are usually greater.

Select Your Type of Cabinetry Business

This really is one of the first steps in a formal business plan. You must select the type, or types, of cabinetry you plan to do. Hit or miss works fine for a time. I've done that, as have thousands of others, and you can do a decent job of making a living that way, taking anything that pops up, as it pops up, with no directed sales effort at anything. Some people I know, and know of, make excellent livings this way. The problems stem from the fact that you may find yourself in over your head, in both knowledge and resources, if you haven't planned for larger jobs or for more complex, smaller jobs. Remodeling is an unlimited market, outdoing even the markets for new homes and additions. The variations in house styles, squareness, owners' desires, and so on seem to me to be much wider than those for owners of newer homes. It may also present you with far more complex problems, sometimes with problems that have nothing approaching a cut-and-dry solution, so you must improvise—which means you must have a very solid foundation in cabinetry methods of almost all kinds.

Gaining Skills

Getting ready to make this first major decision follows a process of learning, and that learning process is critical to success in your chosen area. Cabinetry and cabinetmaking are skills that are both formally and informally taught in this country, with nothing really approaching the apprenticeship system in European countries.

I learned a lot as time passed, and I learned slowly and well when I listened. One of the curses of youth is an inability to listen for any length of time, so my cabinetry education was spotty in some places at the outset. More years ago than I like to think about, in a long-defunct company (Katonah Altar Company, Katonah, New York), my boss thought I was an inept idiot, I'm sure, though he kept explaining and demonstrating, even when he knew I wasn't listening. Some soaked in. It was not enough, but that's my fault.

Thus, a lot of what you have learned and will learn comes down to a desire to listen to someone who already has learned, in whatever manner, to do the work you wish to do. If cabinetmaking is strong in your area, the best way to pick up the knowledge that will allow you to begin on your own is to job-hunt with the top cabinetmakers in your area.

Top Cabinetmakers

Top cabinetmakers aren't difficult to identify. Top cabinetmakers have a reputation to uphold, and they work hard at doing so. This often leads to an added reputation for being slow as well as good. They also may not be favorites of materials supply companies that aren't dedicated to providing top-grade materials. They do not easily adapt to new-fangled ideas, but they do adapt to the better ideas.

New Technology

New technology is not always the greatest way to go. It does pay to stand back and learn how it fits into the jobs you're doing before you jump up and buy one, or more, of the newer tools. Let someone else, someone with more money or less sense, assume the risk of proving the new tools. When the time comes that you realize the new technology is going to save you many, many dollars in labor, and may also give better quality on the job, buy.

Thinking Is the Greatest Skill

Much changes, but much stays the same. While using newer tools that slightly reduce the overall skills needed to produce an accurate cut helps save time on any job, the cabinetmaker still has to know what goes where, and what cuts are needed to make the whats fit the wheres. The actual physical skills take less time to develop, but the thinking skill required to produce a good home or for cabinet installation or other project remains the mainstay of any business you wish to own. If you can't figure it out properly, you can't do it properly, nor can you tell anyone else how to do it properly, so you're going to lose money. All standard residential construction is simply based on the right angle, and that's true as far as it goes, which is most of the way. But you still must decide which part of that angle to use.

What we've got here is an Aristotelian notion of knowing yourself and knowing your skills. Get those two ideas lined up, and you can almost certainly make a go of a small cabinetry business.

Not Easy, Not Simple

No one is going to say getting your business off the ground is going to be either easy or simple. You'll work harder than you've ever worked before, and you'll work longer hours. You'll face more frustrations, in the form of local rules and regulations based on national rules and regulations and on the tax code. The Internal Revenue Service (IRS) often seems specifically designed to drive the small business person completely insane, but the easiest way to deal with that outfit is to keep detailed records of everything you buy that's deductible. (Check with your accountant to see what is and isn't deductible. Your accountant needs to be prodded to give *you* the edge when there is doubt. Too many accountants give the edge to the IRS, though there is nothing illegal about having a difference of opinion with the IRS; so if you and the accountant read the regulation differently, within reason, give some thought to taking the deduction anyway. If you've got a legitimate difference of opinion, you'll have to pay back taxes, but there should be no fines or penalties, though there will be interest at the current rate. That being said, it is also wise to remember that the IRS is monolithic, answers to no one, and can dog your tracks for life, making life more than fairly miserable. So make sure it's a deduction that's worth it if there is a chance of a fight.) Keep track of every single *penny* of income. If you miss a deduction, you lose a few bucks. If you neglect to declare income, you are breaking the law.

Insurance Is Another Essential

Your next biggest thought-provoker is insurance. Although some businesses can afford to skimp on insurance, to skimp on the essentials of liability insurance, disability (workers' compensation), and health insurance, when or as you hire good people, is so risky as to be beyond sensible consideration. Check further details of insurance later on, and add to that a series of talks with your insurance agent or broker, aiming at getting the best possible coverage in all areas for the lowest possible cost. It is still going to be expensive, but there are too many things that could happen on any job and could immediately cost a fortune, small or otherwise, and that you will have to pay if you're not insured.

Start-Up Needs

You'll need to decide how large you plan to start. From that decision stems a further decision on the amount of cash needed for start-up, and reality must be faced here. You write out your business plan, decide how much of your current assets you are going to use to get things going, and then talk to your banker to see what the bank will let you do. You may, or may not, have to modify plans to start larger than most of us do. But a simple, clear business plan with goals for the short term and the long run is close to an imperative in such a situation, combined with a resume of your accomplishments to date.

With knowledge of your craft, whichever version of cabinetry business you try, and reasonable knowledge of your own strengths and limits, you can turn the challenge of operating your own cabinetmaking business into a successful reality.

Zoning and Other Possible Problems

More and more businesses today are being based at home. This can work for a cabinetry business, too, but that depends in large part on the start-up size and the home neighborhood. Office jobs move home, and much else rides the wave of technological trends and uses the phone cable to transmit data and the computer to compile and arrange that data in a change to almost Middle Ages–style working habits. In fact, not since the Hanseatic guilds in Europe in the early Middle Ages have we seen such strong home-based industry.

Small-Shop Business Needs

Small-scale cabinetry for long years has been as much a home-based business as not, though the office in the home and the shop in the home are entirely different parts. But there is more and more to consider each year, and a shop office may well be the best in the long run, if it is at all affordable. You are going to need a shop, and the office must be close by that shop, but could be in the home with the shop out back—if such locations are legal where you live and work and you have separate space for the office. That is to say, the Internal Revenue Service (IRS) isn't going to accept the locus of a business as the edge of your dining room, not if you expect to be able to deduct anything worthwhile in the way of business expenses. Little things such as electricity, heat, and office space become nondeductible items if you do not have a separate space for them.

New IRS Rules

Helped by turn-of-the-millennium technology, small businesses can thrive. Technology can help a cabinetry or cabinetmaking business stay alive and healthy. It cannot run the business for you, but it is amazing how much easier it is to keep books and run an office with even the most basic computer setup than it is without. But, as always, there is a learning curve that is sometimes fairly steep.

Because we're talking here of start-up business, basically done on the proverbial shoestring, I'm not going into massive additions, huge computer purchases, and the laying out of cash for a huge van to cart around lots of gear. My assumption is one or two people, three at most who want to start a business and are capable of hiring one or more helpers to do the work. In the meantime, get the office set up. Setups covered here work well for any small business, whether based in or around the home, or in other quarters.

Setting up an office anywhere is a relatively simple job, especially if you figure no one is going to see the office. After a few months, that will change as customers find your place of business. Stick a desk in a corner, and put a file cabinet next to the desk and a wastebasket on the other side. Flop a phone onto the desktop, and start making calls.

And sometimes you can work it that way. There are a few buts.

First, you'll be driving a vehicle that may or may not fit into your neighborhood scheme of what's parked overnight. Where I live, anything short of an Abrams tank is unremarkable. In fancier suburban areas, simply parking a dirty pickup with a magnetic sign on a door is enough to bring down the ire of the entire community. Given my temperament, I'd move out in about 30 seconds, but that may not be an option. Check around and see if your pickup, van, or other work vehicle, and sometimes some overnight parking of other gear, might create problems for the neighbors. You may or may not care about possible problems, but in some areas, there may be a legal problem with inappropriate vehicles. In any case, losing energy over such petty stuff detracts from business, and from life in general. Check it out first, and talk to those who might object the loudest.

You're probably not in a problem area, for working as a start-up cabinetmaker won't usually pay enough to allow you to live or even loiter long beyond cabinet installation in fancy subdivisions. But keep the thought in mind for later in life, too.

Regardless of location, your shop must offer sufficient parking to make loading and unloading of moderately large vehicles reasonably easy. If the neighborhood will tolerate a small shop but goes bananas over truck traffic, give some thought to another locale.

Zoning

Local regulations preclude some businesses. Because you will be operating a small shop, there may be no problem, other than the possible one above, but check first. Check with the city or county clerk, or whomever they direct you to.

Cabinetmaking, with a wood shop on premises, may well be prohibited in some areas for a variety of reasons—possible release of volatile organic compounds (VOCs), noise, traffic. This is something to consider when you select which type of cabinetry you wish to pursue and where.

Local Provisions—and Restrictions—Vary Widely

The fact is, it is essential to check local regulations, unless your next-door neighbor is doing what you want to do and can answer all your questions, whether that neighbor is a grocery store or your residential neighbor.

In most cases, start the search for information with your county (or town, or city) clerk, who will at least be able to give you a lead as to where information is available, if it is not on hand. You'll find an array of differing requirements, depending on where you live. Many locales allow "customary, or usual, occupations," which can mean various things in various communities.

You must check to be sure, for customary usage varies a lot.

Signs May Be Nixed

And if no other prohibitions exist, you may find yourself prohibited from putting up a sign, or able to put up only a sign that fits specific limits and without lights. Generally, that's not a problem: Garish, strongly lighted signs, even if permitted, probably don't bring a cabinetmaker much business.

You may also find that vehicle parking is limited. Most of us drive trucks, usually pickups, and use vans of varying sizes to cart the finished products to the customer. These are probably commercially licensed for insurance purposes, if not to satisfy the local laws. If we then add a sign, or several signs, to the truck, that may bend the laws past the point where we can get by with parking legally by the shop.

If you're working within such limits, check to see what you can do to get permission to park the vehicle, without signs. If you feel truck signs are an essential part of your business, most office supply stores can offer strip-off signs that apply in seconds and come off just as quickly.

Otherwise, limits on the types of businesses operated from residential areas usually apply to businesses that increase noise, traffic, or both beyond acceptable levels, justifying the *customary home occupation* definition. Most places define a home business use as one that is an accessory use, thus either in an outbuilding, such as a garage, or in a minor percentage of the house. The percentage of the house is never to exceed 30, and often is limited to 20 percent, though sometimes it is set at 25 percent. In many locales, changes to the exterior of the home are restricted as well.

Types of Home Businesses Sometimes Classified as Customary

We're all familiar with the doctor who has an office in the home, though these days the familiarity is more likely traced to old movies and TV shows than reality. Lawyers present another customary home business category, as do artists, writers, music and dance teachers, and similar occupations. Architects and others are also readily allowable, though Chicago has an ordinance that prohibits the use of electronic equipment in the pursuit of business, which effectively, and asininely, outlaws word processing, accounting from the home, writing of most kinds today, desktop publishing, and a host of other businesses. It might also interfere with operating any kind of an office from the home, as telephone answering machines and fax machines are also classified as electronic devices, though the telephone itself doesn't seem to be.

Not far from here (about 180 miles), Fairfax, Virginia, disallows antique shops and funeral homes as customary home business (or any other way), but specifically allows a cabinetmaking shop. The first two are among the quietest businesses there are, though certainly many people feel uneasy living near a funeral home. Antique shops, at least busy ones, can increase traffic in any area, but so does a cabinetmaking shop, and the cabinetry shop is also going to raise the neighborhood noise level considerably. The regulation does illustrate current problems with consistency in the home-based business laws. If you are considering basing a business on your residential property, whether in the home or in an outlying building, check carefully.

Archaic Laws

It is clear why so many people operate home-based businesses without referring to the local authorities. Many artists, writers, architects, word processors, and other people are simply setting up and going about their business without benefit of local legal blessing. The laws are often archaic and do such things as limit the use of outside help: In some areas, you and your family—even if there are a dozen of you—may operate a business from your home; but if you hire a single non-family member, you'll be in violation and can be shut down.

Some areas allow most everything, assuming you get the right permits, while others prohibit even the most usual uses, yet allow others: Some localities specifically prohibit physicians, dentists, lawyers, and clergy from operating from their homes.

Land-Use Guidance Systems

Bedford County, Virginia, had a land-use guidance system (LUGS) not long ago. It was first considered the up-and-coming way of dealing with zoning without having real zoning or the problems zoning sometimes creates. LUGS gives the say to community members, after a business has met specific neighborhood guidelines. Unfortunately, as has been the case in Bedford over the years, the community can be pretty arbitrary in what it refuses and accepts, so that a business that isn't apt to disturb anyone may be refused because four or five neighbors don't like the shape of the business owner's mustache, hairdo, or how she wears her slacks—that's never the expressed reason, of course. It's always something supposedly more meaty, more relevant, though that often turns out to be something that's already been covered and accepted with a similar business nearby. Petty jealousies can play a part and too often a large part with LUGS.

Preliminary Check for Problems

The best advice possible in such a situation is to first do your checking in a very low-key manner. Check with the various zoning and real estate offices as if you are in the very first stages of thinking about starting a business, no matter how close to actually getting set up you are. You might start by saying, "I'm beginning to think of setting up a small building business, using my home lot for a shop building and office. I'd like to check and see how regulations affect that."

If there is any resistance, or if the zoning or LUGS regulations appear weighted too heavily against, hire a local lawyer familiar with the laws and regulations. The investment in the attorney's time and effort will be worthwhile if you are saved the hassle of having your unfolding business shut down for a violation within a month, when you don't know, and can't find out, that such a regulation exists.

If the system, whether standard zoning or LUGS, is set up so that there are specific formal steps, starting with an application and a fee and leading to neighborhood meeting for approval, then you can be sure you need an attorney. An attorney will help close all the loops, making sure you don't miss doing something, or do something you shouldn't, which may create problems later.

There is no way to predict what kind of restrictions you might find in your neighborhood: Note the Fairfax, Virginia, allowance of cabinetmaking. That is great for our purposes, but not so great for other businesses that are less likely to create neighborhood changes but that are disallowed.

Fees, Taxes, and General Local Beatings

Local taxation can also be a nuisance. Here, I pay an annual property tax on office equipment; and if you think it doesn't annoy me, though it's seldom more than $40, then you think wrong. Pay for the property. Pay sales tax on the property after paying income and six other taxes on the money you use to buy it. Then, each and every year, pay a further tax on already taxed property. It should be illegal, but instead is common.

It's up to you to check for applicable taxes on all work, and all personnel, whether sales, income, fee-based, or whatever. Gross business receipts taxes are generally a good argument for moving the business to another headquarters, as are similar taxes applied to stock carried over on January 1 (or whatever date) of a new year.

This state also has a personal property tax, something I did not know when I moved here many years ago. If I'd known, I'd probably not have come, for I find it one of the most irritating forms of taxation. Pay for something once, pay taxes on the money you use to pay for that item, and then pay sales tax overall. Then pay a property tax, again and again and longer. You never own your property. You rent it from the local government.

Licenses

Local regulations may require a false-name license, too. This is also known as *doing business as,* and an alias tax. Your town, county, or other political district probably will, too. The cost here is only $10, and, in fact, it is only an absolute requirement if you use a business name that doesn't include your personal name, though most banks insist on having one in hand before they'll open a business checking account for you. Banks are like little petty governments, making all sorts of regulations you cannot break, or else the bank will fine you or refuse to do business with you. As a cabinetmaker, you *must* have banks, so you can't fight too much over unimportant things.

Various other licenses may be needed. Gross business receipts may be taxed. All this must be checked before you open your doors. You must also check on sales tax liabilities as well as on state and city income tax liabilities, as noted earlier.

Recordkeeping Needs

One tax liability that will create some immediate tax problems is a home office, if it is not impeccably set up. Obviously, you need to keep all receipts, and keep them for many years (7 years is a minimum). But also make sure that your office space, if it is to be deducted, is kept separate from the rest of the house. The IRS requires a door that can be closed, though in real life it may only be closed should the IRS auditor come to your home to check your square footage of office space and similar items listed on Schedule C, or whatever this year's required form will be. That automatically knocks your 50 square foot office on the head if you plan to deduct any of the space.

I no longer bother. I've currently got a closed-off space totaling more than 440 square feet, plus 60 square feet of other room (not counting a storage building that is quite large, with the center third totally used for business). I do not bother to deduct the square footage of the space because it is such a hassle to figure proportions of this and that for a small percentage of actual space used in the house. I'd be entitled to deduct a portion—not as much as I use, but a portion—of my utilities, heat, and some other stuff, but for the $250 or so it might save in taxes, it creates about $5000 in aggravation because of the way the rules are written (not to mention the time consumed in figuring all this out, and more time spent fighting with the IRS about it). I also don't recommend my way of doing this, as it can create its own problems, as well as cause you to lose some money each year. I do deduct most things for my shop space, because I am totally out of business without it.

IRS Publications

What you do is up to you, but read all IRS publications on home offices before you decide to deduct or not deduct, and before you decide how to deduct things. Actually, it's a good idea to review all the forms and literature that the IRS has on small-business taxation. As a start, you're always going to need Form 1040. If you've ever been able to use one of the simpler versions, you can now forget it. You'll also have to have Schedule C, Profit or Loss from Business. Schedule SE becomes imperative, as you are now paying self-employment tax at double your former rate with no employer to kick in one-half. You may need Form 2106, Employee Business Expenses, and you'll certainly need 1040-ES, for your own estimated taxes. For more costly tools and equipment, you need Form 4562, Depreciation and Amortization. There is currently a Form 8829, Expenses for the Business Use of

Home. Of greater immediate interest, call the IRS and ask for Publication 334, Tax Guide for Small Business; Publication 505, Tax Withholding and Estimated Tax; Publication 917, Business Use of a Car; and Publication 525, Taxable and Nontaxable Income.

You'll need more, but those will get you safely started. I'd like to suggest asking an IRS employee what else you might need, but every single time I've asked the IRS for information beyond the printed matter, the IRS has been wrong. Every time. The kicker here is the simple fact that wrong is wrong in the eyes of the IRS, so if you owe more because of the IRS employee's mistake, you still have to pay, and the fines are not relaxed. I suggest you ask your accountant for his or her interpretation when you have questions.

Office Equipment

Office equipment today is a lot different than it used to be, but the jobs it must do are remarkably similar: The primary job of office equipment is to let you keep track of jobs, and of expenses and income, so you're able to keep work flowing smoothly, while income also moves in a way that lets you keep your profits.

The basic office, as noted earlier, requires little more than a filing cabinet, desk, trash basket, telephone, and plans bucket. A chair is handy, but I know people who work from a drafting table and use a stool, so neither desk nor chair is essential. Your desires are paramount here, assuming costs are rational and you really can work out of such a setup. I know other people who place far greater importance on a top coffee maker (me, for one) than on the desk. Some use a table and chair. It's up to you whether you get fancy or go light. The need is for a place to work efficiently in the hours you spend there, making sure the work you conduct in the shop goes smoothly and profitably. The office is the foundation of the successful cabinetry business you're attempting to set up. Efficiency is far more important than any fanciness, but I can solidly state, it's easier to operate out of an office that is both efficient and at least modestly attractive. Ugly doesn't work as well.

While you may have few customers filing through your office, especially in these early days, you have to spend considerable time there yourself. From hard experience, I can promise that you'll work better in an attractive, as neat as possible, clean office. I'm terrible about office housekeeping—filing is my worst job, though I do keep the trash cans emptied and the floor swept. I don't do well with placing books back on shelves and general filing as above. But space can solve those problems.

Spend a little time and energy to get your office to where it suits you. You'll not regret the extra time or the few extra dollars that an attractive office costs.

The Office Computer

Managing of finances can be carried out with a calculator or with a computer, including job estimating, billing, and posting receipts and bills. The main aim is accuracy, but ease and efficiency are also good.

At the start of your business, you may need only a calculator and a Dome book (or some less self-explanatory register) to keep records in. But as time and your recording complexity increase, you're going to wish for other ways.

Regardless of the reason, or rationale, today's bookkeeping needs are strict, and strictly a pain for those of us who do not care for working with columns of figures. I can remember many years ago discovering how much ease the calculator added to doing taxes. I still do taxes with a calculator, but I no longer use the palm-sized machine to do my books.

Kicking in the Computer

Accounts are far more readily and easily kept on a desktop or portable computer: For many good accounting programs, all you do is set up once, enter your material as desired, and get the reports you need to do taxes, figure investments, and stand the stock market on its ear with the mouse cursor gliding around the screen.

Currently, after many years of using Moneycounts, I've begun using Quicken to do my books, and I'm finding it easy to use, except for some minor hassles. (If you can find a computer program other than a game that doesn't present at least minor hassles, drop me a note.) I dropped Moneycounts because I got tired of annual upgrades that added to program complexity without offering me any real benefits. You only need a certain number of financial reports, no matter what, and changing their order and numbers doesn't change that fact, though it can confuse you. Those reports are covered as part of your business plan. They are also part of the everyday chore of doing business, and accurate records need to be kept to get accurate, useful reports. The computer makes keeping those records, and generating the useful reports you get from them, much easier.

And no, you don't need to be a computer expert to deal with these tools. That's right, *tools.* A computer may be a bit more complex than your screwdriver, but it's still nothing but a tool. And it really isn't all that complex. Computers are a pile of superspeedy on/off switches. The complexity comes in arranging the switches, sticking that monitor (think of it as a small TV set) on top, and figuring out a way to arrange the data you put in the machine. As we go along, I'll give you a recommendation for what I think is a reasonable entry-level IBM-compatible personal computer (PC).

Setting Up the Computer

First, I need to explain my biases: I don't use Apple computer products, though I have friends who do. I don't believe Apple's Macintosh truly matches PC-compatible computers for general business work, though for illustration and publications graphics work they still retain a *very* slight edge. The business program edge held by the PC is much wider, and you will be working with those almost exclusively, so the Macintosh is not a good buy. Newer machines will change all that, but so far on a bang-for-your-buck basis, PCs still dominate.

Entry-Level Computer

For basic business work, today's entry-level machine is known as a Pentium computer; actual Pentium chips are made by Intel, but AMD and Cyrix make competing chips that are just as effective. The basic entry-level chip, as this book is written, is a 400-MHz Pentium Celeron (I have no idea at all where these goofy names come from). For those of you not at all familiar with computers, the Pentium is the just about the latest in a long line of central processing unit (CPU) microchips made by Intel. The top model currently is the Pentium III. There will be more by the time this book is out. The Pentium chip, along with what's called a motherboard, is the brain of the computer. All else is extension, though essential to giving the computer any utility.

The case contains the CPU, motherboard, power supply (converts 115 volts to lower-voltage direct current), disk drives, and a couple of lights, plus accessory boards that plug into the motherboard and let you plug into them with your keyboard, mouse, and monitor. On the assumption you've never heard any computer talk, the keyboard is the part of the computer most like a typewriter. The monitor is the screen, basically a very high-definition TV set, without receiving qualities. The mouse is another device that lets you put information into the computer; thus, in computerese, it is an input device (as your keyboard is).

Disk Drives

Disk drives are information storage devices. The main disk is called a fixed, or hard, disk which is permanently mounted in the case and accepts a precise amount of data for storage. Currently, floppy disk drives come in one size, $3^1/_2$-inch-diameter, and are mounted permanently in the case; but they *do not have disks in them* unless you specifically insert those disks. Floppy drives are used to install programs to the hard disk, which then holds the programs and all files created by, or relating to, those programs. The hard disk is the big booby, and needs special thought when you buy the computer.

CD-ROMs are another type of disk drive. In most machines, CDs are not rewritable, so are used to install larger programs, ones that require many floppy disks. There are more complexities than it is worth going into here, but for a start-up machine, you need a read-only 36X- or 40X CD-ROM drive.

I suggest you buy a packaged computer system from a reputable vendor, either locally or by mail. (If you are totally lacking in computer experience, you may be better off buying locally. But a friend of mine in Boston bought his first PC from a mail-order outfit—actually, the manufacturer—and is delighted with the entire setup, though Stan is safely classified as a nontechnical type—the day he put up a single shelf on a wall he had a major celebration.)

Selecting Programs

When selecting the package, you get some installed programs, but probably not exactly what you want. One of the top word processors or word processing suites must be included, as must Windows (at this time, Windows 98 is your best bet, but Windows 2000 is newly released and is a replacement for Windows NT).

The hardware must include a Pentium Celeron 400-MHz or faster processor (faster means a higher number than 400), or a Pentium III 550-MHz. (This is an upward step that currently costs several hundred dollars, so don't go for it without examining your computer needs very carefully. It is not a step up that any business, or computer, beginner really needs.) You want a hard disk that holds at least 12 gigabytes (GB) of data (simply put, that's one heckuva lot of pages).

You need at least a 3½-inch floppy disk drive.

Keyboards and Mouses

Check out some keyboards at a local office supply store or your computer dealers to see what kind feels best to you. You'll be entering a lot of material with this device, so get one that's comfortable, and get a wrist support (to help prevent carpal tunnel syndrome) for both the keyboard and mouse. Do the same checking with the mouse. I don't note a lot of difference in mid-price-range mouses, so it seems you will be best served by going with the one that feels best to you. My current mouse is new and cost all of $8 from a local computer store that buys them in bulk. I bought two.

Monitors

My advice for beginners on monitors goes against the grain of some computer experts: You are going to be offered a 14- or 15-inch monitor (skimpy diagonal measurement, just as with your TV set) with what is called VGA sharpness. Go for at least a sharpness of 1024 lines, with a 0.28 pitch, and get at least a 17-inch monitor. If you can at all afford it, spring for the 19-inch monitor. It will cost about $150 to $250 more than a comparable 17-inch monitor, but the relief it gives your eyes after a long day of crunching numbers or checking out proposals is well worth the extra cost. The best bet is the 19-inch monitor with 1600 $\times$ 1200 lines with a 0.22 or 0.25 pitch. Smaller pitches are acceptable, as are more lines (up to 2200 is currently available, but more costly), but larger pitches are best avoided in any of these three monitor sizes.

The various cables needed are part of the package. A modem will be part of the package, and you may or may not wish to use it: I'm connected to America OnLine, and have also, from time to time, hooked up to other services. Sometimes they're helpful; usually they're interesting. Microsoft runs a small business center that is filled with interesting and useful information.

The computer is a useful tool for modern business, almost an essential one in figuring totals on estimates, assembling larger estimates into neat packages, and for almost any use where a typewriter and adding machine at one time served well enough. Now, you actually find the computer becoming more essential; it is wise to pay some attention. Local community colleges offer courses in basic microcomputer use, and in the use of many of the more popular programs. Take a good look at local offerings. You save an amazing amount of time by having someone teach you shortcuts, instead of trying to take a voyage of self-discovery (which is the time-wasting way I've done most of my computer learning).

And don't spend too much time worrying about being on the cutting edge, up to date, and ready to blast off in all ways with your computer. Almost anything sold these days is obsolete within 6 months, but that doesn't mean it is no longer useful. Get a mid-range (400-MHz or more) computer, and it should serve you faithfully for half a decade, and possibly more.

Insurance Protection for the Small Business

Here is your gateway to the world of protection. That gateway starts at insurance, because if things go wrong and you're uninsured, you no longer have a reason to operate a business, for you'll have nothing to operate it with.

Liability Insurance

It's really hard to say which of your insurance coverages is most important, but general liability coverage in the shop, and on the job site, and for possible later problems when the job is completed, is one of the most important insurance coverages any cabinetry business can have. You *must* have the guidance of a good agent or broker when laying out your liability insurance plan—and the agent or broker can guide you to and through your other needs as a small-business person. My statements here can only be of the most general type, and there is a great deal of variation from state to state that needs to be worked out.

Unfortunately, insurance is extremely important to the small cabinetry shop. Even if you are and plan to remain a one-person shop, you need lots of insurance. It sometimes tends to be the largest single expense in the business, past gathering tools and similar needs.

As always, any tradesperson is best insured with a safe job site, the details of which are not extensively covered in this book: If you're good enough to even think about starting your own cabinetry business, you're good enough to keep a clean shop with safe conditions. But even safe shops sometimes see problems, and injuries occur. These problems can be ruinous if you're not insured. Spreading the risk through insurance is essential, and giving it plenty of thought and time ensures that you get the best available for the most reasonable cost available.

You must have liability insurance, and it's a difficult subject to cover, for the determination of the amount needed is a personal thing, depending on what you feel you have, or may soon have, at risk. Liability comes in many forms, with many aims: Medical coverage liability for premises; damage to property liability; contractor's general protective liability; personal-injury liability; fire liability; collapse caused by excavation liability; damage to underground utilities liability; failure to build to specifications liability. Generally, you want the kind of policy that doesn't specifically exclude much, and one that doesn't exclude *unless* the exclusion is specifically stated. Thus, a long, long talk, or even several long talks, with a very good agent is an essential ingredient to success here.

When you fix on the amount of coverage, take into consideration the possible high costs of legal work when settling claims is required. We all like to think that we're the best and that problems aren't going to arise. Suppose you're doing your first job, a large kitchen in a moderate-size house. Your helpers are inept in hanging, and the result is a section of cabinets, filled with expensive dishes, dropping on an occupant! (If you think that kind of thing doesn't happen, I advise you to check around.) The occupant is not permanently damaged, but suffers a few broken bones.

The result is clearly your fault. The court or a jury will find against you, if the insurance company lets the case get close to court. Bet on a settlement, with lawyers taking a fair chunk of any such award.

Recently, my home neighborhood has seen a lawsuit from a person who dove into the shallow end of a swimming pool and permanently damaged himself. He is suing his pal, the pool owner, the National Association for Pool Manufacturers, the pool manufacturer, and the pool installer for a grand total of $20 million. Why? This home pool had no "Do Not Dive" sign at the shallow end!

Now, not being a lawyer, I say the guy hasn't got a case. I've swum in dozens, maybe hundreds, of home swimming pools, and not one has ever had such a sign. I have never seen such a sign except at rocky areas at some public beaches and in public pools. Nor is it particularly needed for someone who notices that there is a diving board on the opposite end of the pool, and who knows, as does just about everyone over the age of 3 these days, that home pools have shallow ends opposite their deep ends.

But lawyers are a different breed, and one saw a case here, though there is some indication that sobriety might have been missing, at least in local rumor. Sometimes this sort of case seems to depend more on sympathy from the jury than on factual evidence and real law. But in the past such sympathy has paid off in lots of money to plaintiffs.

I have no idea what the resolution will be, but the case was not thrown out of court immediately, as some of us had expected. While you sympathize with someone who is paralyzed, you do not sympathize to the extent that he's supposed to be able to hugely punish others for *his own mistake!* We live in that kind of world, though, so give some thought to just what making a mistake that creates property damage and injury can do to your world. And then think again about the limits of insurance versus the strain, especially early on, of paying for said insurance. It is not the case quoted above that's important, but the attitude it shows: There are far too many people who believe that everything negative that occurs to them is the fault of society, or of some other person in that society. We provide support for that attitude with the litigiousness our legal community thrives on. Thus, insurance is ever more necessary, and is ever more likely to be even more necessary in the future.

Umbrella policies are also available, and they may serve to up your limits considerably at relatively low cost. A great deal depends on the way the umbrella is structured: It is essential to talk over these policies, too, with your agent.

Automobile Insurance

You will need business auto insurance because if your agent discovers you're using that old pickup for work, he'll be forced to yank your family policy coverage. Check the various classifications to make sure you can get into the lowest-priced one possible. Categories are usually commercial, retail, and service, and rates vary considerably. Also aim at getting coverage for nonowned vehicles, if you have any helpers who might be running errands in their own vehicles. This does not cover the driver of said vehicle. It covers you against problems arising should that driver be in an accident while doing your business.

Equipment Theft

This is a great coverage, if you can get it at reasonable cost. Cost changes and varies so much from area to area that there is no way to predict what your costs might be. It basically covers you against quick theft from vehicles and the job site. Try to locate a policy that pays replacement value, rather than cost, though to be honest, tools today do not seem to be increasing in cost as fast as the rest of our expenses. They're expensive, but not nearly as expensive as such items as insurance, materials, and quality labor.

Workers' Compensation

Workers' compensation insurance is meant to provide support for injured workers, no matter who was at fault when the injury occurred, and it also takes up the slack where you're concerned. As an employer, you must, by law, provide medical care, compensation for lost wages, and rehabilitation for any worker injured on any of your projects. With good luck, good safety practices, and good workers, this probably isn't going to amount to more than a few twists, turns, cuts, and sprains most years. But those years where it does add up, it can mount quickly and considerably. You can think of self-insuring, but it is not sensible, practical, or, in many states, legal. There are varieties of accessibility to workers' compensation policies, with some states providing mandatory coverage, while in other states you have to go to private insurers. Some states let you do either.

   This is another insurance where costs are often high. Medical and rehabilitation sophistication adds to prices, as does general inflation, but the biggest factor seems to be unsafe shop or job sites combined with workers more willing to sue the boss. Workers' compensation may cost as much as 20 percent of salaries, and in a few instances may go higher.

You can help reduce costs over time by keeping your shop and job sites safe, and by quickly treating all workers who are hurt, no matter how small the injury. Where possible, shop around for policies, too, for some trades are not as expensive as others, and you may be able to rerate one or two of your employees. As you will note, sometimes you have no option on the coverage and must take what is offered.

The safe work site is most important because it becomes a part of your overall record as a cabinetmaker, and there is usually a modifier for Workers' Compensation and similar insurance coverage based on past claims. Sometimes a company will pay dividends when work safety over a period of time is so good that excess premium money has been paid. Thus, a safe work site and a good safety record can lead to a stable cost, or even a reduced one. A lousy work safety record, which naturally comes from an unsafe work site, will see premiums rise, often very high.

Keeping a safe work site has other priorities than just saving on insurance costs: Your workers are going to be happier if on-the-job injuries are few and far between, even though they must put forth a bit more effort to keep things on the site clean and safe. In general, safe shops make for faster completion of jobs—you don't lose time waiting for cuts to be cleaned and dressed or for ambulances to transport workers to the hospital. You also don't lose time filling out insurance claims forms and any forms that various safety organizations may require. Overall, as well as cost savings in insurance, general humanity is helped, and your wasted time on any job is heavily reduced.

Planning and Financing Your Cabinetmaking Business

Writing a business plan is an excellent method of determining where you stand before you start a business, and looking at where you wish to go, with greater clarity than most people have when they begin a cabinetry business. It's really easy enough to load the old pickup with tools, run a small advertisement, and start to pick up small cabinetry jobs, without doing much else in the way of planning. But you may find yourself a decade down the road doing the same sorts of small jobs, making the same sort of small-job money, scrabbling to get a vehicle that's more reliable and tools that will provide a reasonable job of work without extra effort, while not allowing you to live in a more comfortable style. Cabinetry work doesn't require the same sort of cheery projections put out by other small-business start-ups, for building and remodeling houses is something that most financial sources are up on. Your banker probably has a better idea than you of just how much you can make on any one job, and how tightly controlled the current construction market will force you to be. So a small cabinetry business plan need not have the extended financial and other statements common to overdone entrepreneurial states for less familiar businesses, regardless of whether those other businesses are actually more complex and difficult to run than a cabinetry business.

A simple business plan can allow you to start just that small, make it more certain that you'll succeed at whatever your aims are, and give you a clearer idea of just what those goals are, so you can work more effectively toward reaching them. It need not be complex, and certainly it is best written in simple language. It effectively provides direction where none may exist, and it can simplify your growth as a cabinetmaker—and can truly simplify the chances you have of getting major loans from banks or other investors who might otherwise figure your chances at succeeding are too poor because your planning is not sufficient.

A business plan demands that you be prepared, for you need to think out each and every aspect of the business, providing a blueprint to use in keeping track of where you've been, where you are, and where you're going, while it also serves to attract financial backing. Customers looking at your plan may find themselves more attracted to specific details, and more likely to contract with you than with a cabinetmaker who just flies at things haphazardly (showing customers a business plan is an option).

Your business plan needs to cover ways to finance, operate, and profit from your small cabinetry business. It is also something that is not to be knocked out on a slow Saturday, but it probably won't take more than a couple of weekends to write.

You won't need some aspects of the standard business plan, and you may well never need them. I'm not going to deal with mergers and acquisitions, or the planning for such actions, because in all honesty, I haven't got a clue and don't want one. Most cabinetmakers I know would shudder at the very idea.

You start by assessing the workability of your idea. Go over the information in the preceding chapters and check to see if you've got what a small cabinetry business, no matter its emphasis, requires.

You may need to decide on whether a proprietorship, a partnership, or some form of corporation will work best for your situation. Each has advantages and disadvantages, some of which may be overwhelming.

The most common cabinetmaker's company is the sole proprietorship, and that's probably where you will want to start. Its primary plus is its relative simplicity, with all income, above expenses, classified as personal income and taxed as such. (If you've never worked for yourself before, wait until you get to pay the double whack for Social Security—it's a thriller with no employer to pick up the second half.) At the same time, bookkeeping is simple, and taxation presents no complexities. Nor, in most cases, does licensing, for you usually just need a local fictitious-name license. Depending on your desires, you may or may not be bonded, but you *must* be fully insured when you work as a sole proprietor (that's business-speak for sole owner). Not only do you get to keep all the money you have left after expenses, but also you get to be out in the breeze for any and all possible liabilities. Thus, insurance is imperative, for a single lawsuit can readily bankrupt a small company.

Partnerships are next, and are not my favorite way of doing things, though for some people, and for me on some jobs, they work great. You don't want to grab a partner because he or she is able to do things you can't. If you can't do all the jobs, including keeping the books of a small business, don't begin such a business. If a partner, though, offers complementary skills, whether in office or business management or in the field, then the partnership has a chance of succeeding. According to all my sources, it is wise to never take on a partner for a job that you can hire to be done. In other words, don't become partners with a great bookkeeper. Hire him or her. If you've got a lifelong pal who is a so-so cabinetmaker but who has more money to put in the company than you do, walk on by. That kind of situation can only create major problems. Money for start-up can be a problem, but a skilled cabinetmaker can open up a business with less cash on hand than is the case for many businesses—assuming that cabinetmaker gets into an upward-moving construction market. Simply put, billing for accounts goes out on 30-, 60-, and 90-day notices, and each billing should be enough to cover expenses plus, for that preceding month. Smaller jobs may be worked with the client paying up front a portion of the fee that is large enough to cover the cost of materials.

You still need a sufficient amount of cash on hand to provide you with a stable, livable income for at least 6 months: Starting with less is foolish, for it may take 2 or 3 months of trying before you get your first job of a size to produce decent income. Then you've got to get materials, get a helper working with you, and cover other expenses. Better yet, begin with enough to see you through at least 1 full year.

The size of your working nest egg is determined by the size of your start-up ambitions. If you can live and work on $25,000 for 6 months, all the better; but in many cases more is needed, because helpers have to be paid, tools bought, and materials paid for up front if you haven't arranged credit with local suppliers; and you can bet something is going to go wrong to delay payment on a critically needed completion check before the 6 months are up. The more cash or credit you have up front, before opening day, the better off you will be.

Finances are something I can't even begin to predict for your small cabinetry business because I have no idea whether you wish to start tiny and grow to small or medium, start small and grow to medium or large, or start medium and grow huge (or stay the same size). I strongly suggest you start at least as low on that scale as you can stand. Most of us are better adapted to starting tiny and going from there, because that means selecting primarily one- and two-person jobs, doing them economically and well, and moving on to larger jobs as the cash on hand builds up from the completed, smaller jobs. Generally, people with smaller jobs are less likely to ask for half a dozen estimates before deciding to start. There is also less of a delay in starting many such jobs, so you can turn out a round dozen bathrooms and small kitchens before you even get the contract for a bank lobby.

You must make accurate estimates and do a lot of checking, particularly early in your business life, to make certain you get the best price for the best quality. If you're setting up your business in an area in which you've living and have worked, you'll already have a major jump on this need, especially if you've worked your way up to lead carpenter for a good company (as I strongly recommend you do before starting your own cabinetry business).

Evaluate your business resources. List your assets and skills. What sort of tools do you have? Check your tool needs. Evaluate your skills in the area in which you intend to begin. If you're doing pick-up cabinetry with no real specialization, make sure you have the skills needed for some fairly wide variations in job requirements.

You can make a paper copy of your business plan, and you should, for it's hard to hold all the fine points in mind over time. Do the job neatly, on regular typing paper. (It needs to be typed if you plan to show anyone else any part of the plan. And really it needs to be typed even if only you are reading it, because typing is not only neater, but also easier on the eyes and less apt to be misread.) Make sure your spelling is correct. (If you're not using a computer with a word processor and spell checker, use a desktop dictionary—you *can* find words you can't spell, but the search is harder.) Sentences should be short and simple. You're not striving for stylistic greatness here, but for clarity. While it's not always right, many people who may need to look at your business plan are going to come away with their main impression of you derived from the plan. So make it neat and as accurate as you can, so their main impression will be that of a careful worker with well-thought-out ideas.

Business Plan Needs

Business plans are tools, and they should be regarded as such. Too many of us look at them in near terror, because we have the idea that only an MBA from a fancy school could figure one out. If a business plan is laid out so that only an MBA can understand it, especially for a cabinetmaker, then it is not doing its work. It is a challenge to work up a good business plan, but that plan, and its updates over the years, will help you get a faster start and maintain a faster pace with less confusion. Basically, a business plan sets out your short- and long-term goals for your business and then describes how you're going to get there. In the process, it provides many of the following categories of information, both for your use and for the use of others. (The financial information is a part of the business plan that those who decide to show the plan to clients should *not* show.)

1. Your business
   a. Description of the business (as full as you can make it)
   b. Marketing methods
   c. Local competition
   d. Operating procedures
2. Financial information
   a. Loan applications
   b. Capital equipment list
   c. Balance sheet
   d. Breakeven analysis
   e. Income projections
   f. Cash flow projections

3. Supporting documents

   a. Tax returns

   b. Personal financial statement

   c. Copies of your license and similar legal documents

   d. Copy of your resume

   e. Copies of letters of acceptance from suppliers who will proffer credit

Filling in the Blanks

The above outline simply needs to be expanded along reasonable lines. It is my adaptation of the outline model used by the Small Business Administration (SBA), so it certainly covers most of what lending institutions are going to ask you for. Beyond that, the first section can be separated and used as a marketing tool, if done well.

When you do 1a, describing your business, cover the types of cabinetmaking you will be doing and the reasons why your skills make you especially suited to those types. Look hard and long at the skills you can describe that set you apart from the crowd of cabinetmakers already out there. The longer you think about this part, the more effective it's going to be. If you have a flair for work that provides extra beauty or space or both at lower cost, say so. If you have a flair for providing the lowest possible cost for luxurious surroundings, again, say so. Don't be shy. This is where your business marketing program begins. Next up, at 1b, bring material from the chapter on marketing to this point, and explain where you're going to advertise and why you expect that to be enough. For 1c, cover the local competition, and explain how you believe you will do a different job than they can, or will. Do *not* denigrate the competition. It is not profitable in the long run, for it just builds hard feelings that sometimes never heal. Cover your business operating procedures (estimates, how your labor is hired and paid, etc.). Among other things that the operating procedures section covers is management. If you're in a working partnership, this is the time to say so; explain how different people are going to be doing their different jobs, as partners. If you are a corporation, list the various officers and explain what they'll be doing as the company starts and grows. If, as is most likely, you're going to start as a sole proprietorship, then describe your own business management experience (you have some if you've worked up to foreman, or even if you've only led an installation crew), and how you will apply that to providing the best possible work for your clients. This is the place where you really need to brag about all your abilities, though, not just those of management. Your skills are diverse—they have to be to run any small business today. Cover them all, from managing your time effectively to keeping the books, to reading the plans for the addition or remodeling job or the homes you're going to build, to cutting accurate miters and getting them tightly nailed. At the outset, as the owner, all the jobs will be yours, and you'll need to emphasize the fact that you are more than capable of doing them all. If you have any doubts, this is not the place to air them. If you have trouble coming up with enough material, discuss your experience. Think over jobs you've worked on and go over the things that happened on those jobs, from day-to-day routines to any emergencies you may have helped handle.

It is at this point that you discuss your weaknesses. You're bound to have some, and they're probably less important than you think; but unless you're the world's greatest cabinetmaker at your current age, with no room for improvement, there may be some craft skills that you still need to perfect. Or you may have problems running a computer. Whatever those weaknesses are, describe them, and then explain how you plan to reduce or eliminate them. If computer programs drive you nuts, you might consider a local community college. In a short time, I'll be out the door for a 30-mile drive so I can take my first class in my new word processing program. This is a program I've used daily for a few months, but I'm not getting as much benefit as I might from it, so I'm going to take an advanced course. It's a bargain, as are most community college and adult education courses. Use them to overcome your weaknesses.

You will find, and quickly, that having your own business demands that you educate yourself, often on subjects and in ways you'd have considered way out of your interest range in your free-roving days. If you succeed in business, education is going to be a major support on the way to that success. The day of the rough-and-ready type with gloves in one pocket and a tape measure in another, who could gruffly insist he'd do the "best job you can get—just leave it to me" is long gone, if it ever really existed. I've seen a number of these "leave it to me" types, and I've never seen one who didn't sooner or later get into major binds because of poor workmanship, or imperfect understanding of the job (sometimes on both the cabinetmaker's and buyer's sides). Lawsuits and legal hassles do not contribute anything positive to your workday, so do things properly as well as legally. (The two aren't always the same, but doing a proper job in a cheerful manner, as agreed in the clear contract, at the price specified almost always knocks legal problems in the head.)

Thus, emphasize your strengths, but describe (shortly) your weaknesses, and discuss ways you'll correct those weaknesses.

For section 2, fill out a loan application for part a, and list your capital equipment for part b. A balance sheet may not be necessary, but it is helpful. And a breakeven analysis is probably a time waster for a cabinetmaker, but income and cash flow projections will be a big help for a new business, *if those projections are reasonably accurate.* Do not daydream of castles and kings, but make practical, commonsense estimates of the work you expect to obtain, and how you expect to be paid for that work.

In section 3, your supporting documents are going to start light. You won't have a business tax return until you've been in business over a full fiscal year (simplify your accounting and make your fiscal year the same as the calendar year), or at least into the period that ends on December 31 of the year, and gets to the IRS about April 15 of the following year. Your personal financial statement will pair up with your loan applications. Copies of your license and all legal documents should always be readily available to clients, bankers, insurance agents, and others. A copy of your resume is going to be needed to prove you've got the experience to do the job once you get rolling. And, finally, if you've managed to talk to local warehouses and various suppliers (lumberyards, building supply houses, etc.) and got them to extend you 30-day net credit, find out whether those who have agreed will write a short note to that effect. It simplifies your applications for other credit if you are considered creditworthy already by a number of organizations who will be providing you with materials nearly daily.

Other Business Plan Uses

Business plans require more material to be useful over a long course of time, and for jobs other than getting an immediate line of credit.

(Unless you go in for spec home building, you aren't going to need a huge line of credit to get started—or you shouldn't. If you're so broke you have to have all your start-up money from loans, it's a good idea to rethink the entire start-up.) A business plan lets you set up a timetable for yourself, evaluate your physical and fiscal resources, and helps you set prices.

A business plan also makes it easier to set realistic goals and make practical plans for the future. In fact, a properly done business plan forces you to examine yourself, your resources, and your general plans strongly enough that you can more readily and accurately assess the feasibility of your plan to go into business for yourself. In some cases, a business plan may convince you to back off for the time being. In others, it may convince you to start larger than you'd first planned. Setting tough goals makes sense. Setting unattainable goals just results in long-term problems. Setting goals that are too easy also creates problems with slow business growth. You'll learn to judge the edges more accurately as time goes on, and your projections will grow more accurate. But tough-minded care with a business plan makes original projections fairly workable and useful things to have.

Business Planning Help

If you wish, you can find books in the library to help you with business plan forms, most of which are easily and quickly adaptable to small cabinetmaker businesses. There are also a few business plan programs for home computers, but I've not had time to evaluate any of these to a great extent, so hesitate to recommend any.

The six planners I've managed to get information on can help you summarize your business goals and project profits, but you still have to do an amazing amount of work. Again, you are not going to be preparing a plan that takes a totally unknown business through idea formation to finally incorporation and acceptance on the New York Stock Exchange, so some steps are definitely less important.

In most cases with such businesses, the lender is the final audience considered, and if the plan is not complete down to the last nickel of projected revenue/cost/profit, you're out of the game.

These half-dozen plans (note the 800 numbers following most plans—Business Blueprint requires a toll call) offer slightly different angles on the standard business plan, and my reading evaluation, as against a solid grounding in using the programs, may be something you'll want to supplement with your own research. In fact, I suggest you do.

I've listed these in alphabetical order and given my opinion of each—gently, though, because I have not tested any of the products.

BizPlan Cabinetmaker is from Jian Tools for Sales, and it has been around since 1988. The disk builds Word or WordPerfect files, and data are presented in Lotus 1-2-3 spreadsheet format. The plan is difficult to use and seems aimed at getting too much down on paper in the fear that too little will be presented. This one takes a lot of care in use—you can, of course, delete anything that doesn't work for your new business. It is priced at $129, and the toll-free number is (800) 346-5426.

Business Blueprint is from an outfit called Spreadware, and it works with the Excel 4.0 and later databases. The package is pricey at $149 for a single disk, but there are a dozen charts of text and financial information—far more, really, than a start-up cabinetmaker is going to need. The manual lacks information on the business planning process, the templates are all for Macintosh format, and you will almost certainly find yourself doing more to get lots of printed-out data that aren't going to mean much. Contact Spreadware at (619) 347-2365.

Business Plan Toolkit covers everything from financial analysis to market forecasting, with 46 topics to cover in its text writer area and a large set of spreadsheet templates. The manual is said to be extremely complete, with a great deal of good advice on preparing a business plan. This one may be a lot more than start-up cabinetmakers need, especially at $149.95, but it does a good and complete job more easily than most others. The telephone number is (800) 229-7526.

First Step Business Plan comes to you from the National Business Association in Dallas. This one can't give you a balance sheet (which you almost certainly don't need anyway), but it does provide modules (free to members) for profit or loss statement; cash flow analysis; and a review. This association, with its $5.00 per disk programs, may well be a consideration for some start-up cabinetmakers. The telephone number is (800) 456-0440.

PlanMaker is made by PowerSolutions for Business. The interface is menu-driven and easy to use, with three sample business plans provided as guides. It doesn't include a spreadsheet component, but does have financial tables to help you organize projected operating expenses, amortization of loans, and an overall balance sheet. You pull the figures in from elsewhere, which is okay for cabinetmakers. The split-screen text editor has the instructions in the top half, and the information is entered, by you, in the bottom half. Price is $129; call (800) 955-3337.

Plan Write for Windows includes a full-featured word processor (with a spell checker), a business plan outliner (where has this feature been?!), and a spreadsheet module. You even get a business term glossary, and much advice for various elements of your plan. The final report is good-looking and meets generally defined criteria as closely as, and usually more closely than, the reports of the preceding programs. From business Resource Software, Plan Write costs $129.95; call (800) 423-1228.

A quick and important note on computer programs: Call the above numbers to get further information about the products. Never order a computer program, unless it's an early release of an upgrade, directly from the company that produces it (there are a few exceptions to this rule, but ignoring them won't cost you any money). Order from a local computer retailer or discount office supply house, and you can save up to 45 percent off the retail price, an appreciable difference.

Finishing Up Nonprogram Business Plans

Before any business plan is useful, it must be put in finished form. There are probably as many finished forms as there are business plans, but if you're not using a computerized setup, you'll need to type or word-process your final form. I suggest you beg, borrow, or otherwise jump on a word processor and computer. Pay a few bucks to have it done, if you aren't going to computerize. The range of visual expression in the form of fonts is much greater, and a good laser printer produces pages that look typeset. Typeset appearance is very desirable when you're trying to impress somebody, which is the real use of any business plan, whether the version for clients or the version for bankers and others.

While writing and typing, make sure the finished form is in polished language. By polished, I don't mean bright, shiny, show-off language, but good, workmanlike English, with no misspellings, clumsy sentences, and grammatical errors. If you've got a problem writing that sort of English, hire someone to help. There are many high school teachers out there who welcome the chance to make a few dollars extra doing something they're good at. Not all are, so check carefully.

Remember that long and fancy words don't make much in the way of a favorable impression—they irritate you when other people use them, don't they? Well, then, don't do the same to others. Avoid as much jargon as possible, and leave the buzzwords for others—*power, empowerment, the T* (or any other letter) *word,* and such as among the worst at the moment.

Even if you are good at writing, get a proofreader. I'm always amazed at the number of typographical and other errors I find in my material when I come back to it a week or two later.

Cover Page

A cover page is essential, and because this is a limited edition business plan, I suggest enclosing the entire plan in one of those clear plastic covers as well. The clear cover keeps the business plan from showing premature wear and tear. You still must have a cover page. Go down about $1/2$ inch and type in your first line: Confidential Business Plan. Next, space down about one-third of a page, and type the business name, the address, and, two spaces under those, your telephone and fax numbers. Down three spaces, type your own name. If you wish, you may type a copy number in the lower right corner of the cover page, and it is a good idea to repeat Confidential Business Plan about $1^1/2$ inches from the bottom.

I'm not sure the copy number helps anything but your ability to keep track of the limited number of copies you should turn out. Make a note in a ledger of some sort as to where each copy went and when that person received it.

Table of Contents

Next up, you want a table of contents, which is simply a restatement of your outline, listing the various parts of the business plan. With a word processor, you can readily come back and list page numbers of each section.

Executive Summary

The executive summary is just what it says, a quick summary of the entire document, with the kernel of each section included in short,

Confidential Business Plan

Plane & Saw Cabinetmakers
1415 Beaufort Lane
Sacre Bleu, WH 77793

Telephone: (123) 555-7733
Fax: (123) 555-7737

Owner: **Slim Whittler**

Confidential Business Plan

Copy Number _____

simple words. It is most effective if you can keep the entire summary to no more than one page. The executive summary serves as a pointer, telling the reader whether he or she wants to go on and read in greater detail what you are going to do, and how you plan to do it. It saves time for the person reviewing your business plan (this is also the only part of the business plan I didn't include in the SBA outline).

Business Description

Provide details of the business, from the specifics of ownership—proprietorship, partnership, or corporation—to the number of employees (if any), and on. Tell the reason why the business exists and how it functions, and discuss your strengths, as you see them. If you have a business philosophy, discuss it. Back off technical details of cabinetmaking, but describe clearly the types of cabinetry you do or expect to do shortly. Present a clear word picture of where you stand now, and where you intend and expect to stand in 1 year's time, 2 years' time, and 5 years' time. Do not use the word *hope* to describe your intentions. Write about the present range of your business and its direction. You can discuss your operating hours, but those are seldom important with your kind of work, as you know: You work until the work is done or the light is gone on some days.

Your Financial Plan

This is the point where you get your loan applications together, list capital equipment, make up your balance sheet, and do the income and cash flow projections. If you're aiming to use loans at any point during the first couple of years of your business, this is the most important part of your business plan. Even if you don't directly expect to be using loans, you surely will be needing credit from suppliers, so the importance remains.

The balance sheet is essentially the statement of your financial condition, and I've enclosed a sample here to help you put together an accurate one.

You can use this as you need it, lengthening it if needed and abbreviating where that's a help (though it's unlikely). Start the report with a short narrative of your financial condition (as covered later in the financial chapter). Really, for the start-up cabinetmaker, you are now

---

BALANCE SHEET

ASSETS	Amount	LIABILITIES	Amount
Cash (bank accounts)		Accounts payable	
Accounts receivable		Short-term notes	
Prepaid expenses		Amount due on long-term notes	
Short-term investments		Interest payable	
		Taxes payable	
		Payroll	
Total current assets			
Long-term investments		**Total current liabilities**	
Land			
Buildings (cost)		**Owner's equity** (assets minus liabilities)	
Less depreciation			
Net value		**TOTAL LIABILITIES AND EQUITY**	
Furniture/fixtures			
Less depreciation			
Net value			
Vehicles (cost)			
Less depreciation			
Net value			
Total fixed assets			
Other assets			
TOTAL ASSETS		Date:	

faced with the only part of your financial statement that can be accurate. Everything else is a projection and thus a guess—an educated guess, but still a guess. You may need to state sources of funds here, if you've gotten funds from anywhere other than your own savings. The breakeven analysis is nothing more than an estimate of when that will happen, and it is based on the following two projections: profit or loss, and cash flow.

Sources of Funds

You may wish to add a listing of the sources of your funds as further information for bankers and others who may well want to know what they're going to be lending money against. This is a tool that can be overused, but a reasonable listing of where you got each item or group of items (you surely don't want to list how you paid for each screwdriver and screw), and how much and where you got capitalization for other things, is sometimes as helpful as the overall balance sheet that doesn't include such details.

You can easily see that anyone noting your savings and investments, and the amount you've already invested in getting ready to start your business, must be impressed with your seriousness. Of course,

SOURCES OF FUNDS		
ASSETS	**AMOUNT (COST)**	**SOURCE**
Cash	$31,000.00	Savings
Investments	$9,000.00	Mutual funds
Accounts receivable	$11,000.00	On-going work
Materials and supplies	$5,000.00	On hand
Vehicle	$11,000.00 ($18,250.00)	Bank note
Furniture and fixtures	$850.00	Currently owned
Office equipment	$3,300.00	Currently owned
Tools	$8,340.00	Currently owned

your figures won't agree with those shown on the preceding page, which are totally imaginary, but getting as close as possible to the higher ends here doesn't hurt at all when you go looking for more money for the business. However, in all honesty, this report indicates someone who is in pretty good shape as far as getting through the start-up period of a small business. Having $20,000 or $30,000 or even more above these amounts on hand as a contingency fund against unforeseen costs is a good idea, and serves as a reason to approach an investor with your business plan.

Profit or Loss (Income) Projections

For the business plan, you will want profit or loss projections for a full year, done on a monthly basis, with quarterlies done for the second year and annual projections for the third, fourth, and fifth years. If you start in mid-year, project all months remaining in the year on a monthly basis, and go on for the next year on a monthly basis. A properly done profit or loss projection can be the basis for much other solid information, and can give you a very good idea of just how your business is going to fare.

At the outset, all profit or loss statements are projections, but as you progress in your business, profit or loss quarterly reports provide a superb tracking device for expenses and income, allowing you to figure out where you stand quite quickly. Most computerized bookkeeping programs provide such reports routinely, accurately, and far more quickly and easily than if you had to make them up yourself.

Cash Flow Projections

Cash flow projections can tell you when, as well as if, you can afford some new tool, device, or trip to the outer banks of Hawaii. They are similar to profit or loss statements, but differ in several ways, including the way finances are handled. Your cash flow projections, and later your reports, concern *only* the cash on hand. If it hasn't arrived yet, then it doesn't exist as far as the cash flow reports are concerned; of course, with projections, you must guesstimate how much you're going to receive at any one time. For decades, I've missed on this one vital aspect of a business: I tend too often to believe those fateful words, "The check is in the mail" or "is being cut" or the other bushwah of describing modern payment methods; so my cash flow projections are always far too optimistic. Don't let that happen to you, as it

PROFIT AND LOSS STATEMENT

	January	February	March	First Quarter Totals
Installation revenue				
Cabinetry revenue				
Total revenue				
Cost of goods sold				
Materials and supplies				
Outside labor				
Miscellaneous				
Total cost of goods sold				
GROSS PROFIT				
Expenses				
Wages/salaries				
Payroll deductions				
Advertising				
Vehicle				
Depreciation				
Insurance				
Interest paid				
Legal and professional fees				
Office expenses				
Rent or lease				
Repairs and maintenance				
Supplies				

	January	February	March	First Quarter Totals
Permits and licenses				
Tools				
Travel and entertainment				
Utilities				
Telephone				
Postage				
Dues and publications				
Printing and copying				
Trash pickup				
Miscellaneous				
TOTAL EXPENSES				
NET PROFIT (LOSS)				

PROFIT AND LOSS STATEMENT (*Continued*)

makes for great problems in dealing with your expenses. If anything, learn my lesson early, and learn it well: Take a pessimistic approach to projecting cash flow. Sooner or later, you'll find your lack of faith is justified, because the client you like best is going to be the one with a check that resembles a Ping-Pong ball, or a payment procedure that works along these lines: "You'll have the check within 10 days." You call 15 days later, in mid-month, and are told: "The check will be mailed the 30th." There will almost certainly be at least one further delay, and this will be the person who tells you that your financial problems are not of his causing. Which is true enough, because if you have any sense, you fold your tent on his (or her) work until the money is in hand. My experience tells me that you're apt to give this person extra slack because of a charming personality, but that, sooner or later,

CASH FLOW REPORT

	January	February	March	First Quarter Totals
Cash on hand (1st of month)				
Cash receipts				
Collected receivables				
Other				
Total cash receipts				
Cash expenditures				
Gross wages				
Taxes				
Materials				
Supplies (office, etc.)				
Subcontracts				
Repairs and maintenance				
Advertising				
Vehicle				
Travel and entertainment				
Accounting				
Legal				
Rent				
Telephone				
Utilities				
Printing and copying				

CASH FLOW REPORT (*Continued*)

	January	February	March	First Quarter Totals
Postage				
Shipping (UPS, FedEx, etc.)				
Insurance				
Dues and publications				
Miscellaneous				
Other				
Subtotal				
Capital expenditures				
TOTAL CASH PAID				
CASH POSITION				

you will also face a problem with nonpayment from the same person. Thus, if you have repeat jobs brought in by a chronic late payer, I'd suggest getting all money up front, but under no circumstances get less than 50 percent.

Check Activity before You Start

Check your locality for activity in the construction trades. If a lot of houses are going up, it's easier to get into residential cabinetmaking. Decide what you want and go for it, starting as small as necessary, but as large as practical. You may actually find yourself in one of these infrequent periods when smaller cabinetmakers have a lot more trouble than larger cabinetmakers. Usually, a good, small job or cabinetmaker will do well no matter what the economy is like; but on occasion, periods of time will pass when the person specializing in small cabinetry jobs is getting very little work, because most work in an area is going into bigger construction and large remodeling jobs. And if things are really bad, established cabinetmakers of many sizes may start looking and bidding for the small jobs, just to keep as many of their experienced crews working and together so that they're ready to go when the next upswing arrives.

Examine your location carefully for the most advantageous start-up size, and then you can begin your financial planning, making decisions on where to get the money you know you'll need to live, and pay business bills, for at least 6 months (1 year is better).

Financing Your Start-Up

Decide how you will finance your company. You should have been proving financial responsibility by saving as much money as you could, but the decisions on financing a company start-up, no matter how small, need a lot of thought. Running it out of your own savings until you can start paying yourself a salary is obviously the best way. Running a company start-up from savings means you risk only your own money, not family money, and not money belonging to other people, as is the case when you borrow money to meet short-term goals. At the same time, you must consider practicality, the desirable size of a start-up cabinetry business in your area.

If you feel the time is absolutely perfect for a start-up as a small cabinetmaker, and if you can prove that to yourself and others with real figures, then you want to consider beginning the process of forming your own company even though you're short of cash. Staying short of cash is always a good way to get in a great deal of trouble, but there are avenues open to most responsible people who want to start out bigger than as a one-truck, one-saw, one-project cabinetmaker. If your business plan presents all the facts straightforwardly, and in a neat, precise, and concise manner, then you've got a great shot at getting going with borrowed money.

Business plans help prove your seriousness, and good overall planning, in fact, is essential to success in any business.

Marketing the Small Business

Marketing is nothing more than spreading the word that you're in the cabinetry business and good at what you do. It needs to be treated as a simple providing of information, and not some arcane secret, but there are many methods of doing a more efficient job than most small cabinetry businesses use. Start-up marketing is all-important, but a continuation, a line of notices, is even better. Often, it's possible to get free space and attention in local newspapers and in the trade press, but just as often, you must buy some advertising space to set things up for getting as much work as you desire.

Word-of-mouth advertising works well for many cabinetmakers, and this kind of unpaid advertising is probably the best there is. But the Yellow Pages remains a good place to put your name out, and to let prospective customers know of your specialties, if any. In fact, on any realistic basis of cost versus results, your ad is probably going to do better in the Yellow Pages than in any other single advertising spot. That's not to say you don't need to advertise elsewhere. Advertising frequency and placement is a judgment that you must make over time.

As you can afford it, join the local Chamber of Commerce, Lions, Kiwanis, and similar social-business organizations. You get involved in benefiting your community while also making invaluable contacts for furthering your business.

Certainly, right after you get a simple letterhead, some envelopes, and some business cards printed up, your Yellow Pages ad comes next.

All this means you need to have determined what you're going to call your business and what your specialties are going to be, if any.

Thus, start slowly and build hard. Start listing yourself as a custom cabinetmaker and renovator. Take small jobs if you must—cash flow can be extremely important, but we're covering marketing here—so that you can do excellent work and have something to show the customer with a larger renovation or remodeling job. Take photographs as you go along, particularly of difficult, unusual, or unique aspects of the job and, of course, of the completed work. You can't send prospective clients to each and every completed job to check your work, so good color prints are great as stand-ins for reality.

Building a Portfolio

Build your portfolio just as an artist, photographer, or writer builds a portfolio. You are also a craftsman or craftswoman and have a right to take pride in your artistry and craft. Make sure you have the right! Your portfolio can be enclosed in plastic pages for protection. I suggest getting prints at least 5 inches by 7 inches, with some preference for 8-inch by 10-inch. This does two things. First, it allows the prospective client to see more details; second, it looks more professional. There is a place called 20th Century Photo Products & Accessories [call (800) 767-0777 for a catalog] that sells photo pages in minimum quantities of 50 with pages to fit both 5-inch by 7-inch and 8-inch by 10-inch, and has a low price on a three-ring binder with each 50 pages. You may be able to find the same stuff locally. I never have.

With an ad in the Yellow Pages, your White Pages listing, and the office supplies (letterhead, etc.), you're ready to start a small publicity campaign.

Electronic Media

You'll note that throughout I say nothing about electronic media. That's not a bias. I simply do not see what TV or radio does for small cabinetmakers. It's not going to hurt anything to send your releases to local TV and radio stations. The odds are strongly against their even being read; but should you get a TV news pop, you get a lot of coverage. TV ads are generally too costly for business people just starting out—and for many others.

Writing the Press Release

There are a number of things to remember when you write press releases. First, short is better than long. Second, send the release to the business editor of your local newspaper or newspapers (call the editorial office and ask for the editor's name). Third, make sure there are no misspellings of important materials. Fourth, provide a call-back name and number (your own will do fine). Fifth, don't inundate the editor with material every time you get what seems like a profitable idea.

I didn't include in the five things to remember above one that is all important: The release *must* absolutely be typewritten, and it must be neatly done, with no XXXXs, and no visible White-Out used. If you can't do that yourself, get a friend with a computer and a word processing program to do the release for you. You actually should have a computer yourself and spend the necessary hours getting the basic competence needed with word processing to do such work.

You want to announce that you're starting your business and, at the same time, list any specialties. Your next release might be 6 months later, on completion of your largest job to date, giving a short history of your company and its major accomplishments. (If you specialize in remodeling kitchens, state that as an accomplishment, for example, not the number of jobs done in the past year. If any kitchen has been done with an award-winning designer, architect, or person of special interest, state that. Whatever your specialty, cover the same things.)

For your first release, consider the following form:

ANYTOWN, ANY STATE—Jan XYZ today announced the formation of Cabinetry Unlimited, a proprietorship aimed at providing remodeling through custom cabinetry in old and new homes. The company goal is to provide increased storage, beauty, and general enjoyment of the living areas by using a variety of custom cabinets, in conjunction with other forms of remodeling, so that clients attain an unusual, often unique, appearance and use flow with more moderate costs.

"Cabinetry Unlimited," said Ms. XYZ, "provides wide-spectrum services for those who wish to get the greatest effect from their remodel-

ing, without spending sums of money that might startle the Pentagon."
Preliminary consultations are free.

Cabinetry Unlimited, or CU, is a fully licensed, bondable cabinetry
firm, and is located at 999 Nutly Drive, Anytown. The phone number is
(703) 555-5555.

In essence, you want to use the above form, but change the content
to fit your start-up company. If you're forming a partnership, say so. If
it's a subchapter S corporation, that may be of interest to some, though
probably not. You may well wish to provide a more complete list of
your services and skills. The above is only a minimal suggestion. I do
hope you note it comprises fewer than 130 words, which is a good
goal, though up to 200 words is probably permissible. Your aim is to
get it read and get it published, and making it two or three pages long
makes either goal less likely. Editors are busy and don't want to have
to read a lot of extraneous verbiage, which they'd also only cut before
sending the release to be typeset.

The release sheet may be laid out like this, something easily done if
you are using a computer and less easily done if you're using a typewriter (and only one of many reasons a computer is nearly an essential
item for your business at some very early point):

First, space down about six spaces. Then write **"NEWS RELEASE"**
three times, each time on a new line, at the left margin. Indent the second line for a more pleasing visual effect. Space down three spaces.
Next write "Contact: (Your name)," "Telephone: (Your number)," and
"Address: (Your address)," each on a new line and aligned at the right
margin. Now write the approximate word count of the release on the
next line at the left margin (hint: count the words on each of five lines,
get a per-line average, and multiply that number by the number of
lines). Space down two spaces and write, at the left margin, "DATE-
LINE 9/9/00 (ANYTOWN, ANYSTATE)—" and follow that with the
body of the release. Two spaces below the end of the body and centered, write either "—30—" or "—End—". The latter is probably best.

A second type of news release announces the awarding of a contract for a major job, or the completion of a job that is, in some way,
major enough to get your local newspaper to cover it. With small cabinetmakers, it's difficult to determine what is and isn't off-beat or otherwise strong enough to get mentioned. One thing is certain. If you're
completing a small (but large to you) job weekly, do *not* announce each

and every completion to the newspaper. The paper won't print it, but someone there just might remember you as a pest who sends in a lot of below-average trivia—and there is a great deal that comes to newspapers from professional public relations outfits—who therefore doesn't deserve a reading. Put your name on something and get the editor to develop a habit of round-filing the submission after reading the lead paragraph, and you run the risk of that editor dumping everything with your name on it, without even getting to the lead paragraph. Thus, don't send too many announcements. Even with true major announcements, no small company is going to have enough earth-shattering business news (most won't even make the grass wave) more than once every couple of months.

But when you do deserve a release, get one out quickly, in timely fashion. If you get an award for design excellence, superb craftsmanship, or something else in July, don't wait until August or September to write the release.

And if you don't feel competent to write the releases, don't feel shy about checking out local freelancers to see if there is someone familiar with the construction industry to work up a release for a modest amount of cash (or a moderately reasonable check, by preference, to keep the IRS off both your backs). Newspapers are good places to check for names of freelancers, but you may want to also check with editors of building and construction trade papers and magazines: Their freelancers are often ready and willing to do the work for very little more than an inexperienced local. Today, there is never a real need to meet the writer, or, for that matter, for the writer to be within 1000, or 3000, miles of you and your business. Fax machines transfer finished work instantly, if you don't feel you can wait for the U.S. Post Office. And they're cheap to use: It costs me about 15 cents to send 3 pages about 900 miles. A 3-page letter costs 33 cents for the stamp, plus the envelope, and that 900 miles may take 4 days, though it usually takes 1 or 2. E-mail is even faster and cheaper.

Of course, my letterhead looks a lot better in real life than on fax paper, and yours will, too, but that's about the only real disadvantage of a fax. It is ugly. An old thermal fax does fade—and it fades really fast if you spray bugs over one. I lost a short (thankfully) letter when I had about all I could take with some flies and sprayed the room. When I came back into the room, the letter copy was 95 percent gone!

Here is a press release about an award:

ANYTOWN, ANY STATE—XYZ Company today received an award for most creative use of 2 × 4s on a drywall surface, using nailheads as decorative studs. The award is from the State Association of Bonded and Insured Building Cabinetmakers. XYZ Company was formed by Jim Jackson, in 1999, as a home remodeling business, with its goal of creating the kind of living space desirable to people with growing families, and to people with families that have already left the nest. Greater use of limited space is one feature of XYZ Company work, while more openness in living styles is another.

This is XYZ Company's third (fourth, fifth) award for quality or creativity since the company was founded just 2 years (6 months, whatever) ago. The company is a fully insured, licensed cabinetry contractor for the state of anywhere.

XYZ Company is located at Bumblebee Drive, Anytown, and the phone number is (703) 555-5555.

A release announcing a hire follows:

ANYTOWN, ANY STATE—XYZ Company today announced the hiring of Sylvia S. Sylvia as office controller. Ms. Sylvia has an associate's degree in business administration, and will be working out of the home office. XYZ Company was formed by Jim Jackson, in 2000, as a home cabinetry business, with its goal of creating the kind of living space desirable to people with growing families, and to people with families that have already left the nest. Greater use of limited space is one feature of XYZ Company work.

With this hiring, XYZ Company positions its owner to spend more time in the field, keeping tabs on quality and rapid growth.

The company is a fully insured, licensed contractor for the state of anywhere.

XYZ Company is located at Bumblebee Drive, Anytown, and the phone number is (703) 555-5555.

The above give you bases from which to work, and should simplify your relations with local newspapers, association, and other trade publications. Trade pubs aren't often a great deal of help because their regions of circulation are too large. The reason is simple: Your audience is made up of other cabinetmakers. Association publications may help, but probably they create the same problem.

Marketing, though, is not just a few small ads and simple news releases to local publications. It's an overall effort to use whatever service and skill you have to provide the best-quality work you can give. Of course, the point is to let everyone around know about the great work you do, but first you must do the work.

Your news releases may hit the buttons needed to get a reporter to talk to you about you and your business. And they may not. You may need to be a bit more aggressive here, but you can be almost sure any small town weekly will be interested in running a profile on you and your new business at some point just before, during, or just after your start-up. You may find much the same interest at your local daily paper.

The Public Relations Hook

Find what writers, and editors, call a hook. That's an angle for the story or, as one editor once put it to me, a peg to hang the story on. Do you have an odd college degree for someone who is a cabinetmaker? Do you have specific training that makes your company's emphasis more special than it might for someone without the training? Is your cabinetmaking specialty different enough that it might be the peg for the story? (Do you do special types of wood, special countertops, whatever comes to mind, or are you open to furniture building, etc.?) Dig things out, because publicity for you is publicity for your work, and the more publicity you get, the more work you'll get, assuming quality is consistently high.

Word of Mouth

Marketing is publicity, and publicity comes from many sources, including the most important one, the one the above steps are an attempt to start: *word of mouth.* Good word-of-mouth advertising and publicity can carry a company to a crest. Bad word of mouth can ruin a company. Doing good work is one way to start the tongues wagging in a positive manner, and emphasizing your positive personality traits is another way. Working with civic organizations is another, though when you first start your business, you're not really going to have spare time to do much on the civic front. Join the organizations, of course, but leave heavy participation in their activities, beyond being a fairly regular attendee at meetings when possible, to the time when your business is reasonably well established, and you can take time off to help others without harming your own interests.

Work to be pleasant, even when things are going wrong, and try to turn out the best possible work on all your contracts. You'll note over time that offers come from what seems to be no starting point. Such offers are generated by word-of-mouth advertising, by someone asking about a good cabinetmaker for a particular job and someone else giving your name. It builds, and over time you may well be able to just about quit advertising in any way, except for that single line in the White Pages of your local phone directory. And you may need a cellular phone to keep from missing out on too much while you run from place to place.

Like every other point of the cabinetry business, success in marketing depends on just how much effort you're willing to put into both the marketing and the overall jobs that you do.

Where They Are: Locations of Resources, On- and Off-Line

This is in no sense a full compilation of cabinetmaking sources. There are so many, including local suppliers (your best bet for top-quality wood and hardware at rational prices is usually a local pro supplier), there is probably no way to list them all. This appendix is intended as an aid and a starting point.

In today's computer-deep world, I've included Web sites wherever I can, as well as telephone and fax numbers, and, of course, the general Post Office contact. Where only one or two methods of contact are available, I've listed those.

Always remember, though, that you can quickly and simply do your own Web searches with any of the dozens of available search engines; and because things change more quickly than ever, you will find many more sites listed than I currently can find. Anytime you have trouble locating a company on-line, use variants on its name, preceded by www. and followed by .com. This technique doesn't always work, but it comes up aces a surprising percentage of the time.

Adhesives

3M
(800) 567-1639, ext. 1197
Contact cement

Franklin International
800-669-4583
Titebond adhesives, including several types [Titebond, Titebond II (water-resistant), Titebond II Extend (longer open time), polyurethanes, hide glue, contact cements]

Lutz File & Tool Co.
3929 Virginia Avenue
Cincinnati, OH 45227
(800) 966-3458
www.gorillaglue.com
Probably the first polyurethane wood glue, Gorilla Glue is still considered the best by many.

CAD

KCDw
Cabinetmakers Software
(508) 385-8569
Fax: (508) 385-8467
www.kcdw.com

Hardware

Chair City Supply Co.
P.O. Box 927
Thomasville, NC 27360
(800) 326-2191
Fax: (336) 475-4922
www.chaircitysupply.com
Screws

Doug Mockett & Co., Inc.
Box 3333
Manhattan Beach, CA 90266
(800) 523-1269
Fax: (800) 235-7743
Cabinet hardware

Hafele
(800) 423-3531
Cabinet hardware

Julius Blum, Inc.
Stanley, NC 28164
(800) 438-6788
Hardware for cabinetry, drawer slides, hinges, etc.

Laminates

Chemetal
(800) 807-7341
www.chemetal.com
Metallic laminates

Formica
(800) FORMICA
www.formica.com

Westlund Distributing
(800) 325-6878
Abet Laminati products

Machined wood parts

(Small shops may not wish to set up to do complex doors, drawers, and other items, and most small cabinetmaking shops do not do turnings; all these parts are available, in stock, or custom-made, for many makers.)

Adams Wood Products
974 Forest Drive
Morristown, TN 37814
(423) 587-2942
Fax: (423) 586-2188

Cabinet Factory, Inc.
P.O. Box 1748
La Crosse, WI 54603
(800) 237-1326
Fax: (608) 781-3667
www.cabinetfactory.com

Cab Parts
716 Arrowest Road
Grand Junction, CO 81505
(970) 241-7682
Fax: (970) 241-7689
32-mm cabinet parts

Conestoga
(800) 964-3667
Shaker pattern doors, drawers

DBS (Drawer Box Specialties)
(800) 422-9881

Meridian Products, Inc.
124 Earland Drive
New Holland, PA 17557
(717) 355-7700
Fax: (717) 355-2517
Dovetailed, prefinished drawer boxes

Osborne Wood Products, Inc.
Stock & Custom Wood Turnings
8116 Highway 123 North
Toccoa, GA 30577
(800) 849-8876
www.osbornewood.com

Precision Wood Products
Camden, OH (800) 582-8870
Lamonte, MO (800) 826-5617
Custom cabinet doors

White Oak Custom Woodworking, Ltd.
(800) 205-DOOR
Not just white, oak, but many other
woods, including maple, cherry, walnut

Moisture meters

Delmhorst Insrument Co.
51 Indian Lane East
Towaco, NJ 07082
(800) 222-0638

Electrophysics
Box 1143 Station B
London, ONT CA N6A 5K2

(800) 244-9908
Pin and pinless meters

Lignomat USA Ltd.
P.O. Box 30145
Portland, OR 97294
(800) 227-2105
Fax: (503) 255-1430
www.lignomat.com

Wagner Electronics
(800) 634-9961
www.wwwagner.com
Pinless moisture meters

Tools

Adjustable Clamp Co.
444 N. Ashland Avenue
Chicago, IL 60622
Jorgenson clamps, Pony clamps, much
else in the clamping field

Amana Tool
(800) 445-0077
www.amanatool.com
Top-quality bits, blades

American Tool
www.quick-gripclamp.com
Quick-Grip clamps, along with a wide
variety of other tools. You'll love the
clamps especially for those times when
you're shorthanded.

Apollosprayers, Inc.
1030 Joshua Way
Vista, CA 92083
(800) 578-7606
www.hvlp.com
High-volume low-pressure sprayers

Beisemeyer
(800) 782-1831
www.beisemeyer.com
Industry standard aftermarket fence,
other equipment

Clayton Machine Corp.
  (800) 971-5050
  Oscillating spindle sanders
CMT USA, Inc.
  307-F Pomona Drive
  Greensboro, NC 27407
  (888) CMT-BITS
  Fax: (800) 268-9778
  Saw blades, router and shaper and
  drill bits
Colonial Saw Company
  (800) 252-6355
  www.csaw.com
  Lamello biscuit joiners
Delta Industrial Machinery
  (800) 438-2486
  www.deltawoodworking.com
  Manufacturer of the Unisaw and many,
  many other woodworking tools
De-Sta-Co Industries
  2121 Cole Street
  Birmingham, MI 48009
  (248) 594-5600
  Fax: (800) 682-9686
  www.destaco.com
  Toggle action clamps
DeWalt
  (800) 4DEWALT
  www.dewalt.com
  Manufacturer of many portable and
  other power tools, including a new
  table saw that appears well adapted to
  the small cabinetmaking shop
Excalibur by Sommerville Design
  (800) 357-4118
  Excellent aftermarket table saw fence,
  dust collection and other tools
Fein Power Tools Inc.
  1030 Alcon Street
  Pittsburgh, PA 15220
  (800) 441-9878

Fax: (412) 922-8767
  Superb jigsaw, sanders, superquiet
  vacuum
Festo
  (888) 550-6424
  www.toolguide.net
  Excellent-quality hand power tools
Forrest Mfg. Company, Inc.
  457 River Road
  Clifton, NJ 07014
  (800) 733-7111
  Top-grade saw blades, dado sets
Freeborn Tool Co.
  P.O. Box 6246
  Spokane, WA 99217
  (800) 523-8988
  Shaper-cutters
Freud
  High Point, NC 27263
  (800) 472-7307
  Manufacturer of excellent saw blades,
  shaper-cutters, router bits, drill bits.
Grizzly Industrial
  (800) 523-4777
  Fax: (800) 438-5901
  www.grizzlyindustrial.com
  Importer of power tools, including table
  saws, dust collectors, jointers, planers,
  and much else of use in cabinetmaking
  shop
Jet Equipment & Tools
  2415 West Valley Highway, North
  Auburn, WA 98071
  www.jettools.com
Jesada
  310 Mears Boulevard
  Oldsmar, FL 34677
  (800) 531-5559
  Wide line of router bits and shaper-
  cutters, top-quality saw blades, dado sets

Keller Dovetail System
   1327 I Street
   Petaluma, CA 94952
   (800) 995-2456
   Easiest-to-use dovetail jig system that
   lends itself well to light production work
Laguna Tools
   2265 Laguna Canyon Road
   Laguna Beach, CA 92651
   (800) 234-1976
   Fax: (949) 497-1346
   www.lagunatools.com
   Excellent band saw, many other tools
Leigh Industries, Ltd.
   P.O. Box 357
   Port Coquitlam, BC
   (800) 663-8932
   Remarkably versatile dovetail jig that
   makes finger joints, mortise-and-tenon
   joints, and others
LRH Enterprises
   6961 Valjean Avenue
   Van Nuys, CA 91406
   (818) 782-9226
   Cutters, including special molding types
Makita
   www.makitatools.com
   Manufacturer of wide range of tools
MiniMax
   2475 Satellite Blvd.
   Duluth, GA 30096
   (800) 292-1837
   www.minimax-usa.com
   Excellent line of tools including lathes,
   band saws, jointers, and sanders
Oneida Air Systems, Inc.
   1001 West Fayette Street
   Syracuse, NY 13204
   (315) 476-5151
   www.oneida-air.com
   Dust collection systems

The Original Saw Company
   Britt, IA 50423
   Fax: (515) 843-3869
   www.originalsaw.com
   Large, durable radial arm saws
Porter-Cable
   (800) 487-8665
   www.porter-cable.com
   Extensive and exceptional line of
   portable power tools
Powermatic
   (800) 248-0144
   www.powermatic.com
   The model 66 and all its companions.
   Powermatic has just been bought by Jet,
   so some details are up in the air, but
   assurance is solidly given that model 66
   and top-range tools are going to stay in
   production.
Record Hand Tools
   www.recordtool.com
   Excellent vises, planes, other tools
Sears
   (800) 390-8792
   www.sears.com
   Craftsman power and hand tools
Systimatic/IKS Corp.
   12530 135th Avenue NE
   Kirkland, WA 98034
   (800) 426-0035
   Fax: (425) 821-0804
   Saw blades
Wilke Machinery Company
   3230 Susquehanna Trail
   York, PA 17402
   (800) 235-2100
   www.wilkemach.com
   Bridgewood tools; also carries General
   line, others

Williams & Hussey Machine Co., Inc.
P.O. Box 1149
Wilton, NH 03086
(800) 258-1380
Fax: (603) 654-5446
www.williamsnhussey.com
Planer molder. Open-sided design with blades that take 2 to 3 minutes, tops, to install. Works like a dream.

Woodmaster Tools Inc.
1431 North Topping Avenue
Kansas City, MO 64120
www.woodmastertools.com
Relatively low-cost wide-belt sander

Wood

Blue Ox Hardwoods
P.O. Box 715
Kenmore, NY 14217
(800) 758-0950
The name says it all.

Catskill Mountain Lumber Co.
(800) 828-9663
Hardwoods, white pine

Certainly Wood
11753 Big Tree Road East
Aurora, NY 14052
(716) 655-0206
Fax: (716) 655-3446
Veneers, hardwood

Colonial Hardwoods
(800) 466-5451
More than 100 species of hardwoods in stock

Compton Lumber & Hardware, Inc.
3847 First Avenue South
Seattle, WA 98134
(206) 623-5010
Fax: (206) 340-0851
Hardwoods for the northwest

Constantine's
2050 Eastchester Road
Bronx, NY 10461
(800) 223-8087
Hardwoods

Dieter H. Johnsen, Inc.
P.O. Box 508
Aurora, IN 47001
(812) 926-3944
Fax: (812) 926-1886
Hardwood veneers

Gilmer Wood Co.
2211 NW Saint Helens Road
Portland, OR 97210
(503) 274-1271
Hardwoods, exotics

Groff & Hearne Lumber, Inc.
(800) 342-0001
Fax: (717) 284-2400
Wide variety of hardwoods

Island Hardwoods
P.O. Box 189
Oak Island, NC 28465
(910) 278-1169
Figured hardwood

Landmark Logworks
Route 1, Box 36c
The Plains, VA 22171
(703) 687-4124
Quarter-sawn sycamore and oak, apple, cherry, osage orange, walnut

Menominee Tribal Enterprises
P.O. Box 10
Neopit, WI 54150
(715) 756-2311
Fax: (715) 756-2386
Menominee Tribal Enterprises sells lumber and certified wood products from the Menominee forest certified as sustained yield by Green Cross and Smart Wood.

Niagra Lumber & Wood Products, Inc.
  47 Elm Street
  East Aurora, NY 14052
  (800) 274-0397
  North Appalachian hardwoods
Sandy Pond Hardwoods
  16040 Via Este
  Sonora, CA 95370
  (800) 546-WOOD
  www.sandpond.com
  Lumber and flooring
Steve Wall Lumber Co.
  Box 287
  Mayodan, NC 27027
  (800) 633-4062
  Fax: (336) 427-7588
  www.walllumber.com
  Hardwoods, machinery

Woodstock Products
  Jeffersonville, IN
  (800) 505-8166
  Fax: (812) 288-5935
  www.woodmosaic.com
  Veneers and burls
Willard Brothers Woodcutters
  300 Basin Road
  Trenton, NJ 08619
  (800) 320-6519
  Exotic and domestic hardwood, veneers
  and free-form lumber
Woodtape
  Everett, WA   (800) 426-6362
  Conover, NC   (800) 833-8428
  Mississaugua, Ontario   (800) 461-0061
  Veneer laminate tapes for edges, other
  uses

Finding Out What Wood Looks Like On-Line

The following uniform resource locators (URLs), not all closely related to cabinetmaking (note ukuleles), but all offer a look at woods of various types, can be of help in design work, and are great for just plain having a good time checking out what is out there in the way of woods, including those used for ukuleles. Koa is the wood and Hawaii is the site. Koa is as gorgeous as the islands in which it grows, and it could be a super addition to any cabinetry. As noted, these locations are for provoking thought, while providing information. I hope you both enjoy and learn from them.

By country

http://www.forestworld.com/cgi-bin/countryframe.html

Glossary of terms

http://www.woodweb.com/~treetalk/wow_glossary.html

Names

http://www.forestworld.com/cgi-bin/woodinfo.html

http://www.woodworking.co.uk/html/trees.html

http://www.toolcenter.com/wood/wow.html#anchor1245530

http://www.toolcenter.com/wood/species.html

Pictures and properties

http://www.millerpublishing.com/nw_hardwoods/glossary.html

http://www.windsorplywood.com/worldofwoods/

http://www.wood-worker.com/properties.htm

http://www.windsorplywood.com/worldofwoods/northamerican/

http://www.windsorplywood.com/worldofwoods/tropical/home.html

http://www.curlymaple.com/wood.shtml

http://www.figuredhardwoods.com/photos.html

http://www.ukuleles.com/koawood.html

http://www.pennwood.com/

http://www.woodmosaic.com/sample.html

http://www.erinet.com/hardwood/full.html

http://www.connect.net/hollovar/cardhome.htm

http://www.woodweb.com/~treetalk/wowhome.html

http://www.toolcenter.com/wood/wow.html

http://www.woodweb.com/~treetalk/lapacho.html

http://www.woodweb.com/~treetalk/wowarchives.html

Samples and ID kits

http://www.connect.net/hollovar/cardhome.htm

http://members.wbs.net/homepages/w/i/s/wisconsincrafts.html

Scientific data

http://eco.bio.lmu.edu/WWW_Nat_History/plants/species/plspecie.htm

Tree rings

http://iufro.boku.ac.at/iufro/iufronet/d5/wu50900/li50900.htm

The following URLs can help you find other links to woodworking sites as well as the information provided on the sites listed.

http://www.millerpublishing.com/nw_hardwoods/index.html

http://www.millerpublishing.com/nw_hardwoods/glossary.html

http://www.millerpublishing.com/nw_softwoods/index.html

http://www.wood-worker.com/properties.htm

http://www.woodweb.com/~treetalk/lapacho.html

http://www.woodweb.com/~treetalk/wowarchives.html

For the most part, today, you can drop the *http://* part of the designation and simply start out www. In a few cases, that doesn't work. I like to try it the short way first, especially if I'm typing in the URL.

INDEX

Pages shown in **boldface** have illustrations on them.

ABOUT THE AUTHOR

Charles Self is a professional writer and a contributing editor to *Building & Remodeling News.* He is the past president and former newsletter editor of the National Home & Workshop Writers, and the author of several books on construction topics.